Generalized
Least Squares

Generalized Least Squares

Takeaki Kariya

Kyoto University and Meiji University, Japan

Hiroshi Kurata

University of Tokyo, Japan

John Wiley & Sons, Ltd

Other Wiley Editorial Offices

John Wiley & Sons Inc., 111 River Street, Hoboken, NJ 07030, USA

Jossey-Bass, 989 Market Street, San Francisco, CA 94103-1741, USA

Wiley-VCH Verlag GmbH, Boschstr. 12, D-69469 Weinheim, Germany

John Wiley & Sons Australia Ltd, 33 Park Road, Milton, Queensland 4064, Australia

John Wiley & Sons (Asia) Pte Ltd, 2 Clementi Loop #02-01, Jin Xing Distripark, Singapore 129809

John Wiley & Sons Canada Ltd, 22 Worcester Road, Etobicoke, Ontario, Canada M9W 1L1

Wiley also publishes its books in a variety of electronic formats. Some content that appears
in print may not be available in electronic books.

Library of Congress Cataloging-in-Publication Data

Kariya, Takeaki.
    Generalized least squares / Takeaki Kariya, Hiroshi Kurata.
        p. cm. – (Wiley series in probability and statistics)
    Includes bibliographical references and index.
    ISBN 0-470-86697-7 (alk. paper)
    1. Least squares. I. Kurata, Hiroshi, 1967- II. Title. III. Series.
QA275.K32 2004
511′.42—dc22
                                                            2004047963

British Library Cataloguing in Publication Data

A catalogue record for this book is available from the British Library

ISBN 0-470-86697-7 (PPC)

Produced from LaTeX files supplied by the author and processed by Laserwords Private Limited,
Chennai, India
Printed and bound in Great Britain by TJ International, Padstow, Cornwall
This book is printed on acid-free paper responsibly manufactured from sustainable forestry
in which at least two trees are planted for each one used for paper production.

To my late GLS co-worker Yasuyuki Toyooka and to my wife Shizuko

—Takeaki Kariya

To Akiko, Tomoatsu and the memory of my fathers

—Hiroshi Kurata

Contents

Preface

Regression analysis has been one of the most widely employed and most important statistical methods in applications and has been continually made more sophisticated from various points of view over the last four decades. Among a number of branches of regression analysis, the method of generalized least squares estimation based on the well-known Gauss–Markov theory has been a principal subject, and is still playing an essential role in many theoretical and practical aspects of statistical inference in a general linear regression model. A general linear regression model is typically of a certain covariance structure for the error term, and the examples are not only univariate linear regression models such as serial correlation models, heteroscedastic models and equi-correlated models but also multivariate models such as seemingly unrelated regression (SUR) models, multivariate analysis of variance (MANOVA) models, growth curve models, and so on.

When the problem of estimating the regression coefficients in such a model is considered and when the covariance matrix of the error term is known, as an efficient estimation procedure, we rely on the Gauss–Markov theorem that the Gauss–Markov estimator (GME) is the best linear unbiased estimator. In practice, however, the covariance matrix of the error term is usually unknown and hence the GME is not feasible. In such cases, a generalized least squares estimator (GLSE), which is defined as the GME with the unknown covariance matrix replaced by an appropriate estimator, is widely used owing to its theoretical and practical virtue.

This book attempts to provide a self-contained treatment of the unified theory of the GLSEs with a focus on their finite sample properties. We have made the content and exposition easy to understand for first-year graduate students in statistics, mathematics, econometrics, biometrics and other related fields. One of the key features of the book is a concise and mathematically rigorous description of the material via the *lower and upper bounds approach*, which enables us to evaluate the finite sample efficiency in a general manner.

In general, the efficiency of a GLSE is measured by relative magnitude of its risk (or covariance) matrix to that of the GME. However, since the GLSE is in general a nonlinear function of observations, it is often very difficult to evaluate the risk matrix in an explicit form. Besides, even if it is derived, it is often impractical to use such a result because of its complication. To overcome this difficulty, our book adopts as a main tool the lower and upper bounds approach, which approaches the problem by deriving a sharp lower bound and an effective upper bound for the risk matrix of a GLSE: for this purpose, we begin by showing that the risk matrix of a GLSE is bounded below by the covariance matrix of the GME (*Nonlinear Version of the Gauss–Markov Theorem*); on the basis of this result, we also derive an effective upper bound for the risk matrix of a GLSE relative to the covariance matrix of the GME (*Upper Bound Problems*). This approach has several important advantages: the upper bound provides information on the finite sample efficiency of a GLSE; it has a much simpler form than the risk matrix itself and hence serves as a tractable efficiency measure; furthermore, in some cases, we can obtain the optimal GLSE that has the minimum upper bound among an appropriate class of GLSEs. This book systematically develops the theory with various examples.

The book can be divided into three parts, corresponding respectively to Chapters 1 and 2, Chapters 3 to 6, and Chapters 7 to 9. The first part (Chapters 1 and 2) provides the basics for general linear regression models and GLSEs. In particular, we first give a fairly general definition of a GLSE, and establish its fundamental properties including conditions for unbiasedness and finiteness of second moments. The second part (Chapters 3–6), the main part of this book, is devoted to the detailed description of the lower and upper bounds approach stated above and its applications to serial correlation models, heteroscedastic models and SUR models. First, in Chapter 3, a nonlinear version of the Gauss–Markov theorem is established under fairly mild conditions on the distribution of the error term. Next, in Chapters 4 and 5, we derive several types of effective upper bounds for the risk matrix of a GLSE. Further, in Chapter 6, a uniform bound for the normal approximation to the distribution of a GLSE is obtained. The last part (Chapters 7–9) provides further developments (including mathematical extensions) of the results in the second part. Chapter 7 is devoted to making a further extension of the Gauss–Markov theorem, which is a maximal extension in a sense and leads to a further generalization of the nonlinear Gauss–Markov theorem proved in Chapter 3. In the last two chapters, some complementary topics are discussed. These include concentration inequalities, efficiency under elliptical symmetry, degeneracy of the distribution of a GLSE, and estimation of growth curves.

This book is not intended to be exhaustive, and there are many topics that are not even mentioned. Instead, we have done our best to give a systematic and unified presentation. We believe that reading this book leads to quite a solid understanding of this attractive subject, and hope that it will stimulate further research on the problems that remain.

The authors are indebted to many people who have helped us with this work. Among others, I, Takeaki Kariya, am first of all grateful to Professor Morris L. Eaton, who was my PhD thesis advisor and helped us get in touch with the publishers. I am also grateful to my late coauthor Yasuyuki Toyooka with whom

I published some important results contained in this book. Both of us are thankful to Dr. Hiroshi Tsuda and Professor Yoshihiro Usami for providing some tables and graphs and Ms Yuko Nakamura for arranging our writing procedure. We are also grateful to John Wiley & Sons for support throughout this project. Kariya's portion of this work was partially supported by the COE fund of Institute of Economic Research, Kyoto University.

Takeaki Kariya
Hiroshi Kurata

1

Preliminaries

1.1 Overview

This chapter deals with some basic notions that play indispensable roles in the theory of generalized least squares estimation and should be discussed in this preliminary chapter. Our selection here includes three basic notions: multivariate normal distribution, elliptically symmetric distributions and group invariance. First, in Section 1.2, some fundamental properties shared by the normal distributions are described without proofs. A brief treatment of Wishart distributions is also given. Next, in Section 1.3, we discuss the classes of spherically and elliptically symmetric distributions. These classes can be viewed as an extension of multivariate normal distribution and include various heavier-tailed distributions such as multivariate t and Cauchy distributions as special elements. Section 1.4 provides a minimum collection of notions on the theory of group invariance, which facilitates our unified treatment of generalized least squares estimators (GLSEs). In fact, the theory of spherically and elliptically symmetric distributions is principally based on the notion of group invariance. Moreover, as will be seen in the main body of this book, a GLSE itself possesses various group invariance properties.

1.2 Multivariate Normal and Wishart Distributions

This section provides without proofs some requisite distributional results on the multivariate normal and Wishart distributions.

Multivariate normal distribution. For an n-dimensional random vector y, let $\mathcal{L}(y)$ denote the distribution of y. Let

$$\mu = (\mu_1, \ldots, \mu_n)' \in R^n \text{ and } \Sigma = (\sigma_{ij}) \in \mathcal{S}(n),$$

Generalized Least Squares   Takeaki Kariya and Hiroshi Kurata
© 2004 John Wiley & Sons, Ltd   ISBN: 0-470-86697-7 (PPC)

where $\mathcal{S}(n)$ denotes the set of $n \times n$ positive definite matrices and a' the transposition of vector a or matrix a. We say that y is distributed as an n-dimensional *multivariate normal distribution* $N_n(\mu, \Sigma)$, and express the relation as

$$\mathcal{L}(y) = N_n(\mu, \Sigma), \tag{1.1}$$

if the probability density function (pdf) $f(y)$ of y with respect to the Lebesgue measure on R^n is given by

$$f(y) = \frac{1}{(2\pi)^{n/2}|\Sigma|^{1/2}} \exp\left(-\frac{1}{2}(y - \mu)'\Sigma^{-1}(y - \mu)\right) \quad (y \in R^n). \tag{1.2}$$

When $\mathcal{L}(y) = N_n(\mu, \Sigma)$, the mean vector $E(y)$ and the covariance matrix $\mathrm{Cov}(y)$ are respectively given by

$$E(y) = \mu \quad \text{and} \quad \mathrm{Cov}(y) = \Sigma, \tag{1.3}$$

where

$$\mathrm{Cov}(y) = E\{(y - \mu)(y - \mu)'\}.$$

Hence, we often refer to $N_n(\mu, \Sigma)$ as the normal distribution with mean μ and covariance matrix Σ.

Multivariate normality and linear transformations. Normality is preserved under linear transformations, which is a prominent property of the multivariate normal distribution. More precisely,

Proposition 1.1 *Suppose that* $\mathcal{L}(y) = N_n(\mu, \Sigma)$. *Let A be any* $m \times n$ *matrix such that rank* $A = m$ *and let b be any* $m \times 1$ *vector. Then*

$$\mathcal{L}(Ay + b) = N_m(A\mu + b, A\Sigma A'). \tag{1.4}$$

Thus, when $\mathcal{L}(y) = N_n(\mu, \Sigma)$, all the marginal distributions of y are normal. In particular, partition y as

$$y = \begin{pmatrix} y_1 \\ y_2 \end{pmatrix} \quad \text{with} \quad y_j : n_j \times 1 \quad \text{and} \quad n = n_1 + n_2,$$

and let μ and Σ be correspondingly partitioned as

$$\mu = \begin{pmatrix} \mu_1 \\ \mu_2 \end{pmatrix} \quad \text{and} \quad \Sigma = \begin{pmatrix} \Sigma_{11} & \Sigma_{12} \\ \Sigma_{21} & \Sigma_{22} \end{pmatrix}. \tag{1.5}$$

Then it follows by setting $A = (I_{n_1}, 0) : n_1 \times n$ in Proposition 1.1 that

$$\mathcal{L}(y_1) = N_{n_1}(\mu_1, \Sigma_{11}).$$

Clearly, a similar argument yields $\mathcal{L}(y_2) = N_{n_2}(\mu_2, \Sigma_{22})$. Note here that y_j's are not necessarily independent. In fact,

Proposition 1.2 *If $\mathcal{L}(y) = N_n(\mu, \Sigma)$, then the conditional distribution $\mathcal{L}(y_1|y_2)$ of y_1 given y_2 is given by*

$$\mathcal{L}(y_1|y_2) = N_{n_1}(\mu_1 + \Sigma_{12}\Sigma_{22}^{-1}(y_2 - \mu_2), \Sigma_{11.2}) \qquad (1.6)$$

with

$$\Sigma_{11.2} = \Sigma_{11} - \Sigma_{12}\Sigma_{22}^{-1}\Sigma_{21}.$$

It is important to notice that there is a one-to-one correspondence between (Σ_{11}, Σ_{12}, Σ_{22}) and ($\Sigma_{11.2}, \Theta, \Sigma_{22}$) with $\Theta = \Sigma_{12}\Sigma_{22}^{-1}$. The matrix Θ is often called the *linear regression coefficient* of y_1 on y_2.

As is well known, the condition $\Sigma_{12} = 0$ is equivalent to the independence between y_1 and y_2. In fact, if $\Sigma_{12} = 0$, then we can see from Proposition 1.2 that

$$\mathcal{L}(y_1) = \mathcal{L}(y_1|y_2) \ (= N_{n_1}(\mu_1, \Sigma_{11})),$$

proving the independence between y_1 and y_2. The converse is obvious.

Orthogonal transformations. Consider a class of normal distributions of the form $N_n(0, \sigma^2 I_n)$ with $\sigma^2 > 0$, and suppose that the distribution of a random vector y belongs to this class:

$$\mathcal{L}(y) \in \{N_n(0, \sigma^2 I_n) \mid \sigma^2 > 0\}. \qquad (1.7)$$

Let $\mathcal{O}(n)$ be the group of $n \times n$ orthogonal matrices (see Section 1.4). By using Proposition 1.1, it is shown that the distribution of y remains the same under orthogonal transformations as long as the condition (1.7) is satisfied. Namely, we have

Proposition 1.3 *If $\mathcal{L}(y) = N_n(0, \sigma^2 I_n)$ ($\sigma^2 > 0$), then*

$$\mathcal{L}(\Gamma y) = \mathcal{L}(y) \ \ for \ any \ \Gamma \in \mathcal{O}(n). \qquad (1.8)$$

It is noted that the orthogonal transformation $a \to \Gamma a$ is geometrically either the rotation of a or the reflection of a in R^n. A distribution that satisfies (1.8) will be called a *spherically symmetric distribution* (see Section 1.3). Proposition 1.3 states that $\{N_n(0, \sigma^2 I_n) \mid \sigma^2 > 0\}$ is a subclass of the class of spherically symmetric distributions.

Let $\|A\|$ denote the Euclidean norm of matrix A with

$$\|A\|^2 = tr(A'A),$$

where $tr(\cdot)$ denotes the trace of a matrix $\cdot$. In particular,

$$\|a\|^2 = a'a$$

for a vector a.

Proposition 1.4 *Suppose that $\mathcal{L}(y) \in \{N_n(0, \sigma^2 I_n) \mid \sigma^2 > 0\}$, and let*

$$x \equiv \|y\| \quad and \quad z \equiv y/\|y\| \quad with \quad \|y\|^2 = y'y. \tag{1.9}$$

Then the following three statements hold:

(1) $\mathcal{L}\left(x^2/\sigma^2\right) = \chi_n^2$, *where* χ_n^2 *denotes the* χ^2 *(chi-square) distribution with degrees of freedom n;*

(2) *The vector z is distributed as the uniform distribution on the unit sphere* $\mathcal{U}(n)$ *in* R^n, *where*

$$\mathcal{U}(n) = \{u \in R^n \mid \|u\| = 1\};$$

(3) *The quantities x and z are independent.*

To understand this proposition, several relevant definitions follow. A random variable w is said to be distributed as χ_n^2, if a pdf of w is given by

$$f(w) = \frac{1}{2^{n/2}\Gamma(n/2)} w^{\frac{n}{2}-1} \exp(-w/2) \quad (w > 0), \tag{1.10}$$

where $\Gamma(a)$ is the Gamma function defined by

$$\Gamma(a) = \int_0^\infty t^{a-1} e^{-t} dt \quad (a > 0). \tag{1.11}$$

A random vector z such that $z \in \mathcal{U}(n)$ is said to have a uniform distribution on $\mathcal{U}(n)$ if the distribution $\mathcal{L}(z)$ of z satisfies

$$\mathcal{L}(\Gamma z) = \mathcal{L}(z) \quad \text{for any } \Gamma \in \mathcal{O}(n). \tag{1.12}$$

As will be seen in the next section, statements (2) and (3) of Proposition 1.4 remain valid as long as the distribution of y is spherically symmetric. That is, if y satisfies $\mathcal{L}(\Gamma y) = \mathcal{L}(y)$ for all $\Gamma \in \mathcal{O}(n)$ and if $P(y = 0) = 0$, then $z \equiv y/\|y\|$ is distributed as the uniform distribution on the unit sphere $\mathcal{U}(n)$, and is independent of $x \equiv \|y\|$.

Wishart distribution. Next, we introduce the Wishart distribution, which plays a central role in estimation of the covariance matrix Σ of the multivariate normal distribution $N_n(\mu, \Sigma)$. In this book, the Wishart distribution will appear in the context of estimating a seemingly unrelated regression (SUR) model (see Example 2.4) and a growth curve model (see Chapter 9).

Suppose that p-dimensional random vectors $y_1, \ldots, y_n$ are independently and identically distributed as the normal distribution $N_p(0, \Sigma)$ with $\Sigma \in \mathcal{S}(p)$. We call the distribution of the matrix

$$W = \sum_{j=1}^n y_j y_j'$$

the *Wishart distribution* with parameter matrix Σ and degrees of freedom n, and express it as

$$\mathcal{L}(W) = W_p(\Sigma, n).\tag{1.13}$$

When $n \geq p$, the distribution $W_p(\Sigma, n)$ has a pdf of the form

$$f(W) = \frac{1}{2^{np/2}\Gamma_p(n/2)|\Sigma|^{n/2}}|W|^{\frac{n-p-1}{2}}\exp\left(-\frac{tr(W\Sigma^{-1})}{2}\right),\tag{1.14}$$

which is positive on the set $\mathcal{S}(p)$ of $p \times p$ positive definite matrices. Here $\Gamma_p(a)$ is the multivariate Gamma function defined by

$$\Gamma_p(a) = \pi^{p(p-1)/4}\prod_{j=1}^{p}\Gamma\left(a - \frac{j-1}{2}\right)\quad\left(a > \frac{p-1}{2}\right).\tag{1.15}$$

When $p = 1$, the multivariate Gamma function reduces to the (usual) Gamma function:

$$\Gamma_1(a) = \Gamma(a).$$

If W is distributed as $W_p(\Sigma, n)$, then the mean matrix is given by

$$E(W) = n\Sigma.$$

Hence, we often call $W_p(\Sigma, n)$ the Wishart distribution with mean $n\Sigma$ and degrees of freedom n. Note that when $p = 1$ and $\Sigma = 1$, the pdf $f(W)$ in (1.14) reduces to that of the χ^2 distribution χ_n^2, that is, $W_1(1, n) = \chi_n^2$. More generally, if $\mathcal{L}(w) = W_1(\sigma^2, n)$, then

$$\mathcal{L}(w/\sigma^2) = \chi_n^2.\tag{1.16}$$

(See Problem 1.2.2.)

Wishart-ness and linear transformations. As the normality is preserved under linear transformations, so is the Wishart-ness. To see this, suppose that $\mathcal{L}(W) = W_p(\Sigma, n)$. Then we have

$$\mathcal{L}(W) = \mathcal{L}\left(\sum_{j=1}^{n}y_jy_j'\right),$$

where y_j's are independently and identically distributed as the normal distribution $N_p(0, \Sigma)$. Here, by Proposition 1.1, for an $m \times p$ matrix A such that $rankA = m$, the random vectors $Ay_1, \ldots, Ay_n$ are independent and each Ay_j has $N_p(0, A\Sigma A')$. Hence, the distribution of

$$\sum_{j=1}^{n}Ay_j(Ay_j)' = A\left(\sum_{j=1}^{n}y_jy_j'\right)A'$$

is $W_p(A\Sigma A', n)$. This clearly means that $\mathcal{L}(AWA') = W_p(A\Sigma A', n)$. Thus, we obtain

Proposition 1.5 *If* $\mathcal{L}(W) = W_p(\Sigma, n)$, *then, for any* $A : m \times p$ *such that rank* $A = m$,

$$\mathcal{L}(AWA') = W_p(A\Sigma A', n). \tag{1.17}$$

Partition W and Σ as

$$W = \begin{pmatrix} W_{11} & W_{12} \\ W_{21} & W_{22} \end{pmatrix} \text{ and } \Sigma = \begin{pmatrix} \Sigma_{11} & \Sigma_{12} \\ \Sigma_{21} & \Sigma_{22} \end{pmatrix} \tag{1.18}$$

with $W_{ij} : p_i \times p_j$, $\Sigma_{ij} : p_i \times p_j$ and $p_1 + p_2 = p$. Then, by Proposition 1.5, the marginal distribution of the ith diagonal block W_{ii} of W is $W_{p_i}(\Sigma_{ii}, n)$ $(i = 1, 2)$. A necessary and sufficient condition for independence is given by the following proposition:

Proposition 1.6 *When* $\mathcal{L}(W) = W_p(\Sigma, n)$, *the two matrices* W_{11} *and* W_{22} *are independent if and only if* $\Sigma_{12} = 0$.

In particular, it follows:

Proposition 1.7 *When* $W = (w_{ij})$ *has Wishart distribution* $W_p(I_p, n)$, *the diagonal elements* w_{ii}'s *are independently and identically distributed as* χ_n^2. *And hence,*

$$\mathcal{L}(tr(W)) = \chi_{np}^2. \tag{1.19}$$

Cholesky–Bartlett decomposition. For any $\Sigma \in \mathcal{S}(p)$, the Cholesky decomposition of Σ gives a one-to-one correspondence between Σ and a lower-triangular matrix Θ. To introduce it, let $\mathcal{G}_T^+(p)$ be the group of $p \times p$ lower-triangular matrices with positive diagonal elements:

$$\mathcal{G}_T^+(p) = \{\Theta = (\theta_{ij}) \in \mathcal{G}\ell(p) \mid \theta_{ii} > 0 \ (i = 1, \ldots, p), \ \theta_{ij} = 0 \ (i < j)\},$$
$$\tag{1.20}$$

where $\mathcal{G}\ell(p)$ is the group of $p \times p$ nonsingular matrices (see Section 1.4).

Lemma 1.8 (Cholesky decomposition) *For any positive definite matrix* $\Sigma \in \mathcal{S}(p)$, *there exists a lower-triangular matrix* $\Theta \in \mathcal{G}_T^+(p)$ *such that*

$$\Sigma = \Theta\Theta'. \tag{1.21}$$

Moreover, the matrix $\Theta \in \mathcal{G}_T^+(p)$ *is unique.*

By the following proposition known as the *Bartlett decomposition*, a Wishart matrix with $\Sigma = I_p$ can be decomposed into independent χ^2 variables.

Proposition 1.9 (Bartlett decomposition) *Suppose* $\mathcal{L}(W) = W_p(I_p, n)$ *and let*

$$W = TT'$$

be the Cholesky decomposition in (1.21). Then $T = (t_{ij})$ *satisfies*

(1) $\mathcal{L}(t_{ii}^2) = \chi_{n-i+1}^2$ *for* $i = 1, \ldots, p$;

(2) $\mathcal{L}(t_{ij}) = N(0, 1)$ *and hence* $\mathcal{L}(t_{ij}^2) = \chi_1^2$ *for* $i > j$;

(3) t_{ij}'s $(i \geq j)$ *are independent.*

This proposition will be used in Section 4.4 of Chapter 4, in which an optimal GLSE in the SUR model is derived. See also Problem 1.2.5.

Spectral decomposition. For any symmetric matrix Σ, there exists an orthogonal matrix Γ such that $\Gamma'\Sigma\Gamma$ is diagonal. More specifically,

Lemma 1.10 *Let* Σ *be any* $p \times p$ *symmetric matrix. Then, there exists an orthogonal matrix* $\Gamma \in \mathcal{O}(p)$ *satisfying*

$$\Sigma = \Gamma \Lambda \Gamma' \quad with \quad \Lambda = \begin{pmatrix} \lambda_1 & & 0 \\ & \ddots & \\ 0 & & \lambda_p \end{pmatrix}, \qquad (1.22)$$

where $\lambda_1 \leq \cdots \leq \lambda_p$ *are the ordered latent roots of* Σ.

The above decomposition is called a *spectral decomposition* of Σ. Clearly, when $\lambda_1 < \cdots < \lambda_p$, the jth column vector γ_j of Γ is a latent vector of Σ corresponding to λ_j. If Σ has some multiple latent roots, then the corresponding column vectors form an orthonormal basis of the latent subspace corresponding to the (multiple) latent roots.

Proposition 1.11 *Let* $\mathcal{L}(W) = W_p(I_p, n)$ *and let*

$$W = HLH'$$

be the spectral decomposition of W, *where* $H \in \mathcal{O}(p)$ *and* L *is the diagonal matrix with diagonal elements* $0 \leq l_1 \leq \cdots \leq l_p$. *Then*

(1) $P(0 < l_1 < \cdots < l_p) = 1$;

(2) *A joint pdf of* $l \equiv (l_1, \ldots, l_p)$ *is given by*

$$\frac{\pi^{p^2/2}}{2^{pn/2}\Gamma_p(p/2)\Gamma_p(n/2)} \exp\left(-\frac{1}{2}\sum_{j=1}^{p} l_j\right) \prod_{j=1}^{p} l_j^{(n-p-1)/2} \prod_{i<j} (l_j - l_i),$$

which is positive on the set $\{l \in R^p \mid 0 < l_1 < \cdots < l_p\}$;

(3) *The two random matrices* H *and* L *are independent.*

A comprehensive treatment of the normal and Wishart distributions can be found in the standard textbooks on multivariate analysis such as Rao (1973), Muirhead (1982), Eaton (1983), Anderson (1984), Tong (1990) and Bilodeau and Brenner (1999). The proofs of the results in this section are also given there.

1.3   Elliptically Symmetric Distributions

In this section, the classes of spherically and elliptically symmetric distributions are defined, and their fundamental properties are investigated.

Spherically symmetric distributions. An $n \times 1$ random vector y is said to be distributed as a *spherically symmetric distribution* on R^n, or the distribution of y is called spherically symmetric, if the distribution of y remains the same under orthogonal transformations, namely,

$$\mathcal{L}(\Gamma y) = \mathcal{L}(y) \quad \text{for any } \Gamma \in \mathcal{O}(n), \tag{1.23}$$

where $\mathcal{O}(n)$ denotes the group of $n \times n$ orthogonal matrices. Let $\mathbf{E}_n(0, I_n)$ be the set of all spherically symmetric distributions on R^n. Throughout this book, we write

$$\mathcal{L}(y) \in \mathbf{E}_n(0, I_n), \tag{1.24}$$

when the distribution of y is spherically symmetric.

As is shown in Proposition 1.3, the class $\{N_n(0, \sigma^2 I_n) \mid \sigma^2 > 0\}$ of normal distributions is a typical subclass of $\mathbf{E}_n(0, I_n)$:

$$\{N_n(0, \sigma^2 I_n) \mid \sigma^2 > 0\} \subset \mathbf{E}_n(0, I_n).$$

Hence, it is appropriate to begin with the following proposition, which gives a characterization of the class $\{N_n(0, \sigma^2 I_n) \mid \sigma^2 > 0\}$ in $\mathbf{E}_n(0, I_n)$.

Proposition 1.12 Let $y = (y_1, \dots, y_n)'$ be an $n \times 1$ random vector. Then

$$\mathcal{L}(y) \in \{N_n(0, \sigma^2 I_n) \mid \sigma^2 > 0\} \tag{1.25}$$

holds if and only if the following two conditions simultaneously hold:

(1) $\mathcal{L}(y) \in \mathbf{E}_n(0, I_n)$;

(2) $y_1, \dots, y_n$ *are independent.*

Proof. Note first that $\mathcal{L}(y) \in \mathbf{E}_n(0, I_n)$ holds if and only if the characteristic function of y defined by

$$\psi(t) \equiv E[\exp(it'y)] \quad (t = (t_1, \dots, t_n)' \in R^n) \tag{1.26}$$

satisfies the following condition:

$$\psi(\Gamma't) = \psi(t) \quad \text{for any } \Gamma \in \mathcal{O}(n), \tag{1.27}$$

since $\psi(\Gamma't)$ is the characteristic function of Γy. As will be proved in Example 1.4 in the next section, the above equality holds if and only if there exists a function $\tilde{\psi}$ (on R^1) such that

$$\psi(t) = \tilde{\psi}(t't). \tag{1.28}$$

Suppose that the conditions (1) and (2) hold. Then the characteristic function of y_1, say $\psi_1(t_1)$, is given by letting $t = (t_1, 0, \dots, 0)'$ in $\psi(t)$ in (1.26). Hence from (1.28), the function $\psi_1(t_1)$ is written as

$$\psi_1(t_1) = \tilde{\psi}(t_1^2).$$

Similarly, the characteristic functions of y_j's are written as $\tilde{\psi}(t_j^2)$ $(j = 2, \dots, n)$. Since y_j's are assumed to be independent, the function $\tilde{\psi}$ satisfies

$$\tilde{\psi}(t't) = \prod_{j=1}^{n} \tilde{\psi}(t_j^2) \quad \text{for any } t \in R^n.$$

This equation is known as Hamel's equation, which has a solution of the form $\tilde{\psi}(x) = \exp(ax)$ for some $a \in R^1$. Thus, $\psi(t)$ must be of the form

$$\psi(t) = \exp(at't).$$

Since $\psi(t)$ is a characteristic function, the constant a must satisfy $a \le 0$. This implies that y is normal. The converse is clear. This completes the proof.

When the distribution $\mathcal{L}(y) \in \mathbf{E}_n(0, I_n)$ has a pdf $f(y)$ with respect to the Lebesgue measure on R^n, there exists a function $\tilde{f}$ on $[0, \infty)$ such that

$$f(y) = \tilde{f}(y'y). \tag{1.29}$$

See Example 1.4.

Spherically symmetric distributions with finite moments. Let

$$\mathcal{L}(y) \in \mathbf{E}_n(0, I_n)$$

and suppose that the first and second moments of y are finite. Then the mean vector $\mu \equiv E(y)$ and the covariance matrix $\Sigma \equiv \text{Cov}(y)$ of y take the form

$$\mu = 0 \quad \text{and} \quad \Sigma = \sigma^2 I_n \quad \text{for some } \sigma^2 > 0, \tag{1.30}$$

respectively. In fact, the condition (1.23) implies that $E(\Gamma y) = E(y)$ and Cov $(\Gamma y) = \text{Cov}(y)$ for any $\Gamma \in \mathcal{O}(n)$, or equivalently,

$$\Gamma\mu = \mu \quad \text{and} \quad \Gamma\Sigma\Gamma' = \Sigma \quad \text{for any } \Gamma \in \mathcal{O}(n).$$

This holds if and only if (1.30) holds (see Problem 1.3.1).

In this book, we adopt the two notations, $\mathcal{E}_n(0, \sigma^2 I_n)$ and $\tilde{\mathcal{E}}_n(0, I_n)$, which respectively specify the following two classes of spherically symmetric distributions with finite covariance matrices:

$$\mathcal{E}_n(0, \sigma^2 I_n) = \text{the class of spherically symmetric distributions}$$
$$\text{with mean 0 and covariance matrix } \sigma^2 I_n \tag{1.31}$$

and

$$\tilde{\mathcal{E}}_n(0, I_n) = \bigcup_{\sigma^2 > 0} \mathcal{E}_n(0, \sigma^2 I_n). \tag{1.32}$$

Then the following two consequences are clear:

$$N(0, \sigma^2 I_n) \in \mathcal{E}_n(0, \sigma^2 I_n) \subset \mathbf{E}_n(0, I_n)$$

and

$$\{N_n(0, \sigma^2 I_n) \mid \sigma^2 > 0\} \subset \tilde{\mathcal{E}}_n(0, I_n) \subset \mathbf{E}_n(0, I_n).$$

The uniform distribution on the unit sphere. The statements (2) and (3) of Proposition 1.3 proved for the class $\{N_n(0, \sigma^2 I_n) \mid \sigma^2 > 0\}$ are common properties shared by the distributions in $\mathbf{E}_n(0, I_n)$:

Proposition 1.13 Let $P \equiv \mathcal{L}(y) \in \mathbf{E}_n(0, I_n)$ and suppose that $P(y = 0) = 0$. Then the following two quantities

$$x \equiv \|y\| \quad \text{and} \quad z \equiv y/\|y\| \tag{1.33}$$

are independent, and z is distributed as the uniform distribution on the unit sphere $\mathcal{U}(n)$ in R^n.

Recall that a random vector z is said to have the uniform distribution on $\mathcal{U}(n)$ if

$$\mathcal{L}(\Gamma z) = \mathcal{L}(z) \quad \text{for any } \Gamma \in \mathcal{O}(n).$$

The uniform distribution on $\mathcal{U}(n)$ exists and is unique. For a detailed explanation on the uniform distribution on the unit sphere, see Chapters 6 and 7 of Eaton (1983). See also Problem 1.3.2.

The following corollary, which states that the distribution of $Z(y) \equiv y/\|y\|$ remains the same as long as $\mathcal{L}(y) \in \mathbf{E}_n(0, I_n)$, leads to various consequences, especially in the robustness of statistical procedures in the sense that some properties derived under normality assumption are valid even under spherical symmetry. See, for example, Kariya and Sinha (1989), in which the theory of robustness of multivariate invariant tests is systematically developed. In our book, an application to an SUR model is described in Section 8.3 of Chapter 8.

Corollary 1.14 *The distribution of $z = y/\|y\|$ remains the same as long as $\mathcal{L}(y) \in$ $\mathbf{E}_n(0, I_n)$.*

Proof. Since z is distributed as the uniform distribution on $\mathcal{U}(n)$, and since the uniform distribution is unique, the result follows.

Hence, the mean vector and the covariance matrix of $z = y/\|y\|$ can be easily evaluated by assuming without loss of generality that y is normally distributed.

Corollary 1.15 *If $\mathcal{L}(y) \in \mathbf{E}_n(0, I_n)$, then*

$$E(z) = 0 \quad \text{and} \quad \text{Cov}(z) = \frac{1}{n}I_n. \tag{1.34}$$

Proof. The proof is left as an exercise (see Problem 1.3.3).

Elliptically symmetric distributions. A random vector y is said to be distributed as an *elliptically symmetric distribution* with location $\mu \in R^n$ and scale matrix $\Sigma \in \mathcal{S}(n)$ if $\Sigma^{-1/2}(y - \mu)$ is distributed as a spherically symmetric distribution, or equivalently,

$$\mathcal{L}(\Gamma\Sigma^{-1/2}(y - \mu)) = \mathcal{L}(\Sigma^{-1/2}(y - \mu)) \quad \text{for any } \Gamma \in \mathcal{O}(n). \tag{1.35}$$

This class of distributions is denoted by $\mathbf{E}_n(\mu, \Sigma)$:

$$\mathbf{E}_n(\mu, \Sigma) = \text{the class of elliptically symmetric distributions}$$

$$\text{with location } \mu \text{ and scale matrix } \Sigma. \tag{1.36}$$

To describe the distributions with finite first and second moments, let

$$\mathcal{E}_n(\mu, \sigma^2\Sigma) = \text{the class of elliptically symmetric distributions}$$

$$\text{with mean } \mu \text{ and covariance matrix } \sigma^2\Sigma, \tag{1.37}$$

and

$$\tilde{\mathcal{E}}_n(\mu, \Sigma) = \bigcup_{\sigma^2 > 0} \mathcal{E}_n(\mu, \sigma^2\Sigma). \tag{1.38}$$

Here, it is obvious that

$$\{N_n(\mu, \sigma^2\Sigma) \mid \sigma^2 > 0\} \subset \tilde{\mathcal{E}}_n(\mu, \Sigma) \subset \mathbf{E}_n(\mu, \Sigma).$$

The proposition below gives a characterization of the class $\mathbf{E}_n(\mu, \Sigma)$ by using the characteristic function of y.

Proposition 1.16 *Let $\psi(t)$ be the characteristic function of y:*

$$\psi(t) = E[\exp(it'y)] \quad (t \in R^n). \tag{1.39}$$

Then, $\mathcal{L}(y) \in \mathbf{E}_n(\mu, \Sigma)$ if and only if there exists a function $\tilde{\psi}$ on $[0, \infty)$ such that

$$\psi(t) = \exp(it'\mu) \, \tilde{\psi}(t'\Sigma t). \tag{1.40}$$

Proof. Suppose $\mathcal{L}(y) \in \mathbf{E}_n(\mu, \Sigma)$. Let $y_0 = \Sigma^{-1/2}(y - \mu)$ and hence

$$\mathcal{L}(y_0) \in \mathbf{E}_n(0, I_n).$$

Then the characteristic function of y_0, say $\psi_0(t)$, is of the form

$$\psi_0(t) = \tilde{\psi}(t't) \quad \text{for some function } \tilde{\psi} \text{ on } [0, \infty). \tag{1.41}$$

The function ψ in (1.39) is rewritten as

$$\begin{aligned}
\psi(t) &= \exp(it'\mu) \, E[\exp(it'\Sigma^{1/2}y_0)] \quad \text{(since } y = \Sigma^{1/2}y_0 + \mu) \\
&= \exp(it'\mu) \, \psi_0(\Sigma^{1/2}t) \quad \text{(by definition of } \psi_0) \\
&= \exp(it'\mu) \, \tilde{\psi}(t'\Sigma t) \quad \text{(by (1.41))},
\end{aligned}$$

proving (1.40).

Conversely, suppose (1.40) holds. Then the characteristic function $\psi_0(t)$ of $y_0 = \Sigma^{-1/2}(y - \mu)$ is expressed as

$$\begin{aligned}
\psi_0(t) &\equiv E[\exp(it'y_0)] \\
&= E[\exp(it'\Sigma^{-1/2}y)] \, \exp(-it'\Sigma^{-1/2}\mu) \\
&= \psi(\Sigma^{-1/2}t) \exp(-it'\Sigma^{-1/2}\mu) \\
&= \tilde{\psi}(t't),
\end{aligned}$$

where the assumption (1.40) is used in the last line. This shows that $\mathcal{L}(y_0) \in \mathbf{E}_n(0, I_n)$, which is equivalent to $\mathcal{L}(y) \in \mathbf{E}_n(\mu, \Sigma)$. This completes the proof.

If the distribution $\mathcal{L}(y) \in \mathbf{E}_n(\mu, \Sigma)$ has a pdf $f(y)$ with respect to the Lebesgue measure on R^n, then f takes the form

$$f(y) = |\Sigma|^{-1/2} \tilde{f}((y - \mu)'\Sigma^{-1}(y - \mu)) \tag{1.42}$$

for some $\tilde{f} : [0, \infty) \to [0, \infty)$ such that $\int_{R^n} \tilde{f}(x'x) \, dx = 1$. In particular, when $\mathcal{L}(y) = N_n(\mu, \Sigma)$, the function $\tilde{f}$ is given by

$$\tilde{f}(u) = (2\pi)^{-n/2} \exp(-u/2).$$

Marginal and conditional distributions of elliptically symmetric distributions.
The following result is readily obtained from the definition of $E_n(\mu, \Sigma)$.

Proposition 1.17 *Suppose that* $\mathcal{L}(y) \in E_n(\mu, \Sigma)$ *and let A and b be any* $m \times n$
matrix of rank $A = m$ *and any* $m \times 1$ *vector respectively. Then*

$$\mathcal{L}(Ay + b) \in E_m(A\mu + b, A\Sigma A').$$

Hence, if we partition y, μ and Σ as

$$y = \begin{pmatrix} y_1 \\ y_2 \end{pmatrix}, \quad \mu = \begin{pmatrix} \mu_1 \\ \mu_2 \end{pmatrix} \quad \text{and} \quad \Sigma = \begin{pmatrix} \Sigma_{11} & \Sigma_{12} \\ \Sigma_{21} & \Sigma_{22} \end{pmatrix} \tag{1.43}$$

with $y_i : n_i \times 1$, $\mu_i : n_i \times 1$, $\Sigma_{ij} : n_i \times n_j$ and $n_1 + n_2 = n$, then the following
result holds:

Proposition 1.18 *If* $\mathcal{L}(y) \in E_n(\mu, \Sigma)$, *then the marginal distribution of* y_j *is also
elliptically symmetric:*

$$\mathcal{L}(y_j) \in E_{n_j}(\mu_j, \Sigma_{jj}) \quad (j = 1, 2). \tag{1.44}$$

Moreover, the conditional distribution of y_1 given y_2 is also elliptically symmetric.

Proposition 1.19 *If* $\mathcal{L}(y) \in E_n(\mu, \Sigma)$, *then*

$$\mathcal{L}(y_1|y_2) \in E_{n_1}(\mu_1 + \Sigma_{12}\Sigma_{22}^{-1}(y_2 - \mu_2), \ \Sigma_{11.2}) \tag{1.45}$$

with $\Sigma_{11.2} = \Sigma_{11} - \Sigma_{12}\Sigma_{22}^{-1}\Sigma_{21}$.

Proof. Without essential loss of generality, we assume that $\mu = 0$: $\mathcal{L}(y) \in$
$E_n(0, \Sigma)$. Since there is a one-to-one correspondence between y_2 and $\Sigma_{22}^{-1/2}y_2$,

$$\mathcal{L}(y_1|y_2) = \mathcal{L}(y_1|\Sigma_{22}^{-1/2}y_2)$$

holds, and hence it is sufficient to show that

$$\mathcal{L}(\Gamma\Sigma_{11.2}^{-1/2}w_1|\Sigma_{22}^{-1/2}y_2) = \mathcal{L}(\Sigma_{11.2}^{-1/2}w_1|\Sigma_{22}^{-1/2}y_2) \quad \text{for any } \Gamma \in \mathcal{O}(n_1), \tag{1.46}$$

where $w_1 = y_1 - \Sigma_{12}\Sigma_{22}^{-1}y_2$. By Proposition 1.17,

$$\mathcal{L}(w) \in E_n(0, \Theta) \quad \text{with} \quad \Theta = \begin{pmatrix} \Sigma_{11.2} & 0 \\ 0 & \Sigma_{22} \end{pmatrix},$$

where

$$w = \begin{pmatrix} w_1 \\ w_2 \end{pmatrix}$$

$$= \begin{pmatrix} I_{n_1} & -\Sigma_{12}\Sigma_{22}^{-1} \\ 0 & I_{n_2} \end{pmatrix} \begin{pmatrix} y_1 \\ y_2 \end{pmatrix}$$

$$= \begin{pmatrix} y_1 - \Sigma_{12}\Sigma_{22}^{-1}y_2 \\ y_2 \end{pmatrix}.$$

And thus $\mathcal{L}(x) \in \mathbf{E}_n(0, I_n)$ with $x \equiv \Theta^{-1/2}w$. Hence, it is sufficient to show that

$$\mathcal{L}(x_1|x_2) \in \mathbf{E}_{n_1}(0, I_{n_1}) \quad \text{whenever} \quad \mathcal{L}(x) \in \mathbf{E}_n(0, I_n).$$

Let $\mathbf{P}(\cdot|x_2)$ and P denote the conditional distribution of x_1 given x_2 and the (joint) distribution of $x = (x_1', x_2')'$ respectively. Then, for any Borel measurable set $A_1 \subset R^{n_1}$ and $A_2 \subset R^{n_2}$, and for any $\Gamma \in \mathcal{O}(n_1)$, it holds that

$$\int_{R^{n_1} \times A_2} \mathbf{P}(\Gamma A_1|x_2) P(dx_1, dx_2)$$

$$= \int_{R^{n_1} \times A_2} \chi_{\{x_1 \in \Gamma A_1\}} \, P(dx_1, dx_2)$$

$$= \int_{\Gamma A_1 \times A_2} P(dx_1, dx_2)$$

$$= \int_{A_1 \times A_2} P(dx_1, dx_2)$$

$$= \int_{R^{n_1} \times A_2} \chi_{\{x_1 \in A_1\}} P(dx_1, dx_2)$$

$$= \int_{R^{n_1} \times A_2} \mathbf{P}(A_1|x_2) P(dx_1, dx_2),$$

where χ denotes the indicator function, that is,

$$\chi_{\{x_1 \in A_1\}} = \begin{cases} 1 & \text{if } x_1 \in A_1 \\ 0 & \text{if } x_1 \notin A_1 \end{cases},$$

The first and last equalities are due to the definition of the conditional expectation, and the third equality follows since the distribution of x is spherically symmetric. This implies that the conditional distribution $\mathbf{P}(\cdot|x_2)$ is spherically symmetric a.s. x_2: for any $\Gamma \in \mathcal{O}(n_1)$ and any Borel measurable set $A_1 \subset R^{n_1}$,

$$\mathbf{P}(\Gamma A_1|x_2) = \mathbf{P}(A_1|x_2) \quad \text{a.s. } x_2.$$

This completes the proof.

If $\mathcal{L}(y) \in \mathbf{E}_n(\mu, \Sigma)$ and its first and second moments are finite, then the conditional mean and covariance matrix of y_1 given y_2 are evaluated as

$$E(y_1|y_2) = \mu_1 + \Sigma_{12}\Sigma_{22}^{-1}(y_2 - \mu_2),$$

$$\mathrm{Cov}(y_1|y_2) = g(y_2)\Sigma_{11.2} \tag{1.47}$$

for some function $g : R^{n_2} \to [0, \infty)$, where the conditional covariance matrix is defined by

$$\mathrm{Cov}(y_1|y_2) = E\{(y_1 - E(y_1|y_2))(y_1 - E(y_1|y_2))'|y_2\}.$$

In particular, when $\mathcal{L}(y) \in \mathbf{E}_n(0, I_n)$,

$$E(y_1|y_2) = 0 \quad \text{and} \quad \text{Cov}(y_1|y_2) = E\left(y_1 y_1'|y_2\right) = g(y_2)I_{n_1}. \tag{1.48}$$

More specifically, it can be proved that $g(y_2)$ in the conditional covariance matrix depends on y_2 only through $y_2' y_2$:

$$\text{Cov}(y_1|y_2) = \tilde{g}(y_2' y_2)I_{n_1} \quad \text{for some } \tilde{g} : [0, \infty) \to [0, \infty). \tag{1.49}$$

To see this, suppose that $\mathcal{L}(y)$ has a pdf $f(y)$ with respect to the Lebesgue measure on R^n. Then there exists a function $\tilde{f}$ such that

$$f(y) = \tilde{f}(y'y) = \tilde{f}(y_1' y_1 + y_2' y_2).$$

Hence, the conditional covariance matrix is calculated as

$$\text{Cov}(y_1|y_2) = \int_{R^{n_1}} y_1 y_1' \tilde{f}(y_1' y_1 + y_2' y_2)dy_1 / \int_{R^{n_1}} \tilde{f}(y_1' y_1 + y_2' y_2)dy_1,$$

where the right-hand side of the above equality depends on y_2 only through $y_2' y_2$. For the general case where $\mathcal{L}(y)$ may not have a pdf, see Fang, Kotz and Ng (1990).

Scale mixtures of normal distributions. The class $\mathbf{E}_n(\mu, \Sigma)$ of elliptically symmetric distributions contains various distributions used in practice. A typical example is a *scale mixture* of normal distributions. An $n \times 1$ random vector y is said to have a scale mixture of normal distributions if the distribution of y is expressed as

$$\mathcal{L}(y) = \mathcal{L}(\sqrt{x}w) \tag{1.50}$$

for some random variable x such that $x \geq 0$ a.s. and random vector w satisfies $\mathcal{L}(w) = N_n(0, I_n)$, where x and w are independent.

It is clear that $\mathcal{L}(y) \in \mathbf{E}_n(0, I_n)$ whenever y satisfies (1.50). Thus, various spherically symmetric distributions are produced according to the distribution of x. For simplicity, let us treat the case where x has a pdf $g(x)$ with respect to the Lebesgue measure on R^1. Since $\mathcal{L}(y|x) = N_n(0, xI_n)$, the pdf of y in this case is obtained as

$$f(y) = \frac{1}{(2\pi)^{n/2}} \int_0^\infty x^{-n/2} \exp\left(-\frac{y'y}{2x}\right)g(x)\,dx. \tag{1.51}$$

More generally, if y satisfies

$$\mathcal{L}(\Sigma^{-1/2}(y - \mu)) = \mathcal{L}(\sqrt{x}w),$$

the pdf of y is expressed as

$$f(y) = \frac{1}{(2\pi)^{n/2}|\Sigma|^{1/2}} \int_0^\infty x^{-n/2} \exp\left(-\frac{(y - \mu)'\Sigma^{-1}(y - \mu)}{2x}\right)g(x)\,dx, \tag{1.52}$$

which is an element of $\mathbf{E}_n(\mu, \Sigma)$. The multivariate t distribution with degrees of freedom m is a distribution with pdf

$$f(y) = \frac{\Gamma((n+m)/2)}{\Gamma(m/2)(m\pi)^{n/2}} \left[1 + \frac{(y-\mu)'\Sigma^{-1}(y-\mu)}{m} \right]^{-(n+m)/2},$$

which is produced by setting $\mathcal{L}(m/x) = \chi_m^2$, that is, setting

$$g(x) = \frac{(m/2)^{m/2}}{\Gamma(m/2)} x^{-\frac{m}{2}-1} \exp\left(-\frac{m}{2x}\right),$$

in (1.52). In particular, the multivariate t distribution with degrees of freedom $m = 1$ is known as the *multivariate Cauchy distribution*, which has no moment.

For textbooks that provide a detailed review on the theory of spherically and elliptically symmetric distributions, see, for example, Eaton (1983, 1989), Kariya and Sinha (1989), Fang, Kotz and Ng (1990). The papers by Schoenberg (1938), Kelker (1970), Eaton (1981, 1986), Cambanis, Huang and Simons (1981) are also fundamental.

1.4 Group Invariance

In this section, we provide some basic notions and facts on group invariance, which will be used in various aspects of the theory of generalized least squares estimation. A thorough discussion on this topic will be found in the textbooks given at the end of Section 1.3.

Group. Let $\mathcal{G}$ be a set with a binary operation $\circ : \mathcal{G} \times \mathcal{G} \to \mathcal{G}$. The set $\mathcal{G}$ is called a *group* if $\mathcal{G}$ satisfies the following conditions:

(1) $g_1 \circ (g_2 \circ g_3) = (g_1 \circ g_2) \circ g_3$ holds for any $g_1, g_2, g_3 \in \mathcal{G}$;

(2) There exists an element $e \in \mathcal{G}$ such that $g \circ e = e \circ g = g$ for any $g \in \mathcal{G}$;

(3) For each $g \in \mathcal{G}$, there exists an element $g^{-1} \in \mathcal{G}$ satisfying $g \circ g^{-1} = g^{-1} \circ g = e$.

The elements e and g^{-1} are called the unit of $\mathcal{G}$ and the inverse element of g respectively. Below we write $g_1 \circ g_2$ simply by g_1g_2, when no confusion is caused. Typical examples of the groups are

$\mathcal{G}\ell(n) = $ the group of $n \times n$ nonsingular matrices;

$\mathcal{G}_T^+(n) = $ the group of $n \times n$ nonsingular lower-triangular matrices

with positive diagonal elements;

$\mathcal{O}(n) = $ the group of $n \times n$ orthogonal matrices, which have already appeared in the previous sections. In the three groups mentioned above, the binary operation is the usual matrix multiplication, the unit is the identity matrix I_n, and for each element g, the inverse element is the inverse matrix g^{-1}.

The group $G_T^+(n)$ and $\mathcal{O}(n)$ are subgroups of $G\ell(n)$. Here, a subset $\mathcal{H}$ of a group $\mathcal{G}$ with binary operation $\circ$ is said to be a subgroup of $\mathcal{G}$, if $\mathcal{H}$ is a group with the same binary operation $\circ$.

Group action. Let $\mathcal{G}$ and $\mathcal{X}$ be a group and a set respectively. If there exists a map

$$\mathcal{G} \times \mathcal{X} \to \mathcal{X} : (g, x) \to gx \qquad (1.53)$$

such that

(1) $(g_1 g_2)x = g_1(g_2 x)$ for any $g_1, g_2 \in \mathcal{G}$ and $x \in \mathcal{X}$;

(2) $ex = x$ for any $x \in \mathcal{X}$;

where e is the unit of $\mathcal{G}$, then $\mathcal{G}$ is said to *act on* $\mathcal{X}$ via the group action

$$x \to gx.$$

When $\mathcal{G}$ acts on $\mathcal{X}$, each $g \in \mathcal{G}$ determines a one-to-one and onto transformation T_g on $\mathcal{X}$ by

$$T_g : \mathcal{X} \to \mathcal{X} : x \to gx. \qquad (1.54)$$

Thus, in this case, $\mathcal{G}$ can be viewed as a group of transformations on $\mathcal{X}$ (see Problem 1.4.1).

For each $x \in \mathcal{X}$, the set

$$\mathcal{G}x = \{gx \mid g \in \mathcal{G}\} \qquad (1.55)$$

is called the $\mathcal{G}$-orbit of x. Clearly, the set $\mathcal{X}$ can be expressed as the union of all $\mathcal{G}$-orbits, namely,

$$\mathcal{X} = \bigcup_{x \in \mathcal{X}} \mathcal{G}x. \qquad (1.56)$$

The action of $\mathcal{G}$ on $\mathcal{X}$ is said to be *transitive* if for any $x_1, x_2 \in \mathcal{X}$, there exits an element $g \in \mathcal{G}$ satisfying

$$x_2 = gx_1.$$

In this case, for each $x \in \mathcal{X}$, the set $\mathcal{X}$ can be written as the $\mathcal{G}$-orbit of x, that is, there is only one orbit:

$$\mathcal{X} = \{gx \mid g \in \mathcal{G}\}. \qquad (1.57)$$

Example 1.1 (Action of $\mathcal{G}\ell(n)$ on R^n)   Let $\mathcal{G} \equiv \mathcal{G}\ell(n)$ and $\mathcal{X} \equiv R^n - \{0\}$. Then $\mathcal{G}\ell(n)$ acts on $\mathcal{X}$ via the group action

$$x \to Gx$$

with $G \in \mathcal{G}\ell(n)$ and $x \in R^n$, where Gx is understood as the usual multiplication of matrix and vector. Moreover, the action of $\mathcal{G}\ell(n)$ is transitive on $\mathcal{X}$.

Example 1.2 (Action of $\mathcal{O}(n)$ on R^n)   Let $\mathcal{G} \equiv \mathcal{O}(n)$ and $\mathcal{X} \equiv R^n$. Then $\mathcal{O}(n)$ acts on $\mathcal{X}$ via

$$x \to \Gamma x$$

with $\Gamma \in \mathcal{O}(n)$ and $x \in R^n$. The action of $\mathcal{O}(n)$ on R^n is not transitive. In fact, for each $x_0 \in R^n$, the $\mathcal{O}(n)$-orbit of x_0 is expressed as

$$\mathcal{G}x_0 = \{x \in R^n \mid \|x\| = \|x_0\|\} \quad \text{with } \mathcal{G} = \mathcal{O}(n), \qquad (1.58)$$

which is a sphere in R^n with norm $\|x_0\|$, a proper subset of R^n, and hence $\mathcal{G}x_0 \neq R^n$.

However, if we define $\mathcal{X} \equiv \mathcal{G}x_0$ with arbitrarily fixed $x_0 \in \mathcal{X}$, then clearly the action of $\mathcal{G} \equiv \mathcal{O}(n)$ on $\mathcal{X}$ is transitive. Hence, by letting x_0 be such that $\|x_0\| = 1$, we can see that $\mathcal{O}(n)$ acts transitively on the unit sphere

$$\mathcal{U}(n) = \{u \in R^n \mid \|u\| = 1\}.$$

Invariant functions.   Suppose that a group $\mathcal{G}$ acts on a set $\mathcal{X}$. Let f be a function on $\mathcal{X}$. The function f is said to be *invariant* under $\mathcal{G}$ if f satisfies

$$f(gx) = f(x) \quad \text{for any } g \in \mathcal{G} \text{ and } x \in \mathcal{X}. \qquad (1.59)$$

This condition holds if and only if the function f is constant on each $\mathcal{G}$-orbit, that is, for each $x \in \mathcal{X}$,

$$f(y) = f(x) \quad \text{for any } y \in \mathcal{G}x. \qquad (1.60)$$

If a function m on $\mathcal{X}$ satisfies

(1) (invariance) $m(gx) = m(x)$ for any $g \in \mathcal{G}$ and $x \in \mathcal{X}$;

(2) (maximality) $m(x) = m(y)$ implies $y \in \mathcal{G}x$;

then m is called *a maximal invariant* under $\mathcal{G}$. Note that condition (2) above can be restated as

$$\text{if } m(x) = m(y), \text{ then there exists } g \in \mathcal{G} \text{ such that } y = gx. \qquad (1.61)$$

A maximal invariant is thus an invariant function that distinguishes the orbits.

Proposition 1.20 *A function $f(x)$ is invariant under $\mathcal{G}$ if and only if $f(x)$ is expressed as a function of a maximal invariant m:*

$$f(x) = \tilde{f}(m(x)) \quad for\ some\ function\ \tilde{f}. \tag{1.62}$$

Proof. Suppose that the function f is invariant. Let x and y be such that $m(x) = m(y)$. Then, by the maximality of m, there exists $g \in \mathcal{G}$ satisfying $y = gx$. Hence, by the invariance of f, we have

$$f(y) = f(gx) = f(x),$$

which means that $f(x)$ depends on x only through $m(x)$. Thus, there exists a function $\tilde{f}$ satisfying (1.62). The converse is clear. This completes the proof.

Example 1.3 (Maximal invariant under $\mathcal{G}\ell(n)$)   Let $\mathcal{G} \equiv \mathcal{G}\ell(n)$ and $\mathcal{X} \equiv R^n - \{0\}$. Since the action of $\mathcal{G}\ell(n)$ on $\mathcal{X}$ is transitive, a maximal invariant m on R^n is a constant function: $m(x) = c$ (say). This clearly implies that any invariant function must be constant.

Example 1.4 (Maximal invariant under $\mathcal{O}(n)$)   When $\mathcal{G} \equiv \mathcal{O}(n)$ and $\mathcal{X} \equiv R^n$, a function f on R^n is said to be invariant under $\mathcal{O}(n)$ if

$$f(\Gamma x) = f(x) \quad \text{for any } \Gamma \in \mathcal{O}(n) \text{ and } x \in R^n. \tag{1.63}$$

A maximal invariant under $\mathcal{O}(n)$ is given by

$$m(x) = x'x, \tag{1.64}$$

(see Problem 1.4.4), and hence, every invariant function f can be written as

$$f(x) = \tilde{f}(x'x) \quad \text{for some } \tilde{f} \text{ on } [0, \infty). \tag{1.65}$$

Here, recall that the distribution of an $n \times 1$ random vector y is spherically symmetric if $\mathcal{L}(\Gamma y) = \mathcal{L}(y)$ for any $\Gamma \in \mathcal{O}(n)$. If in addition y has a pdf $f(y)$ with respect to the Lebesque measure on R^n, then f satisfies

$$f(\Gamma y) = f(y) \quad \text{for any } \Gamma \in \mathcal{O}(n) \text{ and } y \in R^n.$$

Hence, the function $f(y)$ can be expressed as $f(y) = \tilde{f}(y'y)$ for some $\tilde{f} : [0, \infty) \to [0, \infty)$. The same result holds for the characteristic function of y. See (1.28), (1.29) and (1.42).

OLS residual vector as a maximal invariant. We conclude this section by showing that in a general linear regression model, the ordinary least squares (OLS) residual vector is a maximal invariant under a group of location transformations. The result will be used in Chapter 2 to establish an essential equivalence between the classes of GLSE and location-equivariant estimators. For relevant definitions, see Chapter 2.

Consider a general linear regression model of the form

$$y = X\beta + \varepsilon \tag{1.66}$$

with

$$y : n \times 1, \quad X : n \times k \text{ and } rank X = k.$$

The Euclidean space $\mathcal{G} \equiv R^k$ can be regarded as a group with binary operation $+$, the usual addition of vectors. Here, the unit is $0 \in R^k$ and the inverse element of $x \in R^k$ is $-x$. Let $\mathcal{Y} \equiv R^n$ be the space of y. Then $\mathcal{G} = R^k$ acts on $\mathcal{Y} = R^n$ via

$$y \to y + Xg \text{ with } y \in \mathcal{Y} \text{ and } g \in \mathcal{G}. \tag{1.67}$$

The group $\mathcal{G} = R^k$ is often called *the group of location transformations* or simply the *translation group*.

A maximal invariant $m(y)$ under $\mathcal{G}$ is given by

$$m(y) = Ny \text{ with } N = I_n - X(X'X)^{-1}X', \tag{1.68}$$

which is nothing but *the OLS residual vector*, when

$$\varepsilon'\varepsilon = (y - X\beta)'(y - X\beta)$$

is minimized with respect to β. To see the invariance of $m(y)$, let $g \in \mathcal{G}$. Then

$$m(y + Xg) = N(y + Xg) = Ny = m(y), \tag{1.69}$$

where the equation

$$NX = 0 \tag{1.70}$$

is used. Here, the matrix N is the orthogonal projection matrix onto $L^{\perp}(X)$ with the null space $L(X)$, where $L(X)$ denotes the linear subspace spanned by the column vectors of X and $L^{\perp}(X)$ is its orthogonally complementary subspace. Next, to show the maximality of $m(y)$, suppose that $m(y) = m(y_*)$ for $y, y_* \in \mathcal{Y}$. Then clearly $N(y - y_*) = 0$ holds, which is in turn equivalent to

$$y - y_* \in L(X).$$

Hence, there exists $g \in \mathcal{G}$ such that $y - y_* = Xg$, proving that $m(y)$ in (1.68) is a maximal invariant under the action of $\mathcal{G} = R^k$ on $\mathcal{Y} = R^n$.

Note that this fact holds no matter what the distribution of the error term ε may be.

1.5 Problems

1.2.1 Suppose that

$$\mathcal{L}(y) = N_n(\mu, \Sigma).$$

The characteristic function and the moment generating function of y are defined by

$$\xi(t) = E[\exp(it'y)] \quad (t \in R^n),$$

and

$$\zeta(t) = E[\exp(t'y)] \quad (t \in R^n),$$

respectively.

(1) Show that ξ is expressed as

$$\xi(t) = \exp\left(i\mu't - \frac{1}{2}t'\Sigma t\right).$$

(2) Show that $\zeta(t)$ exists on R^n and is expressed as

$$\zeta(t) = \exp\left(\mu't + \frac{1}{2}t'\Sigma t\right).$$

The answer will be found in standard textbooks listed at the end of Section 1.2.

1.2.2 Verify (1.16).

1.2.3 Consider the following multivariate linear regression model of the form

$$Y = XB + E,$$

where $Y : n \times p$, $X : n \times k$ with $rank X = k$, and each row ε_j' of E is independently and identically distributed as the normal distribution: $\mathcal{L}(\varepsilon_j) = N_p(0, \Sigma)$. Let a random matrix W be defined as

$$W = Y'NY \quad \text{with} \quad N = I_n - X(X'X)^{-1}X'.$$

Show that the matrix W is distributed as the Wishart distribution:

$$\mathcal{L}(W) = W_p(\Sigma, n - k).$$

See, for example, Theorem 10.1.12 of Muirhead (1982).

1.2.4 When $\mathcal{L}(W) = W_p(I_p, m)$, show that the expectation of W^{-1} is given by

$$E(W^{-1}) = \frac{1}{m - p - 1}I_p.$$

In Problem 3.6 of Muirhead (1982), a more general case is treated.

1.2.5 Suppose that $\mathcal{L}(W) = W_p(I_p, n)$. By using the Bartlett decomposition in Proposition 1.9:

(1) Show that

$$\mathcal{L}(|W|) = \mathcal{L}\left(\prod_{j=1}^{p} \chi^2_{n-j+1}\right),$$

where $|W|$ denotes the determinant of W, and χ^2_a, the random variable distributed as the χ^2 distribution with degrees of freedom a. See, for example, page 100 of Muirhead (1982);

(2) Evaluate the expectation

$$E\left\{\frac{[tr(W)]^2}{4|W|}\right\}$$

when $p = 2$. The answer will be found in the proof of Theorem 4.10 in Chapter 4.

1.3.1 Show that (1.30) is equivalent to the condition

$$\Gamma\mu = \mu \quad \text{and} \quad \Gamma\Sigma\Gamma' = \Sigma \quad \text{for any } \Gamma \in \mathcal{O}(n).$$

1.3.2 Suppose that an $n \times 1$ random vector $y = (y_1, \ldots, y_n)'$ satisfies $\mathcal{L}(y) \in E_n(0, I_n)$ and that y has a pdf of the form

$$f(y) = \tilde{f}(y'y)$$

with respect to the Lebesgue measure on R^n (see (1.29)). Let a transformation of y to polar coordinates be

$$y_1 = r \sin\theta_1 \sin\theta_2 \cdots \sin\theta_{n-2} \sin\theta_{n-1}$$
$$y_2 = r \sin\theta_1 \sin\theta_2 \cdots \sin\theta_{n-2} \cos\theta_{n-1}$$
$$y_3 = r \sin\theta_1 \sin\theta_2 \cdots \sin\theta_{n-3} \cos\theta_{n-2}$$
$$\vdots$$
$$y_{n-1} = r \sin\theta_1 \cos\theta_2$$
$$y_n = r \cos\theta_1,$$

where $r = \|y\| > 0$, $0 < \theta_j \leq \pi$ $(j = 1, \ldots, n-2)$ and $0 < \theta_{n-1} \leq 2\pi$.

(1) Show that the pdf of $(r^2, \theta_1, \ldots, \theta_{n-1})$ is given by

$$\frac{1}{2}(r^2)^{n/2-1} \sin^{n-2}\theta_1 \sin^{n-3}\theta_2 \cdots \sin\theta_{n-2} \tilde{f}(r^2),$$

and thus the quantities r^2 and θ_j's are independent.

(2) Find a pdf of each θ_j $(j = 1, \ldots, n - 1)$, and note that each pdf does not depend on the functional form of $\tilde{f}$.

(3) Find a pdf of $r^2 = \|y\|^2$.

An answer will be found, for example, in Theorem 1.5.5 of Muirhead (1982).

1.3.3 To establish Corollary 1.15, suppose without loss of generality that $\mathcal{L}(y) = N_n(0, I_n)$.

(1) Show that $E(z) = 0$ by using the identity

$$E(y) = E(x)E(z),$$

where $x = \|y\|$.

(2) Show that $\text{Cov}(z) = \frac{1}{n} I_n$ by using the identity

$$\text{Cov}(y) = E(x^2)\text{Cov}(z).$$

1.3.4 Establish Proposition 1.19 under the assumption that $\mathcal{L}(y)$ has a pdf with respect to the Lebesgue measure on R^n.

1.4.1 Show that T_g defined in (1.54) is a one-to-one and onto transformation on $\mathcal{X}$.

1.4.2 Let $\mathcal{G} \equiv \mathcal{G}_T^+(n)$ be the group of $n \times n$ lower-triangular matrices with positive diagonal elements.

(1) Let $\mathcal{X} \equiv \mathcal{G}_T^+(n)$. Show that $\mathcal{G}$ acts on $\mathcal{X}$ via the group action

$$T \to GT \quad \text{with} \quad G \in \mathcal{G} \quad \text{and} \quad T \in \mathcal{X},$$

and that the action is transitive.

(2) Let $\mathcal{X} \equiv \mathcal{S}(n)$ be the set of $n \times n$ positive definite matrices. Show that $\mathcal{G}$ acts on $\mathcal{X}$ via the group action

$$S \to GSG' \quad \text{with} \quad G \in \mathcal{G} \quad \text{and} \quad S \in \mathcal{X},$$

and that the action is transitive.

1.4.3 Let $\mathcal{G} \equiv \mathcal{O}(n)$ and $\mathcal{X} = \mathcal{S}(n)$.

(1) Show that $\mathcal{G}$ acts on $\mathcal{X}$ via the group action

$$S \to \Gamma S \Gamma' \quad \text{with} \quad G \in \mathcal{G} \quad \text{and} \quad S \in \mathcal{X},$$

and that the action is not transitive.

(2) Show that a maximal invariant under $\mathcal{G}$ is the ordered latent roots of S, say, $l_1 \leq \cdots \leq l_n$.

1.4.4 Verify (1.64).

1.4.5 Let $\mathcal{G} \equiv (0, \infty)$, which is a group with binary operation $g_1 \circ g_2 = g_1 g_2$, the usual multiplication of real numbers. The unit of $\mathcal{G}$ is 1, and the inverse element of g is $g^{-1} = 1/g$. The group $\mathcal{G}$ acts on R^n via the group action

$$y \to gy \quad \text{with} \quad g \in \mathcal{G} \text{ and } y \in R^n,$$

and hence $\mathcal{G}$ is often called *a group of scale transformation*.

(1) Show that a maximal invariant $m(y)$ under $\mathcal{G}$ is given by

$$m(y) = y/\|y\|. \tag{1.71}$$

Hence, if a function $f(y)$ is *scale-invariant*, namely,

$$f(gy) = f(y) \quad \text{for any } g > 0 \text{ and } y \in R^n, \tag{1.72}$$

the function f depends on y only through $y/\|y\|$. As will be illustrated in later chapters, several types of scale-invariance will appear in the theory of GLSE. See, for example, Proposition 2.6.

Aside from GLSE, the scale-invariance property (1.72) often implies the robustness of invariant statistics. To see this,

(2) Suppose that $\mathcal{L}(y_1) \in \mathbf{E}_n(0, I_n)$ and $\mathcal{L}(y_2) = N_n(0, I_n)$. Show that for any scale-invariant function f, the following equality holds:

$$\mathcal{L}(f(y_1)) = \mathcal{L}(f(y_2)).$$

Hint: Note that $\mathcal{L}(y_1/\|y_1\|) = \mathcal{L}(y_2/\|y_2\|) =$ the uniform distribution on the unit sphere in R^n.

2

Generalized Least Squares Estimators

2.1 Overview

This chapter is devoted to establishing some basic results on generalized least squares estimators (GLSEs) in a general linear regression model. Here, the general linear regression model is a linear regression model

$$y = X\beta + \varepsilon$$

with general covariance structure of the error term ε, that is, the covariance matrix $\Omega = \sigma^2 \Sigma$ of ε is given by a function of an unknown but estimable parameter θ:

$$\Sigma = \Sigma(\theta).$$

The model includes as its special cases various specific models used in many applied areas. It includes not only univariate linear regression models such as serial correlation model, equi-correlated model and heteroscedastic model, but also multivariate models such as multivariate analysis of variance (MANOVA) model, seemingly unrelated regression (SUR) model, growth curve model and so on. Such models are produced according to the specific structure of the regressor matrix X and the covariance matrix $\Omega = \sigma^2 \Sigma(\theta)$.

In the problem of estimating these models, it is well known in the Gauss–Markov theorem that the Gauss–Markov estimator (GME) of the form

$$b(\Sigma) = (X'\Sigma^{-1}X)^{-1}X'\Sigma^{-1}y$$

is the best linear unbiased estimator (BLUE) of the regression coefficient vector β, when the matrix Σ is known. In most cases, however, the matrix Σ is unknown

Generalized Least Squares   Takeaki Kariya and Hiroshi Kurata
© 2004 John Wiley & Sons, Ltd   ISBN: 0-470-86697-7 (PPC)

and hence, the GME $b(\Sigma)$ is not feasible. In such cases, a GLSE $b(\hat{\Sigma})$, which is defined as the GME with unknown Σ in $b(\Sigma)$ replaced by an estimator $\hat{\Sigma}$, is widely used in practice. In fact, as will be observed later, most of the estimators that have been proposed and applied to real data in the above specific models are special cases of the GLSE $b(\hat{\Sigma})$.

The aim of this chapter is to provide some fundamental results on the GLSEs in a unified manner. The contents of this chapter are summarized as

2.2 General Linear Regression Model

2.3 Generalized Least Squares Estimators

2.4 Finiteness of Moments and Typical GLSEs

2.5 Empirical Example: CO_2 Emission Data

2.6 Empirical Example: Bond Price Data.

In Section 2.2, we define a general linear regression model in a general setup. We also introduce several typical models for applications such as an AR(1) error model, Anderson model, equi-correlated model, heteroscedastic model and SUR model. In Section 2.3, the GLSE is defined on the basis of the Gauss–Markov theorem. The relation between the GLSEs and some other estimators derived from different principles including linear unbiased estimators, location-equivariant estimators and the maximum likelihood estimator is also discussed. Section 2.4 deals with conditions for a GLSE to be unbiased and to have finite second moments. Examples of typical GLSEs and their fundamental properties are also given. In Sections 2.5 and 2.6, we give simple examples of empirical analysis on CO_2 emission data and bond price data by using GLSEs.

2.2   General Linear Regression Model

In this section, a general linear regression model is defined and several specific models that are important in application are introduced.

General linear regression model.   Throughout this book, a *general linear regression model* is defined as

$$y = X\beta + \varepsilon, \tag{2.1}$$

where y is an $n \times 1$ vector and $X : n \times k$ is a known matrix of full rank. Here, the $n \times 1$ error term ε is a random vector with mean 0 and covariance matrix Ω:

$$E(\varepsilon) = 0 \quad \text{and} \quad \text{Cov}(\varepsilon) = E(\varepsilon\varepsilon') = \Omega \in \mathcal{S}(n), \tag{2.2}$$

where $\mathcal{S}(n)$ denotes the set of $n \times n$ positive definite matrices. The finiteness of the second moment in (2.2) is included in the definition of the general linear regression model, meaning that for given $\Omega \in \mathcal{S}(n)$, the class $\mathcal{P}_n(0, \Omega)$ of distributions for ε satisfying (2.2) is combined with the model (2.1). Obviously, the class $\mathcal{P}_n(0, \Omega)$ of distributions with mean 0 and covariance matrix Ω is very broad, and, in particular, it includes the normal distribution $N_n(0, \Omega)$ with mean 0 and covariance matrix Ω.

When it is necessary, the distribution of ε is denoted by P and the expectation of $\cdot$ by $E_P(\cdot)$, or simply by $E(\cdot)$, whenever no confusion is caused. The covariance matrix Ω is usually unknown and is formulated as a function of an unknown but estimable parameter θ:

$$\Omega = \Omega(\theta), \tag{2.3}$$

where the functional form of $\Omega(\cdot)$ is assumed to be known.

It will be observed below that the model (2.1) with the structure (2.3) includes, as its special cases, the serial correlation model, the heteroscedastic model, the SUR model, and so on. Such models are often used in applications. To understand the structure of these models, the details of these models are described below for future references.

Typical models. To begin with, let us consider a family of the models of the following covariance structure:

$$\Omega = \sigma^2 \Sigma(\theta)$$

with

$$\Sigma(\theta)^{-1} = I_n + \lambda_n(\theta)C \quad \text{and} \quad \theta \in \Theta \subset R^1, \tag{2.4}$$

where C is an $n \times n$ known symmetric matrix, $\lambda = \lambda_n = \lambda_n(\theta)$ is a continuous real-valued function on Θ, and the matrix $\Sigma(\theta)$ is positive definite for any $\theta \in \Theta$.

Since C is assumed to be known, it can be set as a diagonal matrix without loss of generality. To see this, let Γ be an orthogonal matrix that diagonalizes C:

$$\Gamma'C\Gamma = \begin{pmatrix} d_1 & & 0 \\ & \ddots & \\ 0 & & d_n \end{pmatrix} \equiv D \quad \text{with} \quad d_1 \leq \cdots \leq d_n. \tag{2.5}$$

Then transforming $y = X\beta + \varepsilon$ to

$$\Gamma'y = \Gamma'X\beta + \Gamma'\varepsilon$$

yields the model with covariance structure $\mathrm{Cov}(\Gamma'\varepsilon) = \sigma^2\Gamma'\Sigma(\theta)\Gamma$, where

$$(\Gamma'\Sigma(\theta)\Gamma)^{-1} = I_n + \lambda_n(\theta)D \tag{2.6}$$

is a diagonal matrix with the jth diagonal element $1 + \lambda_n(\theta)d_j$. Among others, this family includes the Anderson model, equi-correlated model and heteroscedastic model with two distinct variances. We will describe these models more specifically.

Example 2.1 (AR(1) error model and Anderson model) In the linear regression model (2.1) with the condition (2.2), let the error term $\varepsilon = (\varepsilon_1, \ldots, \varepsilon_n)'$ be generated by the following stationary autoregressive process of order 1 (AR(1)):

$$\varepsilon_j = \theta\varepsilon_{j-1} + \xi_j \quad \text{with} \quad |\theta| < 1 \quad (j = 0, \pm1, \pm2, \ldots), \tag{2.7}$$

where ξ_j's satisfy

$$E(\xi_i) = 0, \quad \text{Var}(\xi_i) = \tau^2,$$
$$\text{Cov}(\xi_i, \xi_j) = 0 \ (i \neq j).$$

Then, as is well known, the covariance matrix Ω is expressed as

$$\Omega = \tau^2\Phi \quad \text{with} \quad \Phi = \Phi(\theta) = \frac{1}{1-\theta^2}\left(\theta^{|i-j|}\right), \tag{2.8}$$

and the inverse of Φ is given by

$$\Phi^{-1} = \begin{pmatrix} 1 & -\theta & & & & 0 \\ -\theta & 1+\theta^2 & -\theta & & & \\ & \ddots & \ddots & \ddots & & \\ & & \ddots & \ddots & \ddots & \\ & & & \ddots & 1+\theta^2 & -\theta \\ 0 & & & & -\theta & 1 \end{pmatrix}. \tag{2.9}$$

See Problem 2.2.2. The matrix Φ^{-1} is not of the form (2.4). In fact, it is expressed as

$$\Phi^{-1} = (1-\theta)^2\left[I_n + \lambda(\theta)C + \psi(\theta)B\right], \tag{2.10}$$

where

$$\lambda(\theta) = \frac{\theta}{(1-\theta)^2}, \quad \psi(\theta) = \frac{\theta}{1-\theta},$$

$$C = \begin{pmatrix} 1 & -1 & & & & 0 \\ -1 & 2 & -1 & & & \\ & \ddots & \ddots & \ddots & & \\ & & \ddots & \ddots & \ddots & \\ & & & \ddots & 2 & -1 \\ 0 & & & & -1 & 1 \end{pmatrix}$$

and

$$
B = \begin{pmatrix} 1 & & & 0 \\ & 0 & & \\ & & \ddots & \\ & & 0 & \\ 0 & & & 1 \end{pmatrix}.
$$

The matrix Φ^{-1} is often approximated by replacing the $(1, 1)$th and (n, n)th elements in (2.9) by $1 - \theta + \theta^2$ (Anderson, 1948), which is the same as Φ^{-1} in (2.10) with $\psi(\theta)$ replaced by 0. When it is assumed that the error term has this modified covariance structure, we call the regression model the *Anderson model*.

That is, the Anderson model is the model with covariance structure

$$
\mathrm{Cov}(\varepsilon) = \sigma^2 \Sigma(\theta)
$$

with

$$
\Sigma(\theta)^{-1} = I_n + \lambda(\theta)C \quad \text{and} \quad \theta \in \Theta, \tag{2.11}
$$

where

$$
\sigma^2 = \frac{\tau^2}{(1 - \theta)^2}, \quad \lambda(\theta) = \frac{\theta}{(1 - \theta)^2} \quad \text{and} \quad \Theta = (-1, 1).
$$

Since the latent roots of C are given by (see Problem 2.2.3)

$$
d_j = 2\left[1 - \cos\left(\frac{(j - 1)\pi}{n}\right)\right] \quad (j = 1, \dots, n), \tag{2.12}
$$

the matrix C in (2.11) can be replaced by the diagonal matrix with diagonal elements $d_1, \dots, d_n$:

$$
D = \begin{pmatrix} d_1 & & 0 \\ & \ddots & \\ 0 & & d_n \end{pmatrix}. \tag{2.13}
$$

Since

$$
-\frac{1}{4} < \lambda(\theta) < \infty \quad \text{and} \quad d_1 = 0 < d_2 < \cdots < d_n < 4, \tag{2.14}
$$

$\Sigma(\theta)$ in (2.11) is in fact positive definite on Θ.

Example 2.2 (Equi-correlated model) Suppose that the error term $\varepsilon = (\varepsilon_1, \dots, \varepsilon_n)'$ of the model (2.1) satisfies

$$
\mathrm{Var}(\varepsilon_i) = \tau^2 \quad \text{and} \quad \mathrm{Cov}(\varepsilon_i, \varepsilon_j) = \tau^2\theta \quad (i \neq j), \tag{2.15}
$$

for some $\tau^2 > 0$ and $\theta \in \left(-\frac{1}{n-1}, 1\right) \equiv \Theta$, that is, the covariance matrix Ω is given by

$$\Omega = \tau^2 \begin{pmatrix} 1 & \theta & \cdots & \theta \\ \theta & 1 & & \vdots \\ \vdots & & \ddots & \theta \\ \theta & \cdots & \theta & 1 \end{pmatrix}. \tag{2.16}$$

The model with this covariance structure is called an *equi-correlated model*, since the correlation coefficients between distinct error terms are equally given by θ. This model is a member of the family (2.4), because Ω is of the form

$$\Omega = \sigma^2 \Sigma(\theta) \quad \text{with} \quad \Sigma(\theta)^{-1} = I_n + \lambda_n(\theta)\, 1_n 1_n', \tag{2.17}$$

where

$$\sigma^2 = \tau^2(1 - \theta), \quad \lambda_n(\theta) = -\frac{\theta}{1 + (n-1)\theta}$$

and $1_n = (1, \ldots, 1)' : n \times 1$. Here σ^2 and λ_n are clearly functionally independent.

Since the latent roots of the matrix $1_n 1_n'$ are n and 0s (with multiplicity $n - 1$), applying the argument from (2.5) through (2.6), the model (2.15) is rewritten as

$$\Omega = \sigma^2 \Sigma(\theta) \quad \text{with} \quad \Sigma(\theta)^{-1} = I + \lambda_n(\theta)D, \tag{2.18}$$

where

$$D = \begin{pmatrix} 0 & & & 0 \\ & \ddots & & \\ & & 0 & \\ 0 & & & n \end{pmatrix} \tag{2.19}$$

without any loss of generality. Here, the matrix $\Sigma(\theta)$ is in fact positive definite on Θ, because

$$-\frac{1}{n} < \lambda_n(\theta) < \infty. \tag{2.20}$$

See Problem 2.2.4. In this expression, the equi-correlated model can be regarded as a heteroscedastic model, which we treat next.

Example 2.3 (Heteroscedastic model) The model (2.1) with the following structure

$$y = \begin{pmatrix} y_1 \\ \vdots \\ y_p \end{pmatrix} : n \times 1, \quad X = \begin{pmatrix} X_1 \\ \vdots \\ X_p \end{pmatrix} : n \times k,$$

$$\varepsilon = \begin{pmatrix} \varepsilon_1 \\ \vdots \\ \varepsilon_p \end{pmatrix} : n \times 1 \tag{2.21}$$

and

$$\Omega = \begin{pmatrix} \sigma_1^2 I_{n_1} & & 0 \\ & \ddots & \\ 0 & & \sigma_p^2 I_{n_p} \end{pmatrix} \in \mathcal{S}(n) \tag{2.22}$$

is called a *heteroscedastic model with p distinct variances* or simply, a *p-equation heteroscedastic model*, where

$$y_j : n_j \times 1, \quad X_j : n_j \times k, \quad \varepsilon_j : n_j \times 1 \quad \text{and} \quad n = \sum_{j=1}^{p} n_j.$$

This model consists of p distinct linear regression models

$$y_j = X_j \beta_j + \varepsilon_j \tag{2.23}$$

with

$$E(\varepsilon_j) = 0 \quad \text{and} \quad \text{Cov}(\varepsilon_j) = \sigma_j^2 I_{n_j} \quad (j = 1, \ldots, p),$$

where the coefficient vectors are restricted as

$$\beta_1 = \cdots = \beta_p \equiv \beta. \tag{2.24}$$

Note that when $p = 2$, this model is formally expressed as a member of the family (2.4) with the structure

$$\Omega = \sigma^2 \Sigma(\theta) \quad \text{with} \quad \Sigma(\theta)^{-1} = I_n + \lambda(\theta)D, \tag{2.25}$$

where

$$\sigma^2 = \sigma_1^2, \quad \theta = \sigma_1^2/\sigma_2^2, \quad \lambda(\theta) = \theta - 1 \quad \text{and} \quad D = \begin{pmatrix} 0 & 0 \\ 0 & I_{n_2} \end{pmatrix}.$$

Here $\theta \in \Theta \equiv (0, \infty)$, and $\Sigma(\theta)$ is positive definite on Θ.

Example 2.4 (Seemingly unrelated regression (SUR) model)   When the model $y = X\beta + \varepsilon$ in (2.1) is of the form

$$y = \begin{pmatrix} y_1 \\ \vdots \\ y_p \end{pmatrix} : n \times 1, \quad X = \begin{pmatrix} X_1 & & 0 \\ & \ddots & \\ 0 & & X_p \end{pmatrix} : n \times k,$$

$$\beta = \begin{pmatrix} \beta_1 \\ \vdots \\ \beta_p \end{pmatrix} : k \times 1, \quad \varepsilon = \begin{pmatrix} \varepsilon_1 \\ \vdots \\ \varepsilon_p \end{pmatrix} : n \times 1 \tag{2.26}$$

with covariance structure

$$\Omega = \Sigma \otimes I_m \quad \text{and} \quad \Sigma = (\sigma_{ij}) \in \mathcal{S}(p), \tag{2.27}$$

it is called a *p-equation SUR model*, which was originally formulated by Zellner (1962), where

$$y_j : m \times 1, \ X_j : m \times k_j, \ n = pm, \ k = \sum_{j=1}^{p} k_j$$

and $\otimes$ denotes the Kronecker product. Here, for matrices $P = (p_{ij}) : a \times b$ and $Q = (q_{ij}) : c \times d$, the Kronecker product of P and Q is defined as

$$P \otimes Q = \begin{pmatrix} p_{11}Q & \cdots & p_{1b}Q \\ \vdots & & \vdots \\ p_{a1}Q & \cdots & p_{ab}Q \end{pmatrix} : ac \times bd. \tag{2.28}$$

See Problem 2.2.6.

The model (2.26) is constructed by p different linear regression models with the cross-correlation structure:

$$y_j = X_j \beta_j + \varepsilon_j \tag{2.29}$$

with

$$E(\varepsilon_j) = 0, \quad \text{Cov}(\varepsilon_j) = \sigma_{jj} I_m$$

and

$$\text{Cov}(\varepsilon_i, \varepsilon_j) = \sigma_{ij} I_m \quad (i, j = 1, \dots, p). \tag{2.30}$$

The model (2.26) can also be expressed as a multivariate linear regression model with a restriction on the coefficient matrix:

$$Y = X_* B + E_* \tag{2.31}$$

with

$$E(E_*) = 0, \quad \text{Cov}(E_*) = I_m \otimes \Sigma$$

and

$$B = \begin{pmatrix} \beta_1 & & 0 \\ & \ddots & \\ 0 & & \beta_p \end{pmatrix} : k \times p, \tag{2.32}$$

where

$$Y = (y_1, \ldots, y_p) : m \times p, \quad X_* = (X_1, \ldots, X_p) : m \times k,$$

$$E_* = (\varepsilon_1, \ldots, \varepsilon_p) : m \times p.$$

Note that X_* may not be of full rank. Here, $\mathrm{Cov}(E_*)$ is the covariance matrix of $vec(E_*')$, where for an $n \times m$ matrix $A = (a_1, \ldots, a_m)$ with $a_j : n \times 1$, the quantity $vec(A)$ is defined as

$$vec(A) = \begin{pmatrix} a_1 \\ \vdots \\ a_m \end{pmatrix} : nm \times 1. \tag{2.33}$$

The zeros in (2.32) are regarded as a restriction on the coefficient matrix B in (2.31). If the model (2.31) satisfies

$$k_1 = \cdots = k_p \equiv k_0 \quad \text{and} \quad X_1 = \cdots = X_p \equiv X_0,$$

then it reduces to the familiar multivariate linear regression model

$$Y = X_0 B_0 + E_* \quad \text{with} \quad B_0 = \left(\beta_1, \ldots, \beta_p \right) : k_0 \times p, \tag{2.34}$$

where no restriction is imposed on B_0.

When $p = 2$, the following two special cases are often treated in the literature, because these cases give an analytically tractable structure:

(i) $X_1' X_2 = 0$;

(ii) $L(X_1) \subset L(X_2)$.

Here, $L(A)$ denotes the linear subspace spanned by the column vectors of matrix A. Zellner (1962, 1963) considered the first case in which the regressor matrices X_1 and X_2 are orthogonal: $X_1' X_2 = 0$, while Revankar (1974, 1976) considered the second case in which $L(X_1) \subset L(X_2)$. Such cases are of some interest in the discussion of the efficiency of GLSEs and the OLSE. Some aspects of these simplified SUR models are summarized in Srivastava and Giles (1987).

2.3 Generalized Least Squares Estimators

In this section, we first define a generalized least squares estimator or a GLSE. It is often referred to as a *feasible GLSE* or a *two-stage Aitkin estimator* in the literature. The definition here is based on the well-known Gauss–Markov theorem. We also consider relations among the GLSE and some other important estimators including the linear unbiased estimator, location-equivariant estimator and maximum-likelihood estimator.

Throughout this book, inequalities for matrices should be interpreted in terms of nonnegative definiteness. For example, $A \geq B$ means that A and B are nonnegative definite and $A - B$ is nonnegative definite.

The Gauss–Markov estimator (GME). Consider a general linear regression model of the form

$$y = X\beta + \varepsilon \quad \text{with} \quad P \equiv \mathcal{L}(\varepsilon) \in \mathcal{P}_n(0, \sigma^2 \Sigma), \tag{2.35}$$

where

$$y : n \times 1, \quad X : n \times k \quad \text{and} \quad rank\, X = k.$$

First suppose that Σ is known. In this setup, the Gauss–Markov theorem plays a fundamental role in the problem of estimating the coefficient vector β. To state the theorem formally, let $\mathcal{C}_0$ be the class of linear unbiased estimators of β, that is,

$$\mathcal{C}_0 = \{\hat{\beta} = Cy \mid C \text{ is a } k \times n \text{ matrix such that } CX = I_k\}. \tag{2.36}$$

(See Problem 2.3.1.) Here, recall that an estimator $\hat{\beta}$ is called *linear* if it is of the form $\hat{\beta} = Cy$ for a $k \times n$ matrix C. If a linear estimator is unbiased, it is called *linear unbiased*.

Theorem 2.1 (Gauss–Markov theorem) *The estimator of the form*

$$b(\Sigma) = (X'\Sigma^{-1}X)^{-1}X'\Sigma^{-1}y, \tag{2.37}$$

which we call the Gauss–Markov estimator (GME) throughout this book, is the BLUE of β, that is, the GME is the unique estimator that satisfies

$$\text{Cov}(b(\Sigma)) \leq \text{Cov}(\hat{\beta}) \tag{2.38}$$

for any $\hat{\beta} \in \mathcal{C}_0$ and $P \in \mathcal{P}_n(0, \sigma^2\Sigma)$. The covariance matrix of $b(\Sigma)$ is given by

$$\text{Cov}(b(\Sigma)) = \sigma^2(X'\Sigma^{-1}X)^{-1}.$$

Proof. Decompose a linear unbiased estimator $\hat{\beta} = Cy$ in $\mathcal{C}_0$ as

$$\hat{\beta} - \beta = \left[b(\Sigma) - \beta\right] + \left[\hat{\beta} - b(\Sigma)\right] \tag{2.39}$$

$$= (X'\Sigma^{-1}X)^{-1}X'\Sigma^{-1}\varepsilon + \left[C - (X'\Sigma^{-1}X)^{-1}X'\Sigma^{-1}\right]\varepsilon.$$

Since the two terms on the right-hand side of (2.39) are mutually uncorrelated (see Problem 2.3.2), the covariance matrix of $\hat{\beta}$ is evaluated as

$$\text{Cov}(\hat{\beta}) = \text{Cov}(b(\Sigma)) + E[(\hat{\beta} - b(\Sigma))(\hat{\beta} - b(\Sigma))']$$

$$= \sigma^2(X'\Sigma^{-1}X)^{-1}$$

$$+ \sigma^2[C - (X'\Sigma^{-1}X)^{-1}X'\Sigma^{-1}]\Sigma[C - (X'\Sigma^{-1}X)^{-1}X'\Sigma^{-1}]'$$

$$\tag{2.40}$$

$$\geq \sigma^2(X'\Sigma^{-1}X)^{-1},$$

since the second term of (2.40) is nonnegative definite. The equality holds if and only if $C = (X'\Sigma^{-1}X)^{-1}X'\Sigma^{-1}$. This completes the proof.

It is noted that the distribution P of ε is arbitrary so long as $\mathrm{Cov}_P(\varepsilon) = \sigma^2 \Sigma$. When $\Sigma = I_n$ holds, the GME in (2.37) reduces to the *ordinary least squares estimator* (OLSE)

$$b(I_n) = (X'X)^{-1}X'y. \tag{2.41}$$

In other words, the OLSE is the BLUE, when $\Sigma = I_n$.

Generalized least squares estimators (GLSEs). In applications, Σ is generally unknown and hence the GME is not feasible, as it stands. In this case, a natural estimator of β is a GME with unknown Σ in $b(\Sigma)$ replaced by an estimator $\hat{\Sigma}$, which we shall call a **GLSE**. More precisely, an estimator of the form

$$b(\hat{\Sigma}) = (X'\hat{\Sigma}^{-1}X)^{-1}X'\hat{\Sigma}^{-1}y \tag{2.42}$$

is called a GLSE if $\hat{\Sigma}$ is almost surely positive definite and is a function of the OLS residual vector e, where

$$e = Ny \quad \text{with} \quad N = I_n - X(X'X)^{-1}X'. \tag{2.43}$$

Let Z be an $n \times (n - k)$ matrix such that

$$N = ZZ', \quad Z'Z = I_{n-k} \quad \text{and} \quad X'Z = 0, \tag{2.44}$$

that is, the set of $(n - k)$ columns of the matrix Z forms an orthonormal basis of the orthogonally complementary subspace of the column space $L(X)$ of X. In the sequel, for a given X, we pick a matrix Z satisfying (2.44) and fix it throughout.

A GLSE defined in (2.42) is in general highly nonlinear in y and hence it is generally difficult to investigate its finite sample properties. However, as will be seen in the next proposition, the class of GLSEs can be viewed as an extension of the class C_0 of linear unbiased estimators. In fact, any GLSE $b(\hat{\Sigma})$ is expressed as

$$b(\hat{\Sigma}) = C(e)y \quad \text{with} \quad C(e) = (X'\hat{\Sigma}^{-1}X)^{-1}X'\hat{\Sigma}^{-1}$$

and the matrix-valued function $C(\cdot)$ satisfies $C(e)X = I_k$ for any e. Clearly, a linear unbiased estimator is obtained by letting C be a constant function. More precisely, let

$$C_1 = \text{the class of all GLSEs of the form (2.42).}$$

The following proposition given by Kariya and Toyooka (1985) provides a characterization of the class C_1.

Proposition 2.2 *The class C_1 is expressed as*

$$C_1 = \{\hat{\beta} \equiv C(e)y \mid C(\cdot) \text{ is a } k \times n \text{ matrix-valued measurable}$$

$$\text{function on } R^n \text{ satisfying } C(\cdot)X = I_k\}. \tag{2.45}$$

More specifically, any estimator of the form $\hat{\beta} = C(e)y$ *satisfying* $C(e)X = I_k$ *is a GLSE* $b(\hat{\Sigma})$ *with*

$$\hat{\Sigma}^{-1} = C(e)'C(e) + N. \tag{2.46}$$

Proof. For any estimator of the form $\hat{\beta} = C(e)y$ satisfying $C(e)X = I_k$, let $\hat{\Sigma}^{-1} = C(e)'C(e) + N$. Substituting this into $b(\hat{\Sigma})$ and noting that $C(e)X = I_k$ and $X'N = 0$ proves the equality $b(\hat{\Sigma}) = C(e)y$. The nonsingularity of $\hat{\Sigma}^{-1}$ follows because

$$\hat{\Sigma}^{-1} = (C(e)', Z) \begin{pmatrix} C(e) \\ Z' \end{pmatrix},$$

and

$$\begin{pmatrix} C(e) \\ Z' \end{pmatrix} (X, Z) = \begin{pmatrix} I_k & C(e)Z \\ 0 & I_{n-k} \end{pmatrix}, \tag{2.47}$$

where the matrix (X, Z) is clearly nonsingular. This completes the proof.

Clearly, from Proposition 2.2, the class $\mathcal{C}_1$ of GLSEs includes the class $\mathcal{C}_0$ in (2.36).

Proposition 2.3   *The relation* $\mathcal{C}_0 \subset \mathcal{C}_1$ *holds. More specifically, any linear unbiased estimator* $\hat{\beta} = Cy \in \mathcal{C}_0$ *is a GLSE* $b(\Psi)$ *with*

$$\Psi^{-1} = C'C + N.$$

Proof. Let $C(e)$ be a constant function: $C(e) \equiv C$. Then the result follows.

Note that the class $\mathcal{C}_1$ includes some biased estimators. However, it will be found soon that most of the reasonable GLSEs, such as the ones introduced in the next section, are unbiased. Note also that the class $\mathcal{C}_1$ in (2.45) does not depend on $\sigma^2\Sigma = \text{Cov}(\varepsilon)$, and hence, some of the results in Sections 2.3 and 2.4 hold even if the model is incorrectly specified.

When Σ is known to be a function of the unknown but estimable vector θ, say

$$\Sigma = \Sigma(\theta),$$

it is often the case that θ is first estimated and the GLSE of the form

$$b(\Sigma(\hat{\theta})) = (X'\Sigma(\hat{\theta})^{-1}X)^{-1}X'\Sigma(\hat{\theta})^{-1}y \tag{2.48}$$

is used, where $\hat{\theta} = \hat{\theta}(e)$ is an estimator of θ based on the OLS residual vector e. Clearly, such a GLSE is in $\mathcal{C}_1$.

Next, we consider relations among the GLSEs, location-equivariant estimators and maximum likelihood estimators (MLEs), and show that these estimators are GLSEs in our sense.

Relation between GLSEs and location-equivariant estimators. To define a location-equivariant estimator, consider the following transformation defined on the space R^n of y as

$$y \rightarrow y + Xg \quad \text{with} \quad g \in R^k. \tag{2.49}$$

In other words, as was exposited in Section 1.4 of Chapter 1, the group $\mathcal{G} \equiv R^k$ acts on the set $\mathcal{X} \equiv R^n$ via the group action (2.49). This action induces the following action on the space R^k of β:

$$\beta \rightarrow \beta + g, \tag{2.50}$$

that is, by the invariance principle, since the mean of y is $X\beta$, observing $y + Xg$ is regarded to be equivalent to considering $\beta + g$ in the parameter space R^k. Thus, it will be natural to limit consideration to estimators $\hat{\beta} = \hat{\beta}(y)$ satisfying the following condition

$$\hat{\beta}(y + Xg) = \hat{\beta}(y) + g \quad \text{for any} \ g \in R^k, \tag{2.51}$$

which is called a *location-equivariant* estimator of β. For a general theory on equivariant estimation, the readers may be referred to Ferguson (1967), Eaton (1983) and Lehmann (1983).

Let $\mathcal{C}_2$ be the class of location-equivariant estimators:

$$\mathcal{C}_2 = \{\hat{\beta} \mid \hat{\beta} \text{ satisfies } (2.51)\}.$$

Although the notion of the location-equivariant estimators appears to be quite different from that of the GLSEs, the class $\mathcal{C}_2$ is in fact essentially equivalent to the class $\mathcal{C}_1$ of GLSEs. To clarify this, we begin with the following proposition.

Proposition 2.4 *The class $\mathcal{C}_2$ is characterized as*

$$\mathcal{C}_2 = \{\hat{\beta}(y) \equiv b(I_n) + d(e) \mid d \text{ is a } k \times 1 \text{ vector-valued}$$
$$\text{measurable function on } R^n\}. \tag{2.52}$$

Proof. First letting $g = -b(I_n)$ in (2.51), the estimator $\hat{\beta}(y)$ must be of the form

$$\hat{\beta}(y) = b(I_n) + d(e) \tag{2.53}$$

for some measurable function d, where

$$b(I_n) = (X'X)^{-1}X'y$$

is the OLSE. Conversely, the estimator $\hat{\beta}(y)$ of the form (2.53) satisfies the condition (2.51). This completes the proof.

Here, note that the residual $e \equiv e(y)$ as a function of y is a maximal invariant under $\mathcal{G} = R^k$ (see Section 1.4 of Chapter 1 for the definition). In fact, $e(y)$ is invariant under the transformation (2.49) and $e(y) = e(y_*)$ implies $y = y_* + Xg$ for some $g \in R^k$.

To state the relation between the classes $\mathcal{C}_1$ and $\mathcal{C}_2$, let χ_A denote the indicator function of set A. Further, let

$$\tilde{\mathcal{C}}_1 = \{\tilde{\beta} = \hat{\beta} + a \, \chi_{\{e=0\}} \mid \hat{\beta} \in \mathcal{C}_1, \, a \in R^k\}, \tag{2.54}$$

which satisfies $\tilde{\mathcal{C}}_1 \supset \mathcal{C}_1$. Clearly, an estimator $\tilde{\beta}$ in $\tilde{\mathcal{C}}_1$ is identically equal to an estimator $\hat{\beta}$ in $\mathcal{C}_1$ except on the set $\{e = 0\}$. The following result is due to Kariya and Kurata (2002).

Proposition 2.5 *The class $\mathcal{C}_2$ in (2.52) of location-equivariant estimators is equal to $\tilde{\mathcal{C}}_1$, that is,*

$$\tilde{\mathcal{C}}_1 = \mathcal{C}_2. \tag{2.55}$$

In particular, if the distribution P of ε satisfies $P(e = 0) = 0$, then

$$\mathcal{C}_1 = \mathcal{C}_2 \quad a.s. \tag{2.56}$$

and hence a location-equivariant estimator is a GLSE and vice versa.

Here, for two sets A and B, $A = B$ a.s. means that for any $a \in A$, there exists $b \in B$ such that $a = b$ a.s. and conversely.

Proof. $\tilde{\mathcal{C}}_1 \subset \mathcal{C}_2$ follows because for any $\tilde{\beta}(y) = C(e)y + a \, \chi_{\{e=0\}} \in \tilde{\mathcal{C}}_1$,

$$\begin{aligned}
\tilde{\beta}(y) &= C(e)[X(X'X)^{-1}X' + N]y + a \, \chi_{\{e=0\}} \\
&= b(I_n) + [C(e)e + a \, \chi_{\{e=0\}}] \\
&\equiv b(I_n) + d(e) \quad \text{(say)},
\end{aligned} \tag{2.57}$$

which is in $\mathcal{C}_2$. In the first line of (2.57), the identity $X(X'X)^{-1}X' + N = I_n$ is used.

On the other hand, $\tilde{\mathcal{C}}_1 \supset \mathcal{C}_2$ follows because for any $\hat{\beta}(y) = b(I_n) + d(e) \in \mathcal{C}_2$,

$$\begin{aligned}
\hat{\beta}(y) &= (X'X)^{-1}X'y + d(e)[\chi_{\{e=0\}} + \chi_{\{e\neq0\}}] \\
&= [(X'X)^{-1}X' + \chi_{\{e\neq0\}}d(e)(e'e)^{-1}e']y + d(e) \, \chi_{\{e=0\}} \\
&= [(X'X)^{-1}X' + \chi_{\{e\neq0\}}d(e)(e'e)^{-1}e']y + d(0) \, \chi_{\{e=0\}}.
\end{aligned} \tag{2.58}$$

Thus, by letting

$$C(e) = (X'X)^{-1}X' + \chi_{\{e\neq0\}}d(e)(e'e)^{-1}e' \quad \text{and} \quad d(0) = a,$$

the estimator $\hat{\beta}(y)$ is rewritten as $\hat{\beta}(y) = C(e)y + a\chi_{\{e=0\}}$, which is in $\tilde{\mathcal{C}}_1$. Note that the equality $e'y = e'e$ is used in the second line of (2.58).

If $P(e = 0) = 0$, then $\tilde{\mathcal{C}}_1 = \mathcal{C}_1$ a.s. holds, from which (2.56) follows. This completes the proof.

In particular, if P has a probability density function with respect to the Lebesgue measure on R^n, then clearly $P(e = 0) = 0$ holds, implying $C_1 = C_2$ a.s. It is noted that when a regression model (2.1) is considered with the assumption (2.2), a possibility that $P(\varepsilon = 0) > 0$ for some distribution $P \in \mathcal{P}_n(0, \sigma^2 \Sigma)$ is not excluded. Here, note that $P(\varepsilon = 0) \le P(e = 0)$.

Relation between GLSEs and MLEs. Finally, it is sketched that when ε is normally distributed, the MLE of β is always a GLSE of the form $b(\Sigma(\hat{\theta}))$ and that $\hat{\theta} = \hat{\theta}(e)$ is an even function of the residual vector e, that is, $\hat{\theta}(e) = \hat{\theta}(-e)$. The result is due to Magnus (1978) and Kariya and Toyooka (1985). Suppose that the distribution of ε is the normal distribution

$$N_n(0, \sigma^2 \Sigma(\theta)) \quad \text{with} \quad \theta = (\theta_1, \ldots, \theta_d)' \in R^d, \tag{2.59}$$

and that the MLE of $(\beta, \sigma^2, \theta)$ exists. Then the MLE is a solution of the following log-likelihood equation:

$$\frac{\partial}{\partial \beta} L(\beta, \sigma^2, \theta) = 0, \quad \frac{\partial}{\partial \sigma^2} L(\beta, \sigma^2, \theta) = 0, \quad \frac{\partial}{\partial \theta} L(\beta, \sigma^2, \theta) = 0, \tag{2.60}$$

where L is the log-likelihood function:

$$L(\beta, \sigma^2, \theta) = -\frac{n}{2} \log(2\pi) - \frac{n}{2} \log(\sigma^2) - \frac{1}{2} \log(|\Sigma(\theta)|)$$

$$- \frac{1}{2\sigma^2} (y - X\beta)' \Sigma(\theta)^{-1} (y - X\beta). \tag{2.61}$$

The value of $(\beta, \sigma^2, \theta)$, which maximizes $L(\beta, \sigma^2, \theta)$, is given by a solution of the following equation:

$$\beta = b(\Sigma(\theta)), \quad \sigma^2 = \hat{\sigma}^2(\theta), \quad \frac{\partial}{\partial \theta} Q(\theta) = 0, \tag{2.62}$$

where

$$\hat{\sigma}^2(\theta) = \frac{1}{n} [y - Xb(\Sigma(\theta))]' \Sigma(\theta)^{-1} [y - Xb(\Sigma(\theta))]$$

and

$$Q(\theta) = -\frac{n}{2} \log(2\pi) - \frac{n}{2} \log(\hat{\sigma}^2(\theta)) - \frac{1}{2} \log |\Sigma(\theta)| - \frac{n}{2}.$$

Here, the function $\Sigma(\theta)$ is assumed to be differentiable with respect to θ. See Problem 2.3.3.

Note that

$$y - Xb(\Sigma) = [I_n - X(X'\Sigma^{-1}X)^{-1}X'\Sigma^{-1}]y$$

$$= [I_n - X(X'\Sigma^{-1}X)^{-1}X'\Sigma^{-1}][X(X'X)^{-1}X' + N]y$$

$$= [I_n - X(X'\Sigma^{-1}X)^{-1}X'\Sigma^{-1}]e, \tag{2.63}$$

where $\Sigma = \Sigma(\theta)$, and the matrix identity

$$X(X'X)^{-1}X' + N = I_n$$

is used in the second line. This implies that if the equation $\partial Q(\theta)/\partial \theta_j = 0$ ($j = 1, \ldots, d$) has a solution $\hat{\theta}$, it must be a function of e. Thus, the MLE is a GLSE in our sense.

Furthermore, the log-likelihood equation with (2.63) implies that $\hat{\theta}$ is an even function of e, which in turn implies that the MLE is unbiased. In fact, as is shown in the next section, if $\hat{\theta}(e)$ is an even function, then the GLSE $b(\Sigma(\hat{\theta}(e)))$ is unbiased as long as its first moment is finite.

2.4   Finiteness of Moments and Typical GLSEs

In this section, we introduce certain GLSEs in the specific models stated in Section 2.2. We also derive tractable sufficient conditions for the GLSEs to be unbiased and to have a finite second moment. The results derived here serve as a basis for later chapters, when the efficiency of the GLSEs in terms of risk matrix or covariance matrix is considered.

Example 2.5 (Anderson model)   Let us consider the Anderson model (2.11) in Example 2.1, which is restated as

$$y = X\beta + \varepsilon \quad \text{with} \quad \mathcal{L}(\varepsilon) \in \mathcal{P}_n(0, \sigma^2\Sigma), \tag{2.64}$$

where X is an $n \times k$ known matrix of full rank and Σ is of the form

$$\Sigma^{-1} = \Sigma(\theta)^{-1} = I_n + \lambda(\theta)C \quad (\theta \in \Theta).$$

Relevant definitions are given in Example 2.1.

To define a GLSE for this model, assume $P(e = 0) = 0$, where $e = (e_1, \ldots, e_n)'$ is the OLS residual vector:

$$e = Ny \quad \text{with} \quad N = I_n - X(X'X)^{-1}X'. \tag{2.65}$$

Then a typical estimator of θ is given by

$$\hat{\theta} \equiv \hat{\theta}(e) = \frac{\sum_{j=2}^{n} e_j e_{j-1}}{\sum_{j=1}^{n} e_j^2} = \frac{e'Ke}{e'e}, \tag{2.66}$$

where the matrix K is given by

$$K = \frac{1}{2}\begin{pmatrix} 0 & 1 & & & & 0 \\ 1 & 0 & 1 & & & \\ & \ddots & \ddots & \ddots & & \\ & & \ddots & \ddots & \ddots & \\ & & & 1 & 0 & 1 \\ 0 & & & & 1 & 0 \end{pmatrix} : n \times n. \tag{2.67}$$

This naturally leads to the GLSE

$$b(\hat{\Sigma}) = C(e)y \quad \text{with} \quad C(e) = (X'\hat{\Sigma}^{-1}X)^{-1}X'\hat{\Sigma}^{-1}, \tag{2.68}$$

where

$$\hat{\Sigma}^{-1} = \Sigma(\hat{\theta})^{-1} = I_n + \lambda(\hat{\theta})C \quad \text{with} \quad \lambda(\hat{\theta}) = \frac{\hat{\theta}}{(1-\hat{\theta})^2}. \tag{2.69}$$

This GLSE is well defined, that is, the estimator $\hat{\Sigma}$ is almost surely positive definite, since applying the Cauchy–Schwarz inequality

$$(a'b)^2 \le \|a\|^2 \|b\|^2 \quad \text{for any } a, b \in R^n \tag{2.70}$$

to the right-hand side of (2.66) yields

$$\hat{\theta}(e) \in \Theta = (-1, 1) \text{ a.s.}$$

Note that the estimator $\hat{\theta}(e)$ is a continuous function of e and satisfies

(i) $\hat{\theta}(-e) = \hat{\theta}(e)$;

(ii) $\hat{\theta}(ae) = \hat{\theta}(e)$ for any $a > 0$.

This implies that $C(e)$ is a continuous function satisfying

(i') $C(-e) = C(e)$;

(ii') $C(ae) = C(e)$ for any $a > 0$.

As will be seen soon, these conditions imply that the GLSE in (2.68) is an unbiased estimator with a finite second moment.

General theory. Motivated by this example, we derive a condition for which a GLSE is unbiased and has a finite second moment. Consider the general linear regression model

$$y = X\beta + \varepsilon \quad \text{with} \quad P \equiv \mathcal{L}(\varepsilon) \in \mathcal{P}_n(0, \Omega), \tag{2.71}$$

where

$$y : n \times 1, \quad X : n \times k, \quad rank\, X = k,$$

and

$$\Omega = \sigma^2 \Sigma \in \mathcal{S}(n).$$

Here, $\mathcal{L}(\cdot)$ denotes the distribution of $\cdot$ and $\mathcal{S}(n)$ the set of $n \times n$ positive definite matrices.

Proposition 2.6 *Let $\hat{\beta}$ be an estimator of the form $\hat{\beta} = C(e)y$ satisfying $C(e)X = I_k$, where e is the OLS residual vector.*

(1) *Suppose that*

$$\mathcal{L}(\varepsilon) = \mathcal{L}(-\varepsilon)$$

and the function C is an even function in the sense that

$$C(-e) = C(e), \qquad (2.72)$$

then the estimator $\hat{\beta} = C(e)y$ is unbiased as long as its first moment is finite.

(2) *Suppose that*

$$P(e = 0) = 0$$

and that the function C is continuous and is scale-invariant in the sense that

$$C(ae) = C(e) \;\; \text{for any} \;\; a > 0, \qquad (2.73)$$

then the estimator $\hat{\beta} = C(e)y$ has a finite second moment.

Proof. Note that the condition (2.73) is equivalent to the one

$$C(e) = C(e/\|e\|), \qquad (2.74)$$

where the norm $\|x\|$ of a vector x is defined as $\|x\| = \sqrt{x'x}$.

For unbiasedness of $\hat{\beta} \in C_1$, since $\hat{\beta} = \beta + C(e)\varepsilon$, it suffices to show that

$$E[C(e)\varepsilon] = 0. \qquad (2.75)$$

This holds since

$$E[C(e)\varepsilon] = E[C(-e)(-\varepsilon)] = -E[C(e)\varepsilon], \qquad (2.76)$$

where the first equality follows from the assumption that $\mathcal{L}(\varepsilon) = \mathcal{L}(-\varepsilon)$ and the second, from the condition (2.72).

Next, to verify (2), let $C(e) = (c_{ij}(e))$ and $y = (y_1, \ldots, y_n)'$. Then clearly, the (i, j)th element of $E(\hat{\beta}\hat{\beta}')$ is given by

$$E\left[\left(\sum_{u=1}^{n} c_{iu}(e)y_u\right)\left(\sum_{v=1}^{n} c_{jv}(e)y_v\right)\right] \qquad (2.77)$$

By (2.74), the function C can be regarded as a function on the compact set $\{e \in R^n | \;\|e\| = 1\}$. Since any continuous function defined on a compact (i.e., bounded and closed) set is bounded, all the elements $c_{ij}(e)$'s of $C(e)$ are bounded. Let $|c_{ij}(e)| < M$. Then we have

$$E\left[\left(\sum_{u=1}^{n} c_{iu}(e)y_u\right)^2\right] \leq E\left[\left(\sum_{u=1}^{n} c_{iu}^2(e)\right)\left(\sum_{u=1}^{n} y_u^2\right)\right]$$

$$\leq nM^2 E(y'y)$$

$$= nM^2[\sigma^2 tr(\Sigma) + \beta'X'X\beta] \ (< \infty), \qquad (2.78)$$

where the first inequality follows from Cauchy–Schwarz inequality in (2.70). Thus, the (i, i)th element of $E(\hat{\beta}\hat{\beta}')$ is finite.

As for the off-diagonal elements, combining the result above with the following version of the Cauchy–Schwarz inequality:

$$[E(AB)]^2 \leq E(A^2)E(B^2), \qquad (2.79)$$

where A and B are arbitrary random variables with finite second moments, shows that the (i, j)th element with $i \neq j$ is also finite. This completes the proof.

The result (1) of Proposition 2.6 is due to Kariya and Toyooka (1985) and Eaton (1985), and (2) is given by Kariya and Toyooka (1985) and Toyooka and Kariya (1995). See Andrews (1986) in which a more general result of (1) is given.

Further examples. We introduce several typical GLSEs with proof of the finiteness of the moments and the unbiasedness. To apply Proposition 2.6, we note that a GLSE $b(\Psi)$ is scale-invariant in the sense that

$$B(a\Psi) = B(\Psi) \quad \text{for any } a > 0, \qquad (2.80)$$

where

$$B(\Psi) = (X'\Psi^{-1}X)^{-1}X'\Psi^{-1}.$$

This scale-invariance property will be frequently used to establish the condition (2.73).

Example 2.6 (Heteroscedastic model) Let us consider the p-equation heteroscedastic model, which is a model (2.71) with the following structure:

$$y = \begin{pmatrix} y_1 \\ \vdots \\ y_p \end{pmatrix} : n \times 1, \quad X = \begin{pmatrix} X_1 \\ \vdots \\ X_p \end{pmatrix} : n \times k,$$

$$\varepsilon = \begin{pmatrix} \varepsilon_1 \\ \vdots \\ \varepsilon_p \end{pmatrix} : n \times 1 \qquad (2.81)$$

and

$$\Omega = \Omega(\theta) = \begin{pmatrix} \theta_1 I_{n_1} & & 0 \\ & \ddots & \\ 0 & & \theta_p I_{n_p} \end{pmatrix} \in \mathcal{S}(n), \tag{2.82}$$

where

$$n = \sum_{j=1}^{p} n_j, \quad y_j : n_j \times 1, \quad X_j : n_j \times k, \quad \varepsilon_j : n_j \times 1.$$

To define a GLSE, suppose $P(e = 0) = 0$, where e is the OLS residual vector:

$$e = Ny \quad \text{with} \quad N = I_n - X(X'X)^{-1}X'. \tag{2.83}$$

A typical GLSE treated here is of the form

$$b(\hat{\Omega}) = C(e)y \quad \text{with} \quad C(e) = (X'\hat{\Omega}^{-1}X)^{-1}X'\hat{\Omega}^{-1}, \tag{2.84}$$

where

$$\hat{\Omega} \equiv \Omega(\hat{\theta}) = \begin{pmatrix} \hat{\theta}_1 I_{n_1} & & 0 \\ & \ddots & \\ 0 & & \hat{\theta}_p I_{n_p} \end{pmatrix}, \tag{2.85}$$

and $\hat{\theta}$ is an estimator of θ of the form

$$\hat{\theta} \equiv \hat{\theta}(e) = (\hat{\theta}_1, \ldots, \hat{\theta}_p)' = (\hat{\theta}_1(e), \ldots, \hat{\theta}_p(e))' : p \times 1, \tag{2.86}$$

where $\hat{\theta}_j$'s are assumed to be almost surely positive.

We shall introduce two specific GLSEs in accordance with different estimation procedures of variances θ_j's. Recall that the present model can be viewed as a collection of the homoscedastic models

$$y = X_j\beta_j + \varepsilon_j, \quad E(\varepsilon_j) = 0 \quad \text{and} \quad \text{Var}(\varepsilon_j) = \theta_j I_{n_j}$$

with the following restriction

$$\beta_1 = \cdots = \beta_p \equiv \beta. \tag{2.87}$$

Then the variances θ_j's can be estimated with or without the restriction (2.87). With the restriction, the coefficient vector β is common to all the submodels and hence θ_j's are estimated by using the OLSE $b(I_n)$ applied to the full model. The estimators $\hat{\theta}_j$'s thus obtained are called the *restricted estimators*. Without the restriction (2.87), the variances θ_j's are estimated by the residual of each submodel.

Such estimators are called the *unrestricted estimators*. More specifically, let the OLS residual vector e in (2.83) be decomposed as

$$e = \begin{pmatrix} e_1 \\ \vdots \\ e_p \end{pmatrix} : n \times 1 \quad \text{with} \quad e_j : n_j \times 1. \tag{2.88}$$

The *restricted GLSE* is a GLSE $b(\Omega(\hat{\theta}))$ with $\hat{\theta} = \hat{\theta}^R$ for $\hat{\theta}$ in (2.86), where

$$\hat{\theta}^R = \hat{\theta}^R(e) = (\hat{\theta}_1^R, \ldots, \hat{\theta}_p^R)' \tag{2.89}$$

with

$$\hat{\theta}_j^R = \hat{\theta}_j^R(e) = e_j' e_j / n_j. \tag{2.90}$$

The restricted GLSE is well defined, since the $\hat{\theta}_j$'s are almost surely positive. Further, it has a finite second moment. To see this, note first that $\hat{\theta}^R(e)$ is a continuous function satisfying

$$\hat{\theta}^R(ae) = a^2 \hat{\theta}^R(e) \quad \text{for any } a > 0. \tag{2.91}$$

Hence, $\tilde{\Omega}(e) \equiv \Omega(\hat{\theta}^R(e))$ is also continuous in e and satisfies

$$\tilde{\Omega}(ae) = a^2 \tilde{\Omega}(e). \tag{2.92}$$

This further implies that the function

$$C(e) = [X'\tilde{\Omega}(e)^{-1}X]^{-1}X'\tilde{\Omega}(e)^{-1}$$

is continuous and satisfies the condition (2.73) of Proposition 2.6:

$$\begin{aligned} C(ae) &= [X'\tilde{\Omega}(ae)^{-1}X]^{-1}X'\tilde{\Omega}(ae)^{-1} \\ &= [X'a^{-2}\tilde{\Omega}(e)^{-1}X]^{-1}X'a^{-2}\tilde{\Omega}(e)^{-1} \\ &= [X'\tilde{\Omega}(e)^{-1}X]^{-1}X'\tilde{\Omega}(e)^{-1} \\ &= C(e), \end{aligned} \tag{2.93}$$

where the third equality follows from the scale-invariance property (2.80).

Furthermore, if $\mathcal{L}(\varepsilon) = \mathcal{L}(-\varepsilon)$, the restricted GLSE is unbiased. To see this, note that $\hat{\theta}^R$ satisfies

$$\hat{\theta}^R(e) = \hat{\theta}^R(-e). \tag{2.94}$$

This clearly implies that $C(e) = C(-e)$ and hence the condition (2.72) of Proposition 2.6 is satisfied.

On the other hand, the *unrestricted GLSE* is a GLSE $b(\Omega(\hat{\theta}))$ with $\hat{\theta} = \hat{\theta}^U$, where

$$\hat{\theta}^U = \hat{\theta}^U(e) = (\hat{\theta}_1^U, \ldots, \hat{\theta}_p^U)' \tag{2.95}$$

with

$$\hat{\theta}_j^U = \hat{\theta}_j^U(e) = e_j' N_j e_j / (n_j - r_j). \tag{2.96}$$

Here, $r_j = rank\, X_j$ and N_j is the orthogonal projection matrix defined as

$$N_j = I_{n_j} - X_j(X_j'X_j)^+ X_j', \tag{2.97}$$

where A^+ denotes the Moore–Penrose generalized inverse of a matrix A. It can be easily seen by the same arguments as in the restricted GLSE that the unrestricted GLSE has a finite second moment and is unbiased when $\mathcal{L}(\varepsilon) = \mathcal{L}(-\varepsilon)$.

The difference between $\hat{\theta}^R$ and $\hat{\theta}^U$ is found through

Proposition 2.7

$$\hat{\theta}_j^R = [y_j - X_j b(I_n)]'[y_j - X_j b(I_n)]/n_j \tag{2.98}$$

and

$$\hat{\theta}_j^U = [y_j - X_j \hat{\beta}_j]'[y_j - X_j \hat{\beta}_j]/(n_j - r_j), \tag{2.99}$$

where $\hat{\beta}_j$ is the OLSE calculated under the submodel $y_j = X_j \beta + \varepsilon_j$:

$$\hat{\beta}_j = (X_j'X_j)^+ X_j' y_j \quad (j = 1, \ldots, p). \tag{2.100}$$

To establish this result, it is convenient to use the following lemma in which the relation between the OLS residual vector $e = (e_1', \ldots, e_p')'$ in (2.83) and the equation-wise residual vectors

$$\hat{\varepsilon}_j = y_j - X_j \hat{\beta}_j = N_j y_j \quad (j = 1, \ldots, p) \tag{2.101}$$

are clarified.

Lemma 2.8 *The vector $e_j : p \times 1$ can be decomposed in terms of $\hat{\beta}_j$'s and $\hat{\varepsilon}_j$'s as*

$$e_j = \hat{\varepsilon}_j - X_j(X'X)^{-1} \sum_{i=1}^p X_i'X_i(\hat{\beta}_i - \hat{\beta}_j) \tag{2.102}$$

Proof. The proof is the calculation: Let $M_j = X_j(X_j'X_j)^+ X_j'$ and so $I_{n_j} = M_j + N_j$. Then we have

$$e_j = y_j - X_j(X'X)^{-1}X'y$$

$$= y_j - X_j(X'X)^{-1}\left(\sum_{i=1}^p X_i'y_i\right)$$

$$= (M_j + N_j)y_j - X_j(X'X)^{-1}\left[\sum_{i=1}^{p} X_i'(M_i + N_i)y_i\right]$$

$$= (X_j\hat{\beta}_j + \hat{\varepsilon}_j) - X_j(X'X)^{-1}\left(\sum_{i=1}^{p} X_i'X_i\hat{\beta}_i\right)$$

$$= \hat{\varepsilon}_j - X_j(X'X)^{-1}\left[\sum_{i=1}^{p} X_i'X_i(\hat{\beta}_i - \hat{\beta}_j)\right],$$

where the last equality is due to

$$X_j\hat{\beta}_j = X_j(X'X)^{-1}\left[\sum_{i=1}^{p} X_i'X_i\hat{\beta}_j\right].$$

This completes the proof of Lemma 2.8.

Proposition 2.7 readily follows since

$$\hat{\varepsilon}_j = N_j e_j.$$

Example 2.7 (SUR model)   Let us consider the p-equation SUR model introduced in Example 2.4, which is described as $y = X\beta + \varepsilon$ with

$$y = \begin{pmatrix} y_1 \\ \vdots \\ y_p \end{pmatrix} : n \times 1, \quad X = \begin{pmatrix} X_1 & & 0 \\ & \ddots & \\ 0 & & X_p \end{pmatrix} : n \times k,$$

$$\beta = \begin{pmatrix} \beta_1 \\ \vdots \\ \beta_p \end{pmatrix} : k \times 1, \quad \varepsilon = \begin{pmatrix} \varepsilon_1 \\ \vdots \\ \varepsilon_p \end{pmatrix} : n \times 1 \tag{2.103}$$

and the covariance structure is given by

$$\Omega = \Sigma \otimes I_m \quad \text{and} \quad \Sigma = (\sigma_{ij}) \in \mathcal{S}(p), \tag{2.104}$$

where

$$y_j : m \times 1, \ X_j : m \times k_j, \ n = pm \ \text{and} \ k = \sum_{j=1}^{p} k_j.$$

As has been seen in Example 2.4, this model is equivalent to the multivariate linear regression model $Y = X_*B + E_*$ as in (2.31) with prior zero restriction (2.32) on B.

To define a GLSE, we assume $P(e = 0) = 0$, where e is the OLS residual vector, that is,

$$e = Ny \quad \text{with} \quad N = I_n - X(X'X)^{-1}X'. \tag{2.105}$$

We consider the GLSE of the form

$$b(\hat{\Sigma} \otimes I_m) = C(e)y \tag{2.106}$$

with

$$C(e) = [X'(\hat{\Sigma}^{-1} \otimes I_m)X]^{-1}X'(\hat{\Sigma}^{-1} \otimes I_m),$$

where $\hat{\Sigma} = \hat{\Sigma}(e)$ is an estimator of Σ depending only on e. The typical GLSEs introduced here are the *restricted Zellner estimator* (RZE) and the *unrestricted Zellner estimator* (UZE)(Zellner, 1962, 1963). To describe these estimators, let the OLS residual vector e be decomposed as

$$e = (e_1', \ldots, e_p')' \quad \text{with} \quad e_j : m \times 1. \tag{2.107}$$

Here, the structure of the regressor matrix X in (2.103) enables us to rewrite e_j simply as

$$e_j = N_j y_j \quad \text{with} \quad N_j = I_m - X_j(X_j'X_j)^{-1}X_j', \tag{2.108}$$

which is the OLS residual obtained from the jth regression model $y_j = X_j\beta_j + \varepsilon_j$.

The RZE is defined as a GLSE $b(\hat{\Sigma} \otimes I_m)$ with

$$\hat{\Sigma} = \hat{\Sigma}(e) = (e_i'e_j/m). \tag{2.109}$$

Note that $e_i'e_j/m$ is an estimator of σ_{ij} based on the OLS residual vector in (2.108). Since $\hat{\Sigma}(e)$ is a continuous function of e such that

$$\hat{\Sigma}(e) = \hat{\Sigma}(-e)$$

and

$$\hat{\Sigma}(ae) = a^2\hat{\Sigma}(e) \text{ for any } a > 0, \tag{2.110}$$

by the same arguments as in Example 2.6, the RZE has a finite second moment and is unbiased when $\mathcal{L}(\varepsilon) = \mathcal{L}(-\varepsilon)$.

On the other hand, the UZE is a GLSE $b(\hat{\Sigma} \otimes I_m)$ with $\hat{\Sigma} = S/q$, where

$$S = S(e) = \left(e_i'N_*e_j\right),$$

$$N_* = I_m - X_*(X_*'X_*)^{+}X_*',$$

$$r = rank X_* \quad \text{and} \quad q = m - r. \tag{2.111}$$

Here, the $m \times k$ matrix X_* is defined in (2.31). Note that $e_i' N_* e_j / q$ is an estimator of σ_{ij} based on the multivariate OLS residual $\hat{E} = N_* Y$ in the model $Y = X_* B + E_*$. In fact, S can be rewritten as

$$S = \hat{E}' \hat{E} = Y' N_* Y = E_*' N_* E_* \qquad (2.112)$$

(see Problem 2.4.5). The term "unrestricted" is used because σ_{ij} is estimated from the multivariate regression model by ignoring the zero restrictions on B in (2.32), while the term "restricted" is used because the estimator $\hat{\Sigma}$ in (2.109) utilizes the restriction. Using the scale-invariance property in (2.80), we see that

$$b\left((S/q) \otimes I_m\right) = b(S \otimes I_m). \qquad (2.113)$$

Throughout this book, we use the right-hand side of (2.113) as the definition of the UZE, since it is simpler than the expression on the left-hand side. The UZE $b(S \otimes I_m)$ also has a finite second moment and is unbiased when $\mathcal{L}(\varepsilon) = \mathcal{L}(-\varepsilon)$.

Note also that when the error term ε is normally distributed: $\mathcal{L}(\varepsilon) = N_n(0, \Sigma \otimes I_m)$, the matrix S has the Wishart distribution $W_p(\Sigma, q)$ with mean $q\Sigma$ and degrees of freedom q:

$$\mathcal{L}(S) = W_p(\Sigma, q). \qquad (2.114)$$

2.5 Empirical Example: CO_2 Emission Data

This section demonstrates with real data an example of GLSEs in an AR(1) error model and Anderson model. The analysis here is in line with Nawata (2001) where a causal relation between GNP (gross national product) and CO_2 (carbon dioxide) emission is discussed from various points of view. We use the same data set given in his book. For practical techniques required in regression analysis, consult, for example, Sen and Srivastava (1990).

CO_2 emission and GNP. Table 2.1 gives the data on the volume of CO_2 emission in million ton, and the GNP in trillion yen (deflated by 1990 price) in Japan from 1970 to 1996. Of course, variations in the GNP are here regarded as a causal variable of those in the CO_2 emission volume.

We specify the relation as

$$CO_2 = \alpha_1 \, (GNP)^{\alpha_2}. \qquad (2.115)$$

In this relation, the quantity α_2 can be understood as being the elasticity of CO_2 relative to GNP as in Figure 2.1. In general, when a relation between two variables x and y is expressed as $y = f(x)$, the elasticity of y relative to x is defined by

$$\begin{aligned}
\text{elasticity} &= \lim_{\Delta \to 0} \left\{ \frac{f(x + \Delta) - f(x)}{f(x)} \right\} \bigg/ \left\{ \frac{(x + \Delta) - x}{x} \right\} \\
&= \frac{f'(x)x}{f(x)}.
\end{aligned} \qquad (2.116)$$

Table 2.1 CO_2 emission data of Japan.

Year	GNP	Volume of CO_2 Emission	Year	GNP	Volume of CO_2 Emission
1970	187	739	1984	329	934
1971	196	765	1985	344	909
1972	213	822	1986	354	907
1973	230	916	1987	370	897
1974	227	892	1988	393	984
1975	234	854	1989	412	1013
1976	244	875	1990	433	1071
1977	254	919	1991	449	1093
1978	268	916	1992	455	1105
1979	283	951	1993	456	1080
1980	290	920	1994	459	1131
1981	299	903	1995	466	1137
1982	309	879	1996	485	1168
1983	316	866			
	Trillion yen	Million ton		Trillion yen	Million ton

Source: Nawata (2001) with permission.

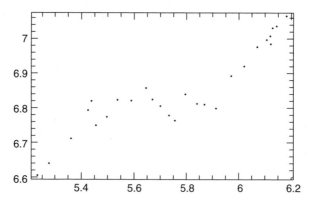

Figure 2.1   Scatter plot of (GNP, CO_2) for Japanese data (in log-scale).

Therefore, the elasticity measures percentage change of y for 1% change of x. In (2.115), this is equal to α_2 as $f(x) = \alpha_1 x^{\alpha_2}$. The model in (2.115) is a model with constant elasticity. The equation (2.115) is clearly equivalent to

$$\log(CO_2) = \beta_1 + \beta_2 \log(GNP), \tag{2.117}$$

where $\beta_1 = \log(\alpha_1)$ and $\beta_2 = \alpha_2$. Hence, this model is expressed as a linear regression model:

$$y = X\beta + \varepsilon \tag{2.118}$$

with $X : n \times k$, $n = 27$, $k = 2$,

$$y = \begin{pmatrix} \log(CO2_1) \\ \vdots \\ \log(CO2_{27}) \end{pmatrix} = \begin{pmatrix} \log(739) \\ \vdots \\ \log(1168) \end{pmatrix} : 27 \times 1,$$

$$X = \begin{pmatrix} 1 & \log(GNP_1) \\ \vdots & \vdots \\ 1 & \log(GNP_{27}) \end{pmatrix} = \begin{pmatrix} 1 & \log(187) \\ \vdots & \vdots \\ 1 & \log(485) \end{pmatrix} : 27 \times 2,$$

$$\beta = \begin{pmatrix} \beta_1 \\ \beta_2 \end{pmatrix} : 2 \times 1, \quad \varepsilon = \begin{pmatrix} \varepsilon_1 \\ \vdots \\ \varepsilon_{27} \end{pmatrix} : 27 \times 1,$$

where $CO2_j$ and GNP_j denote the values of CO_2 and GNP at year $1969 + j$ ($j = 1, \ldots, 27$) respectively. (We do not use the notation $CO2_j$ for simplicity.)

The model is estimated by the OLSE

$$b(I_{27}) = (X'X)^{-1}X'y$$

as

$$\log(CO_2) = 4.754 + 0.364 \, \log(GNP), \quad R^2 = 0.8142, \quad s = 0.0521.$$

Thus, the elasticity β_2 in question is estimated by 0.364. Here, the quantity R^2 is a measure for goodness of fit defined by

$$R^2 = 1 - \frac{e'e}{y'N_0 y} \quad \text{with} \quad N_0 = I_{27} - 1_{27}(1'_{27}1_{27})^{-1}1'_{27},$$

and s denotes the standard error of regression:

$$s = \sqrt{\frac{1}{27-2}e'e},$$

where e is the OLS residual vector

$$e = \begin{pmatrix} e_1 \\ \vdots \\ e_{27} \end{pmatrix} = [I_{27} - X(X'X)^{-1}X']y.$$

The statistics s and s^2 are measures of variation of the error term ε. In fact, s^2 is the uniformly minimum variance unbiased estimator of σ^2, if the error terms ε_j's are independently and identically distributed as the normal distribution $N(0, \sigma^2)$, that is,

$$\mathcal{L}(\varepsilon) = N_{27}(0, \sigma^2 I_{27}). \tag{2.119}$$

GLSE in an AR(1) error model. Although computing the OLSE is a basic procedure to take in the analysis of data, the OLSE may not be efficient, as will be discussed in Chapters 3, 4 and 5 of this book. As a diagnostic checking for the assumption of error terms, it is often the case that the Durbin–Watson test statistic is computed for time-series data. The Durbin–Watson test statistic DW is in our case calculated as

$$DW = \frac{\sum_{j=2}^{27}(e_j - e_{j-1})^2}{\sum_{j=1}^{27} e_j^2} = 0.5080,$$

which suggests the presence of positive serial correlation of AR(1) type among the error terms ε_j's. Hence, we specify that ε_j's follow an AR(1) process

$$\varepsilon_j = \theta \varepsilon_{j-1} + \xi_j \quad \text{with} \quad |\theta| < 1, \tag{2.120}$$

where ξ_j's are supposed to be independently and identically distributed as $N(0, \tau^2)$. In this specification, the distribution of the error term ε is expressed as $\mathcal{L}(\varepsilon) = N_{27}(0, \sigma_*^2 \Sigma_*(\theta))$, where

$$\sigma_*^2 = \frac{\tau^2}{(1-\theta)^2},$$

$$\Sigma_*(\theta)^{-1} = I_{27} + \lambda(\theta)C + \psi(\theta)B \tag{2.121}$$

with

$$\lambda(\theta) = \frac{\theta}{(1-\theta)^2} \quad \text{and} \quad \psi(\theta) = \frac{\theta}{1-\theta}.$$

See Example 2.1 for the definitions of the matrices C and B. A typical estimator of θ is given by

$$\hat{\theta} = \frac{\sum_{j=2}^{27} e_j e_{j-1}}{\sum_{j=1}^{27} e_j^2}, \tag{2.122}$$

which is calculated as $\hat{\theta} = 0.7001$. This value also suggests an application of the GLSE for the estimation of β, and the GLSE

$$b(\Sigma_*(\hat{\theta})) = (X'\Sigma_*(\hat{\theta})^{-1}X)^{-1}X'\Sigma_*(\hat{\theta})^{-1}y \tag{2.123}$$

is $b(\Sigma_*(\hat{\theta})) = (4.398, 0.425)'$. Hence, the model is estimated as

$$\log(CO_2) = 4.398 + 0.425 \ \log(GNP).$$

The estimate suggests that 1% increase in GNP causes 0.425% increase in CO_2 emission. The quantities $\lambda(\theta)$ and $\psi(\theta)$ are also evaluated by

$$\lambda(\hat{\theta}) = 7.7814 \quad \text{and} \quad \psi(\hat{\theta}) = 2.3340.$$

GLSE in Anderson model. On the other hand, the Anderson model, which is the model with $\psi(\theta)$ in (2.121) replaced by zero, may be a possible model for this data. Hence, assume that $\mathcal{L}(\varepsilon) = N_{27}(0, \sigma^2 \Sigma(\theta))$, where

$$\sigma^2 = \frac{\tau^2}{(1-\theta)^2},$$

$$\Sigma(\theta)^{-1} = I_{27} + \lambda(\theta)C. \tag{2.124}$$

Under this model, the GLSE $b(\Sigma(\hat{\theta}))$ is calculated as

$$b(\Sigma(\hat{\theta})) = (X'\Sigma(\hat{\theta})^{-1}X)^{-1}X'\Sigma(\hat{\theta})^{-1}y$$

$$= (4.521, 0.404)', \tag{2.125}$$

from which we obtain

$$\log(CO_2) = 4.521 + 0.404 \, \log(GNP). \tag{2.126}$$

Thus, the elasticity of CO_2 relative to GNP in this model is evaluated by 0.404, which is slightly lower than the case of AR(1) error model. An interesting result is found here. In the Anderson model, in Figure 2.2, the GLSE $b(\Sigma(\hat{\theta}))$ and the OLSE $b(I_{27})$ satisfy

$$\frac{1}{27}1'_{27}Xb(\Sigma(\hat{\theta})) = 6.8495 \tag{2.127}$$

and

$$\frac{1}{27}1'_{27}Xb(I_{27}) = 6.8495, \tag{2.128}$$

where the right-hand sides of the two equations are equal to

$$\bar{y} = \frac{1}{27}\sum_{j=1}^{27}y_j = 6.8495.$$

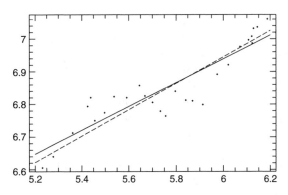

Figure 2.2   Two regression lines: the line obtained from the OLSE and the one from the GLSE in the Anderson model (in dotted line).

The results in (2.127) and (2.128) show that the distribution of the difference between $b(\Sigma(\hat{\theta}))$ and $b(I_{27})$ is degenerate. This fact will be fully investigated in Section 8.4 of Chapter 8.

Multiple linear regression models. As is well known, a serial correlation problem is often solved by adding some appropriate explanatory variables. In Nawata (2001), various models including polynomial regression models such as

$$\log(CO_2) = \beta_1 + \beta_2 \log(GNP) + \beta_3 [\log(GNP)]^2 + \beta_4 [\log(GNP)]^3,$$

and the models with dummy variables such as

$$\log(CO_2) = \beta_1 + \beta_2 D + \beta_3 \log(GNP) + \beta_4 D \log(GNP),$$

are estimated and compared in terms of AIC (Akaike Information Criterion). Here, $D = (d_1, \ldots, d_{27})'$ is a dummy variable defined by

$$d_j = \begin{cases} 0 & (j = 1, \ldots, 11) \\ 1 & (j = 12, \ldots, 27), \end{cases}$$

which detects whether the intercept and/or the slope of regression line has changed after the year 1980. The models mentioned above will be treated again in Section 4.6 in Chapter 4 in the context of estimation of the SUR model.

Table 2.2 CO_2 emission data of USA.

Year	GNP	Volume of CO_2 Emission	Year	GNP	Volume of CO_2 Emission
1970	3494	4221	1984	5050	4356
1971	3597	4247	1985	5201	4426
1972	3782	4440	1986	5342	4511
1973	3987	4614	1987	5481	4631
1974	3976	4451	1988	5691	4895
1975	3951	4267	1989	5878	4921
1976	4148	4559	1990	5954	4824
1977	4328	4575	1991	5889	4799
1978	4541	4654	1992	6050	4856
1979	4670	4671	1993	6204	5023
1980	4642	4575	1994	6423	5137
1981	4712	4414	1995	6585	5162
1982	4611	4196	1996	6817	5301
1983	4767	4211			
	Billion dollars	Million ton		Billion dollars	Million ton

Source: Nawata (2001) with permission.

In Table 2.2, the data on the amount of CO_2 emission and the GNP in the USA is also shown. Readers may try to analyze the data by using the GLS estimation procedure (see Problem 2.5.1).

2.6 Empirical Example: Bond Price Data

In this section, as an example of empirical analysis with GLSEs, we treat a linear regression model $y = X\beta + \varepsilon$ that models the cross-sectional relation between the prices $y_1, \ldots, y_n$ of n bonds observed in the market. The covariance matrix $\mathrm{Cov}(\varepsilon) = \sigma^2 \Sigma(\theta)$ of the error term ε is supposed to be a function of the unknown one-dimensional parameter $\theta \in (-1, 1)$. The matrix $\Sigma(\theta)$ models a heteroscedatic and serially correlated structure that is caused by different timepoints of cash flows and different maturity of bonds. The model is estimated by a GLSE $b(\Sigma(\hat{\theta}))$, which is obtained as a solution of minimization of the function $\psi(\beta, \theta) = (y - X\beta)'\Sigma(\theta)^{-1}(y - X\beta)$.

Covariance structure with one-dimensional parameter. We begin by providing a rough sketch of the model considered in this section. A complete description of the model will be given later.

The model considered here is, in short, a general linear regression model $y = X\beta + \varepsilon$ with covariance structure

$$\mathrm{Cov}(\varepsilon) = \sigma^2 \Sigma(\theta),$$

where $\Sigma(\theta) = (\sigma_{ij}(\theta))$ and

$$\sigma_{ij}(\theta) = \begin{cases} a_{ii} f_{ii} & (i = j) \\ \theta a_{ij} f_{ij} & (i \neq j). \end{cases}$$

Here, θ is an unknown one-dimensional parameter of interest such that $-1 < \theta < 1$, and a_{ij}'s and f_{ij}'s are the quantities that can be calculated from the attributes or characteristics of the n bonds in question. As will be explained below, the definition of the quantities a_{ij}'s and f_{ij}'s is based on the nature of heteroscedasticity and correlation shared by the prices of n bonds that are simultaneously and stochastically determined in the bonds market.

Although θ is one-dimensional, we cannot diagonalize the matrix Σ since an orthogonal matrix diagonalizing Σ in general depends on θ. Furthermore, we cannot form an estimator $\hat{\theta}$ of θ through an intuitive or constructive approach as in the specific models treated in the previous sections. Hence, in estimating the model, we need to obtain a GLSE $b(\Sigma(\hat{\theta}))$ by minimizing

$$\psi(\beta, \theta) = (y - X\beta)'\Sigma(\theta)^{-1}(y - X\beta)$$

with respect to β and θ, or equivalently, by minimizing

$$\tilde{\psi}(\theta) \equiv \psi(b(\Sigma(\theta)), \theta) = (y - Xb(\Sigma(\theta)))'\Sigma(\theta)^{-1}(y - Xb(\Sigma(\theta)))$$

with respect to θ. See Problem 2.2.1.

Basic nature of bonds and discount function. To fix the idea, let P be the present price (at time 0) of a bond that generates the cash flow $C(s)$ at a (future) timepoint s. $C(\cdot)$ is called *a cash flow function as a function of s*. Suppose there exist M timepoints $(0<)s_1 < \cdots < s_M$ at which the cash flows are generated. Thus, the cash flows are given by $C(s_1), \ldots, C(s_M)$. The last timepoint s_M is called *maturity*.

However, it does not hold that the present value of the bond is equal to

$$P = \sum_{j=1}^{M} C(s_j),$$

since $C(s_j)$s are generated in future timepoints. Instead, each cash flow $C(s_j)$ should be discounted according to the time interval $|s_j - 0| = s_j$. In other words, the present value (at 0) of the cash flow occurring at s_j is discounted by a discount factor $D(s_j)$ as $C(s_j)D(s_j)$. Hence, the value of a bond with cash flow $\{C(s_j) \mid j = 1, \ldots, M\}$ is equal to

$$P = \sum_{j=1}^{M} C(s_j)D(s_j),$$

where $D(\cdot)$ is called *a discount function as a function of a timepoint* $\cdots$.

Usually, the discount function $D(\cdot)$ is regarded as an unobservable random quantity that depends on characteristics (attributes), say Z, of the bond as well as the current state of the market. It is noted that the random variables $\{D(s_j) \mid j = 1, \ldots, M\}$ given by $D(\cdot)$ are correlated in general. Hence, to obtain a model for P, such conditions should be taken into the structure of the model.

Furthermore, there exist many bonds in a market and these prices are simultaneously and stochastically determined in the market. This implies that possible correlation among the prices of bonds should also be embedded into the structure of Σ.

Background of the model. Kariya (1993) introduced a bond pricing model for individual coupon bonds with different cash flows and different attributes. A main feature of the model is that a random realization of each individual price is viewed as equivalent to a random realization of the random discount function discounting cash flows from each individual bond, where individual attributes are included in the modeling. This feature differentiates this model from a traditional model such as in McCulloch (1971, 1975) where the discount function for cash flows is common to all the bonds.

To describe the model, suppose that there are n bonds to be modeled at t, so that these n bond prices are observed at t. Without loss of generality, let $t = 0$. The ith bond is then characterized as $(C_i(\cdot), Z_i, P_i)$, where $C_i(\cdot)$ is the cash flow function known at 0, Z_i is the set of attributes and P_i is its price at 0. Let

$$0 < s(i)_1 < s(i)_2 < \cdots < s(i)_{M(i)} \tag{2.129}$$

be the timepoints in future at which the ith bond generates the cash flows (coupons and principal). In other words, the cash flow function $C_i(s)$ is zero except for these points $s = s(i)_j$ $(j = 1, \ldots, M(i))$ and the values of the cash flows are

$$\{C_i(s(i)_j) \mid j = 1, \ldots, M(i)\}. \tag{2.130}$$

If the ith bond is a typical bond such as a government bond, $C(s(i)_{M(i)}) = 100$ (principal). Considering bonds with no default such as government bonds, we take

$$Z_{i1} = \text{coupon rate of } i\text{th bond},$$

$$Z_{i2} = \text{term to maturity } s(i)_{M(i)} \text{ of } i\text{th bond} \tag{2.131}$$

as attributes for each bond, though attributes such as default risk, and so on, can be included.

All the prices of n bonds are realized stochastically and simultaneously in the market with the attributes being taken into account. The prices thus formed in the market are the discounted present values of the future cash flows that are stochastically discounted. Thus, let

$$D_i(s): \ 0 \le s \le s(i)_{M(i)} \tag{2.132}$$

represent the stochastic discount function of the ith bond so that the price is identified with

$$P_i = \sum_{j=1}^{M(i)} C_i(s(i)_j) D(s(i)_j). \tag{2.133}$$

In this expression, the $M(i)$ discount factors $\{D(s(i)_j)\}$ are unobservable random variables, while the cash flows $\{C_i(s(i)_j)\}$ are known at 0.

In specifying $D_i(s)$, we need to consider

(1) heteroscedastic property of the prices $\{P_i \mid i = 1, \ldots, n\}$ that the shorter the maturity of a bond is, the smaller the variance tends to be;

(2) separation of market variations common to all the bond prices from individual variations specific to each individual bond;

(3) correlations among discounts $\{D(s(i)_j)\}$'s at different timepoints with each bond and correlations among different bond prices;

(4) parsimonious parameterization.

Bond pricing model. To give a model pricing, with all the bonds with these points taken into account, let

$$0 < s_{a1} < s_{a2} < \cdots < s_{aM} \tag{2.134}$$

with

$$M = \max\{M(i)|i = 1, \ldots, n\} \quad \text{and} \quad s_{aM} = \max\{s_{M(i)}|i = 1, \ldots, n\}$$

denote all the combined timepoints at which cash flows are generated at least by one of the n bonds. In this notation, the ith bond price is expressed as

$$P_i = \sum_{j=1}^{M} C_i(s_{aj})D(s_{aj}), \tag{2.135}$$

where $C_i(s_{aj}) = 0$ unless s_{aj} is a timepoint for one of the cash flows of the ith bond. For the stochastic discount function, the mean function is assumed to be specified as a polynomial with coefficients depending on the bond attributes:

$$\mu_i(s) \equiv E[D_i(s)] = 1 + \delta_1(z_i)s + \cdots + \delta_p s^p, \tag{2.136}$$

where

$$\delta_j(z_i) = \delta_{j1}z_{i1} + \cdots + \delta_{jq}z_{iq}. \tag{2.137}$$

Here, note that δ_j's are common parameters to all the bonds. The random deviations of the discount function from the mean

$$v_i(s) = D_i(s) - \mu_i(s), \quad s \in \{s_{aj}|j = 1, \ldots, M\} \tag{2.138}$$

are in general correlated not only within those of the ith bonds but also among those of different bonds. Considering the above four points, Kariya (1993) specified them as

$$\text{Cov}(v_i(s_{aj}), \ v_k(s_{ar})) = \lambda_{ik}\phi_{ik.jr} \tag{2.139}$$

with

$$\phi_{ik.jr} = \exp(-|s_{aj} - s_{ar}|),$$
$$\lambda_{ik} = \begin{cases} a_{ii} & (i = k) \\ \theta a_{ik} & (i \neq k), \end{cases} \tag{2.140}$$

where

$$a_{ik} = \exp(-|s_{M(i)} - s_{M(k)}|). \tag{2.141}$$

Using these specifications for the random discounts, the whole model for n bond prices is expressed as a general linear regression model

$$y = X\beta + \varepsilon \quad \text{with} \quad \mathcal{L}(\varepsilon) \in \mathcal{P}_n(0, \sigma^2\Sigma), \tag{2.142}$$

where

$$y = \begin{pmatrix} y_1 \\ \vdots \\ y_n \end{pmatrix} : n \times 1 \text{ with } y_i = P_i - \sum_{j=1}^{M} C_i(s_{aj}),$$

$$\beta = \begin{pmatrix} \delta_1 \\ \vdots \\ \delta_p \end{pmatrix} : pq \times 1 \text{ with } \delta_i = \begin{pmatrix} \delta_{i1} \\ \vdots \\ \delta_{iq} \end{pmatrix} : q \times 1,$$

and the matrix X of explanatory variables are defined by

$$X = \begin{pmatrix} X_1' \\ \vdots \\ X_n' \end{pmatrix} : n \times pq$$

with

$$X_i = \begin{pmatrix} u_{i1} \\ \vdots \\ u_{ip} \end{pmatrix} : pq \times 1 \text{ and } u_{ir} = \begin{pmatrix} u_{i1r} \\ \vdots \\ u_{iqr} \end{pmatrix} : q \times 1.$$

The quantity u_{ikr} is defined by

$$u_{ikr} = \sum_{j=1}^{M} z_{iks}{}^r_{aj} C_i(s_{aj}).$$

As for the error term ε, let

$$\varepsilon = \begin{pmatrix} \varepsilon_1 \\ \vdots \\ \varepsilon_n \end{pmatrix} : n \times 1$$

and let

$$\Sigma = (\sigma_{ik}) : n \times n \text{ with } \sigma_{ik} = \begin{cases} a_{ii} f_{ii} & (i = k) \\ \theta a_{ik} f_{ik} & (i \neq k), \end{cases}$$

where

$$f_{ik} = C_i' \Phi_{ik} C_k.$$

Here, Φ_{ik} is an $M \times M$ matrix with the (j, r)th element being $\phi_{ik.jr}$:

$$\Phi_{ik} = (\phi_{ik.jr}) : M \times M$$

and the vector C_i is defined by

$$C_i = \begin{pmatrix} C_i(s_{a1}) \\ \vdots \\ C_i(s_{aM}) \end{pmatrix} : M \times 1.$$

In other words, in the pricing model in (2.142), the covariance structure is expressed as

$$\Sigma = \Sigma(\theta) \quad \text{with} \quad -1 < \theta < 1,$$

where θ is an unknown parameter in Σ. Hence, the GLSE

$$b(\Sigma(\hat{\theta})) = (X'\Sigma(\hat{\theta})^{-1}X)^{-1}X'\Sigma(\hat{\theta})^{-1}y \tag{2.143}$$

is obtained as the estimator minimizing

$$\psi(\beta, \theta) = (y - X\beta)'\Sigma(\theta)^{-1}(y - X\beta) \tag{2.144}$$

with respect to (β, θ). In this specification, the points (1) through (4) made before have been taken care of as follows. First of all, the covariances of v_i's in (2.139) are those of $D_i(s_{aj})$'s and the covariance structure determines that of the bond prices P_i's through (2.135), which is the covariance structure of ε_i's in (2.142). The covariance matrix in (2.142) is heteroscedastic as well as correlated, because the variances and covariances of y_i's depend not only on those of $D_i(s_{aj})$'s but also on the cash flows $C_i(s_{aj})$'s. Point (1) is reflected in the specification of (2.141), which is an expression of the fact that the shorter the maturity of a bond is, the smaller the variance tends to be, because a bond with shorter maturity generates less cash flow, leading to a smaller variance. The second point (2) is taken care of by the mean structure that the parameters common to all the bonds are separated from the attributes of individual bonds. Point (3) is clear from the specification of the covariances of $D_i(s_{aj})$ and $D_k(s_{ar})$ in (2.139). Finally, a parameterization was carried out in the specification of the model because there are only five parameters for more than 100 bond prices.

Empirical result. In Kariya and Tsuda (1994), the Japanese Government bonds are priced by the above model, which they call the *cross-sectional model* (CSM). The model is cross-sectionally fitted with data of monthly individual bond prices (at the end of each month).

As attributes, coupon rate and term to maturity are taken for z_{i1} and z_{i2}. The sample size of each model varies from 33 to 85. The model seems to perform very well. In fact, almost all the residual standard deviations of the 120 CSMs over the period from 1983.1 to 1992.12 are less than 1 yen where the face value of a Japanese Government bond is 100 yen and the initial life of each bond is 10 years. In addition, in each model, the residuals of the individual bond prices are less than

Table 2.3 Standard deviations for the CSM (yen).

Date	Number of Bonds	θ	δ_{11} (t-value)	δ_{21} (t-value)	δ_{12} (t-value)	δ_{22} (t-value)	Standard Deviation
84.12	49	0.30	−0.007073 (−3.086)	0.001516 (2.563)	−0.012829 (−2.162)	0.000333 (0.696)	0.668
85.01	50	0.44	−0.006452 (−2.041)	0.001326 (1.868)	−0.013181 (−1.640)	0.000472 (0.733)	0.854
89.12	70	0.33	−0.007219 (−6.414)	0.001060 (3.652)	−0.008762 (−4.944)	0.000344 (1.995)	0.338
90.01	70	0.10	−0.007564 (−9.102)	0.001236 (4.793)	−0.011002 (−7.721)	0.000469 (3.783)	0.312

1 yen except for a few prices. In the analysis, the polynomial of the mean discount function is chosen as

$$\overline{D}_i(s) = 1 + (\delta_{11}z_{i1} + \delta_{12}z_{i2})s + (\delta_{21}z_{i1} + \delta_{22}z_{i2})s^2. \tag{2.145}$$

Table 2.3 gives the standard deviations of the CSMs and the five parameters involved in each CSM. All the standard deviations of the models are smaller than 1 yen and hence, less than 1% relative to the face value 100. Hence, the model shows accuracy and will be useful. In fact, it has been used by the Nissay Life Co., where the model is extended to a dynamic model.

To check the validity of the CSM as a pricing model for individual bonds, we draw in Figure 2.3 the two graphs of the realized individual prices and their estimated

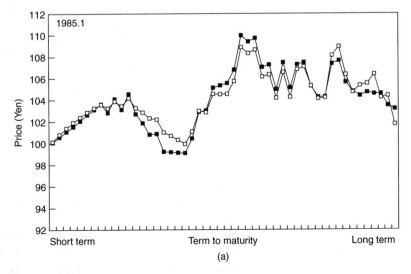

Figure 2.3 Realized individual prices and their estimated values in the 85.1 and 90.1 models.

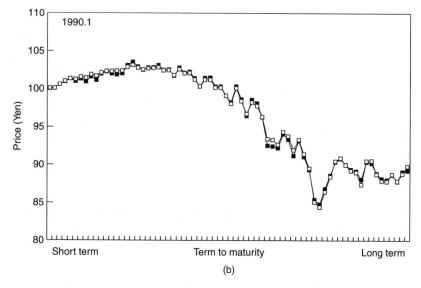

Figure 2.3 *continued*

Table 2.4   Price of bonds, coupon rate and term to maturity.

Data Number	Coupon Rate	Term to Maturity	Price of Bond	Data Number	coupon Rate	Term to Maturity	Price of Bond
1	7.7	0.05	100.05	36	6.0	5.88	98.20
2	8.0	0.05	100.06	37	6.5	5.88	100.34
3	8.7	0.30	100.59	38	6.1	5.97	98.61
4	8.5	0.55	100.95	39	5.7	5.97	96.43
5	8.5	0.80	101.37	40	6.1	6.39	98.55
6	8.0	0.80	100.99	41	6.0	6.39	98.09
7	8.0	1.05	101.33	42	5.7	6.39	96.26
8	7.6	1.05	100.93	43	5.1	6.39	92.51
9	8.0	1.30	101.61	44	5.1	6.47	92.44
10	7.6	1.30	101.13	45	5.1	6.89	92.20
11	8.0	1.55	101.99	46	5.4	6.89	94.05
12	8.0	1.80	102.28	47	5.3	6.89	93.35
13	7.7	2.05	102.02	48	5.0	6.97	91.16
14	7.5	2.39	101.89	49	5.3	7.38	93.17
15	7.5	2.47	102.03	50	5.0	7.38	91.06
16	8.0	2.47	103.10	51	4.7	7.38	89.34
17	8.0	2.89	103.54	52	4.0	7.38	85.35
18	7.7	2.89	102.94	53	3.9	7.38	84.78
19	7.5	2.97	102.51	54	4.3	7.64	86.83

Table 2.4 *continued*

Data Number	Coupon Rate	Term to Maturity	Price of Bond	Data Number	coupon Rate	Term to Maturity	Price of Bond
20	7.5	3.38	102.80	55	4.6	7.64	88.58
21	7.5	3.47	102.85	56	4.9	7.64	90.35
22	7.5	3.88	103.13	57	5.0	7.88	90.78
23	7.3	3.88	102.51	58	4.9	8.13	89.98
24	7.3	3.97	102.55	59	4.8	8.13	89.36
25	7.0	4.38	101.74	60	4.8	8.38	89.17
26	7.3	4.47	102.80	61	4.6	8.38	88.05
27	7.1	4.47	102.11	62	5.0	8.64	90.35
28	7.1	4.88	102.26	63	5.0	8.88	90.17
29	6.8	4.88	101.41	64	4.8	8.88	88.84
30	6.5	4.97	100.30	65	4.7	8.88	88.17
31	6.8	4.97	101.43	66	4.7	9.13	87.99
32	6.8	5.38	101.52	67	4.8	9.13	88.67
33	6.5	5.38	100.32	68	4.7	9.38	87.79
34	6.5	5.47	100.32	69	4.8	9.38	89.07
35	6.2	5.47	99.10	70	4.9	9.38	89.21

values in the 85.1 model with $\hat{\sigma}_{85.1} = 0.854$ and the 90.1 model with $\hat{\sigma}_{90.1} = 0.312$. The horizontal axis of Figure 2.3 denotes the maturities of individual bonds. The graphs show that the individual model values are rather close to the observed bond prices. From these observations, it may be concluded that the CSM is a good model for pricing individual Japanese Government bonds and for bond trading in practice. In the Table 2.4, the data used in the estimation of bond pricing model at 90.1 is summarized.

2.7 Problems

2.2.1 In a general linear regression model $y = X\beta + \varepsilon$ with $X : n \times k$ and *rank* $X = k$, show that

(1) The OLSE $b(I_n) = (X'X)^{-1}X'y$ minimizes

$$S(\beta) = (y - X\beta)'(y - X\beta)$$

with respect to $\beta \in R^k$;

(2) The GLSE $b(\Psi) = (X'\Psi^{-1}X)^{-1}X'\Psi^{-1}y$ with $\Psi \in \mathcal{S}(n)$ minimizes

$$S(\beta; \Psi) = (y - X\beta)'\Psi^{-1}(y - X\beta)$$

with respect to $\beta \in R^k$.

(3) Let $\Sigma(\theta) \in S(n)$ be a differentiable function of $\theta \in R^d$ with $d < n$. Show that the function

$$S(\beta, \theta) = (y - X\beta)'\Sigma(\theta)^{-1}(y - X\beta)$$

is minimized by

$$(\beta, \theta) = \left(b(\Sigma(\hat{\theta})), \hat{\theta}\right),$$

where $\hat{\theta}$ is a solution of minimization of

$$T(\theta) \equiv S\left(b(\Sigma(\theta)), \theta\right) = (y - Xb(\Sigma(\theta)))' \Sigma(\theta)^{-1} (y - Xb(\Sigma(\theta))) .$$

Note that the above results hold without distinction of the distribution of the error term ε.

2.2.2 Show that Φ^{-1} in (2.9) is in fact the inverse matrix of Φ.

2.2.3 Verify (2.12). The answer will be found in Theorem 6.5.4 of Anderson (1971), in which the latent roots of the matrix

$$C_* = \frac{1}{2} \begin{pmatrix} 1 & 1 & & & & & 0 \\ 1 & 0 & 1 & & & & \\ & & \ddots & \ddots & \ddots & & \\ & & & \ddots & \ddots & \ddots & \\ & & & & \ddots & 0 & 1 \\ 0 & & & & & 1 & 1 \end{pmatrix}$$

are derived as

$$\cos\left(\frac{(j-1)\pi}{n}\right) \quad (j = 1, \dots, n).$$

The latent roots of C in question is readily obtained from

$$C = 2I_n - 2C_*.$$

The latent vectors are also given there.

2.2.4 Prove the following statements:

(1) The latent roots of $1_n 1_n'$ are 0 (with multiplicity $n - 1$) and n;

(2) The matrix $I_n + \alpha 1_n 1_n'$ is positive definite if and only if $\alpha > -1/n$;

(3) The matrix $(1 - \theta)I_n + \theta 1_n 1_n'$ is positive definite if and only if $\theta \in (-1/(n - 1), 1)$;

(4) $(I_n + \alpha 1_n 1_n')^{-1} = I_n - \frac{\alpha}{1+n\alpha} 1_n 1_n'$ for $\alpha > -1/n$.

2.2.5 Verify (2.25).

2.2.6 Establish the following formulas:

(1) $(A \otimes B)(C \otimes D) = AC \otimes BD$;

(2) $(A \otimes B)' = A' \otimes B'$;

(3) $(A \otimes B)^{-1} = A^{-1} \otimes B^{-1}$ when A and B are nonsingular;

(4) $vec(ABC) = (C' \otimes A)vec(B)$.

See Section 2.2 of Muirhead (1982).

2.3.1 Show that the class $\mathcal{C}_0$ is actually the class of linear unbiased estimator of β, that is, an estimator of the form Cy is unbiased if and only if $CX = I_k$.

2.3.2 Show that the two terms of the right-hand side of (2.39) are mutually uncorrelated. That is,

$$(X'\Sigma^{-1}X)^{-1}X'\Sigma^{-1}E(\varepsilon\varepsilon')\Big[C - (X'\Sigma^{-1}X)^{-1}X'\Sigma^{-1}\Big]' = 0.$$

2.3.3 For the log-likelihood equation in (2.60):

(1) By using the identity $y = Xb(\Sigma(\theta)) + [y - Xb(\Sigma(\theta))]$, show that

$$L(\beta, \sigma^2, \theta) \leq L\Big[b(\Sigma(\theta)), \sigma^2, \theta\Big] \equiv L_1(\sigma^2, \theta)$$

for any $\beta \in R^k$, $\sigma^2 > 0$ and $\theta \in R^d$;

(2) By differentiating L_1 with respect to σ^2, show that

$$L_1(\sigma^2, \theta) \leq L_1(\hat{\sigma}^2(\theta), \theta)$$

for any $\sigma^2 > 0$ and $\theta \in R^d$;

(3) Confirm that $L_1(\hat{\sigma}^2(\theta), \theta) = Q(\theta)$.

2.4.1 By using the Cauchy–Schwarz inequality (2.70), show that the estimator $\tilde{\theta}(e)$ in (2.66) is in $(-1, 1)$ for any $e \in R^n$.

2.4.2 The Anderson model with covariance matrix (2.11) is an approximation to a regression model with covariance matrix (2.8) for AR(1) errors. Hence, in application, a GLSE under the AR(1) structure is preferred. To obtain a GLSE in this case, assume $\mathcal{L}(\varepsilon) = N_n(0, \tau^2\Phi(\theta))$ with $\Phi(\theta)$ in (2.8) and derive the estimating equation in (2.62) for the MLE. Show that the MLE has a finite second moment.

2.4.3 Consider the heteroscedastic model in Example 2.6. Let

$$\hat{\theta} = (\hat{\theta}_1, \dots, \hat{\theta}_p)'$$

be an estimator of θ. Show that the GLSE $b(\hat{\Omega})$ with $\hat{\Omega} = \Omega(\hat{\theta})$ is rewritten as a weighted sum of $\hat{\beta}_j$'s:

$$b(\hat{\Omega}) = \left(\sum_{j=1}^{p} \hat{\theta}_j^{-2} X_j' X_j \right)^{-1} \sum_{j=1}^{p} \hat{\theta}_j^{-2} X_j' X_j \hat{\beta}_j.$$

2.4.4 In the heteroscedastic model in Example 2.6,

(1) Show that the statistic $(\hat{\beta}_1, \dots, \hat{\beta}_p; \hat{\varepsilon}_1, \dots, \hat{\varepsilon}_p)$ is a minimal sufficient statistic if $\mathcal{L}(\varepsilon) = N_n(0, \Omega)$.

(2) Let $p = 2$ for simplicity. Show that $\hat{\theta}^R = (\hat{\theta}_1^R, \hat{\theta}_2^R)'$ can be rewritten as a function of the above-mentioned minimal sufficient statistic as

$$\hat{\theta}_1^R = \left\{ \hat{\varepsilon}_1' \hat{\varepsilon}_1 + (\hat{\beta}_2 - \hat{\beta}_1)' Q (\hat{\beta}_2 - \hat{\beta}_1) \right\} / n_1$$

with $Q = (X_2' X_2)(X'X)^{-1} X_1' X_1 (X'X)^{-1} (X_2' X_2)$. Here, $\hat{\theta}_2^R$ is obtained by interchanging the suffix.

(3) Evaluate $E(\hat{\theta}_j^R)$.

Some related results will be found in Section 5.4. For the definition of minimal sufficiency, see, for example, Lehmann (1983).

2.4.5 Establish (2.112).

2.4.6 (Moore–Penrose generalized inverse) The Moore–Penrose generalized inverse matrix $A^+ : n \times m$ of a matrix $A : m \times n$ is defined as the unique matrix satisfying $AA^+A = A$, $A^+AA^+ = A^+$, $(AA^+)' = AA^+$ and $(A^+A)' = A^+A$.

(1) Show that $(A')^+ = (A^+)'$, $(A'A)^+ = A^+ (A^+)'$, rank A = rank A^+ = rank (AA^+) and $(AA^+)^2 = AA^+$.

(2) Show that A^+ is unique.

(3) Give an explicit form of A^+ when A is symmetric.

See, for example, Rao and Mitra (1971).

2.5.1 For the CO_2 emission data in Table 2.2, fit the model with AR(1) error and the Anderson model, and estimate the models by the GLSEs treated in Section 2.5.

3

Nonlinear Versions of the Gauss–Markov Theorem

3.1 Overview

In Chapter 2, we discussed some fundamental aspects of generalized least squares estimators (GLSEs) in the general linear regression model

$$y = X\beta + \varepsilon. \tag{3.1}$$

In this chapter, we treat a prediction problem in regression and establish a nonlinear version of the Gauss–Markov theorem in both prediction and estimation.

First, to formulate a prediction problem, which includes an estimation problem as its special case, we generalize the model (3.1) to

$$\begin{pmatrix} y \\ y_0 \end{pmatrix} = \begin{pmatrix} X \\ X_0 \end{pmatrix} \beta + \begin{pmatrix} \zeta \\ \zeta_0 \end{pmatrix}, \tag{3.2}$$

and consider the problem of predicting a future value of y_0. Here it is assumed that

$$E\left(\begin{pmatrix} \zeta \\ \zeta_0 \end{pmatrix} \right) = \begin{pmatrix} 0 \\ 0 \end{pmatrix}$$

and

$$\mathrm{Cov}\left(\begin{pmatrix} \zeta \\ \zeta_0 \end{pmatrix} \right) = \sigma^2 \begin{pmatrix} \Sigma & \Delta' \\ \Delta & \Sigma_0 \end{pmatrix} \equiv \Phi. \tag{3.3}$$

In (3.2), the following three quantities

$$y : n \times 1, \quad X : n \times k \quad \text{and} \quad X_0 : m \times k$$

Generalized Least Squares   Takeaki Kariya and Hiroshi Kurata
© 2004 John Wiley & Sons, Ltd   ISBN: 0-470-86697-7 (PPC)

are observable variables, and

$$y_0 : m \times 1$$

is to be predicted. It is further assumed that

$$rank X = k \quad \text{and} \quad \Sigma \in \mathcal{S}(n),$$

where $\mathcal{S}(n)$ denotes the set of $n \times n$ positive definite matrices. Clearly, Σ_0 is an $m \times m$ nonnegative definite matrix, $\sigma^2 > 0$ and $\Delta : m \times n$.

In this setup, the problem of estimating β in (3.1) is embedded into the problem of predicting y_0 in (3.2). In fact, letting

$$m = k, \quad X_0 = I_k, \quad \Delta = 0 \quad \text{and} \quad \Sigma_0 = 0 \tag{3.4}$$

yields $y_0 = \beta$ and hence, when (3.4) holds, predicting y_0 in (3.2) is equivalent to estimating β in (3.1).

Typical models for (3.1) and (3.2) include serial correlation models, heteroscedastic models, equi-correlated models, seemingly unrelated regression (SUR) models and so on, which have been discussed in Chapter 2. Furthermore, in the model (3.2), the problem of estimating missing observations can be treated by regarding y_0 as a missing observation vector. In this case, the matrix Φ in (3.3) is usually assumed to be positive definite.

Under the model (3.2), we establish a nonlinear version of the Gauss–Markov theorem for generalized least squares predictors (GLSPs) in a prediction problem, a particular case of which gives a nonlinear version of the Gauss–Markov theorem in estimation. More specifically, the organization of this chapter is as follows:

3.2 Generalized Least Squares Predictors

3.3 A Nonlinear Version of the Gauss–Markov Theorem in Prediction

3.4 A Nonlinear Version of the Gauss–Markov Theorem in Estimation

3.5 An Application to GLSEs with Iterated Residuals.

In Section 3.2, we first define a GLSP under the model (3.2) and then describe its basic properties. Section 3.3 is devoted to establishing a nonlinear version of the Gauss–Markov theorem in prediction, which states that the risk matrix of a GLSP is bounded below by that of the Gauss–Markov predictor (GMP). The results are applied to the estimation problem in Section 3.4, and a lower bound for the risk matrices of the GLSEs is obtained. In Section 3.5, a further application to iterated GLSEs is presented.

3.2 Generalized Least Squares Predictors

In this section, we define a GLSP and describe its fundamental properties.

The Gauss–Markov theorem in prediction. To begin with, by transforming

$$\zeta = \varepsilon \quad \text{and} \quad \zeta_0 = \Delta\Sigma^{-1}\varepsilon + \varepsilon_0$$

in (3.2), the model is rewritten as

$$\begin{pmatrix} y \\ y_0 \end{pmatrix} = \begin{pmatrix} X \\ X_0 \end{pmatrix}\beta + \begin{pmatrix} I_n & 0 \\ \Delta\Sigma^{-1} & I_m \end{pmatrix}\begin{pmatrix} \varepsilon \\ \varepsilon_0 \end{pmatrix} \tag{3.5}$$

with

$$E\left(\begin{pmatrix} \varepsilon \\ \varepsilon_0 \end{pmatrix}\right) = \begin{pmatrix} 0 \\ 0 \end{pmatrix},$$

$$\text{Cov}\left(\begin{pmatrix} \varepsilon \\ \varepsilon_0 \end{pmatrix}\right) = \sigma^2\begin{pmatrix} \Sigma & 0 \\ 0 & \Sigma_0 - \Delta\Sigma^{-1}\Delta' \end{pmatrix} \equiv \Omega. \tag{3.6}$$

Thus, we can discuss the model in terms of $(\varepsilon', \varepsilon_0')'$. To state the distributional assumption on the error term in a compact form, let $P \equiv \mathcal{L}(\cdot)$ denote the distribution of $\cdot$ and let

$$\mathcal{P}_{n+m}(\mu, \Psi) = \text{the set of distributions on } R^{n+m} \text{ with mean } \mu$$

$$\text{and covariance matrix } \Psi. \tag{3.7}$$

Thus, the assumption (3.6) on the vector $(\varepsilon', \varepsilon_0')'$ is simply denoted by

$$P = \mathcal{L}\left(\begin{pmatrix} \varepsilon \\ \varepsilon_0 \end{pmatrix}\right) \in \mathcal{P}_{n+m}(0, \Omega). \tag{3.8}$$

With this preparation, let us first establish the Gauss–Markov theorem in prediction in model (3.5). A predictor

$$\hat{y}_0 = \hat{y}_0(y) : R^n \to R^m$$

of y_0 is a vector-valued function defined on R^n, the space of the observable vector y, to R^m, the space of the predicted vector y_0. A predictor $\hat{y}_0$ is said to be unbiased if

$$E_P(\hat{y}_0) = E_P(y_0) \; (= X_0\beta), \tag{3.9}$$

where the suffix P denotes the distribution under which the expectation is taken, and is omitted when no confusion is caused. A predictor $\hat{y}_0$ is called *linear* if it takes the form

$$\hat{y}_0 = Cy,$$

where C is an $m \times n$ nonrandom matrix. Further, a predictor is called *linear unbiased* if it is linear and unbiased. As is easily seen, a linear predictor $\hat{y}_0 = Cy$ is unbiased (i.e., $\hat{y}_0$ is linear unbiased) if and only if the matrix C satisfies

$$CX = X_0.$$

See Problem 3.2.1. Thus, the class $C_0(X_0)$ of linear unbiased predictor of y_0 is described as

$$C_0(X_0) = \{\hat{y}_0 = Cy \mid C \text{ is an } m \times n \text{ matrix such that } CX = X_0\}. \qquad (3.10)$$

In this section, the efficiency of a GLSP $\hat{y}_0$ of y_0 is measured by the risk matrix defined by

$$R_P(\hat{y}_0, y_0) = E_P[(\hat{y}_0 - y_0)(\hat{y}_0 - y_0)']. \qquad (3.11)$$

Goldberger (1962) extended the Gauss–Markov theorem to the case in prediction.

Theorem 3.1 (Gauss–Markov theorem in prediction) *Suppose that the matrices Σ and Δ are known. The predictor of the form*

$$\tilde{y}_0(\Sigma, \Delta) = X_0 b(\Sigma) + \Delta \Sigma^{-1}[y - Xb(\Sigma)] \qquad (3.12)$$

is the best linear unbiased predictor (BLUP) of y_0 in the sense that

$$R(\tilde{y}_0(\Sigma, \Delta), \ y_0) \leq R(\hat{y}_0, y_0) \qquad (3.13)$$

holds for any $\hat{y}_0 \in C_0(X_0)$. Here, the quantity $b(\Sigma)$ in (3.12) is the Gauss–Markov estimator (GME) of β:

$$b(\Sigma) = (X'\Sigma^{-1}X)^{-1}X'\Sigma^{-1}y.$$

Proof. To see that $\tilde{y}_0(\Sigma, \Delta)$ is a linear unbiased predictor, rewrite it as

$$\tilde{y}_0(\Sigma, \Delta) = C_* y \qquad (3.14)$$

with

$$C_* = X_0(X'\Sigma^{-1}X)^{-1}X'\Sigma^{-1}$$
$$+ \Delta\Sigma^{-1}[I_n - X(X'\Sigma^{-1}X)^{-1}X'\Sigma^{-1}].$$

Then the matrix C_* clearly satisfies $C_*X = X_0$.

Next, to establish the inequality (3.13), fix $\hat{y}_0 = Cy \in C_0(X_0)$ arbitrarily, and decompose it as

$$\hat{y}_0 - y_0 = [\tilde{y}_0(\Sigma, \Delta) - y_0] + [\hat{y}_0 - \tilde{y}_0(\Sigma, \Delta)]$$
$$= Y_1 + Y_2 \quad \text{(say)}. \qquad (3.15)$$

Then, the two terms in the right-hand side of (3.15) are written as

$$Y_1 = \tilde{y}_0(\Sigma, \Delta) - y_0$$
$$= (X_0 - \Delta\Sigma^{-1}X)(X'\Sigma^{-1}X)^{-1}X'\Sigma^{-1}\varepsilon - \varepsilon_0 \qquad (3.16)$$

and

$$Y_2 = \hat{y}_0 - \tilde{y}_0(\Sigma, \Delta)$$
$$= (C - \Delta\Sigma^{-1})[I_n - X(X'\Sigma^{-1}X)^{-1}X'\Sigma^{-1}]\varepsilon \qquad (3.17)$$

respectively. Let

$$V_{ij} = E(Y_i Y_j') \quad (i, j = 1, 2).$$

Then

$$R(\tilde{y}_0(\Sigma, \Delta), y_0) = V_{11}$$

and

$$R(\hat{y}_0, y_0) = V_{11} + V_{12} + V_{21} + V_{22}.$$

Here, it is shown by direct calculation that

$$V_{12} = 0 = V_{21}'.$$

See Problem 3.2.2. Hence, we obtain

$$R(\hat{y}_0, y_0) = V_{11} + V_{22} \geq V_{11} = R(\tilde{y}_0(\Sigma, \Delta), y_0).$$

Here, V_{11} and V_{22} are given by

$$V_{11} = \sigma^2[(X_0 - \Delta\Sigma^{-1}X)(X'\Sigma^{-1}X)^{-1}(X_0 - \Delta\Sigma^{-1}X)'$$
$$+ (\Sigma_0 - \Delta\Sigma^{-1}\Delta')] \qquad (3.18)$$

and

$$V_{22} = \sigma^2(C - \Delta\Sigma^{-1})[\Sigma - X(X'\Sigma^{-1}X)^{-1}X'](C - \Delta\Sigma^{-1})', \qquad (3.19)$$

respectively. This completes the proof.

We call the predictor $\tilde{y}_0(\Sigma, \Delta)$ in (3.12) *the Gauss–Markov predictor (GMP)*.

Note that the Gauss–Markov theorem in estimation established in Chapter 2 is a special case of this result. In fact, when the condition (3.4) holds, the predictand y_0 becomes β and the GMP $\tilde{y}_0(\Sigma, \Delta)$ in (3.12) reduces to the GME $b(\Sigma)$. Correspondingly, the class $C_0(X_0) = C_0(I_n)$ reduces to the class C_0 of linear unbiased estimators. Hence, for any $\hat{\beta} \in C_0$,

$$R(\hat{\beta}, \beta) \equiv E[(\hat{\beta} - \beta)(\hat{\beta} - \beta)']$$
$$\geq E[(b(\Sigma) - \beta)(b(\Sigma) - \beta)']$$
$$\equiv R(b(\Sigma), \beta) = \text{Cov}(b(\Sigma))$$
$$= \sigma^2(X'\Sigma^{-1}X)^{-1}. \qquad (3.20)$$

Generalized least squares predictors. Next let us consider GLSPs. When Σ and Δ are unknown, a natural way of defining a predictor of y_0 is to replace the unknown Σ and Δ in the GMP $\tilde{y}_0(\Sigma, \Delta)$ by their estimators, $\hat{\Sigma}$ and $\hat{\Delta}$. This leads to the definition of a *generalized least squares predictor or GLSP*.

More specifically, a predictor of the form

$$\tilde{y}_0(\hat{\Sigma}, \hat{\Delta}) = X_0 b(\hat{\Sigma}) + \hat{\Delta}\hat{\Sigma}^{-1}[y - Xb(\hat{\Sigma})] \tag{3.21}$$

is called a GLSP if the estimators $\hat{\Sigma}$ and $\hat{\Delta}$ are functions of the ordinary least squares (OLS) residual vector:

$$e = Ny \quad \text{with} \quad N = I_n - X(X'X)^{-1}X', \tag{3.22}$$

where $b(\hat{\Sigma})$ in (3.21) is a GLSE of β:

$$b(\hat{\Sigma}) = (X'\hat{\Sigma}^{-1}X)^{-1}X'\hat{\Sigma}^{-1}y. \tag{3.23}$$

A GLSP $\tilde{y}_0(\hat{\Sigma}, \hat{\Delta})$ is clearly rewritten as

$$\tilde{y}_0(\hat{\Sigma}, \hat{\Delta}) = C(e)y$$

by letting

$$C(e) = X_0(X'\hat{\Sigma}^{-1}X)^{-1}X'\hat{\Sigma}^{-1}$$
$$+ \hat{\Delta}\hat{\Sigma}^{-1}[I_n - X(X'\hat{\Sigma}^{-1}X)^{-1}X'\hat{\Sigma}^{-1}]. \tag{3.24}$$

The function $C(e)$ satisfies $C(e)X = X_0$. This property characterizes the class of the GLSPs. More precisely,

Proposition 3.2 *Let $C_1(X_0)$ be the class of GLSPs (3.21). Then it is expressed as*

$$C_1(X_0) = \{\hat{y}_0 = C(e)y \mid C(\cdot) \text{ is an } m \times n \text{ matrix-valued measurable}$$

$$\text{function satisfying } C(\cdot)X = X_0\}. \tag{3.25}$$

Proof. For any $\hat{y}_0 = C(e)y$ satisfying $C(e)X = X_0$, choose $\hat{\Sigma}(e)$ and $\hat{\Delta}(e)$ such that

$$\hat{\Delta}(e)\hat{\Sigma}(e)^{-1} = C(e). \tag{3.26}$$

(A possible choice is $\hat{\Sigma}(e) = I_n$ and $\hat{\Delta}(e) = C(e)$.) Then for such $\hat{\Sigma}$ and $\hat{\Delta}$, $\tilde{y}_0(\hat{\Sigma}, \hat{\Delta})$ becomes

$$\tilde{y}_0(\hat{\Sigma}, \hat{\Delta}) = X_0(X'\hat{\Sigma}^{-1}X)^{-1}X'\hat{\Sigma}^{-1}y$$

$$+ C(e)[I_n - X(X'\hat{\Sigma}^{-1}X)^{-1}X'\hat{\Sigma}^{-1}]y$$

$$= C(e)y, \tag{3.27}$$

since $C(e)X = X_0$. This completes the proof.

It readily follows from (3.10) and (3.25) that

$$\mathcal{C}_0(X_0) \subset \mathcal{C}_1(X_0),$$

that is, any linear unbiased predictor of y_0 is a GLSP in our sense.

When Σ and Δ are functions of unknown parameter θ, say $\Sigma = \Sigma(\theta)$ and $\Delta = \Delta(\theta)$, the GLSP of the form

$$\tilde{y}_0(\hat{\Sigma}, \hat{\Delta}) = X_0 b(\hat{\Sigma}) + \hat{\Delta}\hat{\Sigma}^{-1}[y - Xb(\hat{\Sigma})] \tag{3.28}$$

is usually used, where

$$\hat{\Sigma} = \Sigma(\hat{\theta}), \quad \hat{\Delta} = \Delta(\hat{\theta})$$

and $\hat{\theta} = \hat{\theta}(e)$ is an estimator of θ based on the OLS residual vector e. Obviously, such a GLSP also belongs to the class $\mathcal{C}_1(X_0)$.

Conditions for unbiasedness and finiteness of moments. The results stated in Proposition 2.6 can be extended to the prediction problems as

Proposition 3.3 *Let $\hat{y}_0$ be a GLSP of the form $\hat{y}_0 = C(e)y$ satisfying $C(e)X = X_0$.*

(1) *Suppose that $\mathcal{L}(\varepsilon) = \mathcal{L}(-\varepsilon)$ and that $C(e)$ is an even function of e, then the predictor $\hat{y}_0 = C(e)y$ is unbiased as long as the expectation is finite.*

(2) *Suppose that the function $C(e)$ is continuous and satisfies $C(ae) = C(e)$ for any $a > 0$, then the predictor $\hat{y}_0 = C(e)y$ has a finite second moment.*

Proof. The proof is quite similar to that of Proposition 2.6 and omitted.

Since Σ and Δ are estimated, it is conjectured that the GLSP $\tilde{y}_0(\hat{\Sigma}, \hat{\Delta})$ is less efficient than the GMP $\tilde{y}_0(\Sigma, \Delta)$, that is, for any GLSP $\tilde{y}_0(\hat{\Sigma}, \hat{\Delta})$ in $\mathcal{C}_1(X_0)$, the risk matrix $R(\tilde{y}_0(\hat{\Sigma}, \hat{\Delta}), y_0)$ is expected to be bounded below by that of the GMP $\tilde{y}_0(\Sigma, \Delta)$:

$$R(\tilde{y}_0(\Sigma, \Delta), y_0) \leq R(\tilde{y}_0(\hat{\Sigma}, \hat{\Delta}), y_0). \tag{3.29}$$

The discussion in the next section concerns the inequality (3.29). We will establish a *nonlinear version of the Gauss–Markov theorem* under a mild assumption on the distribution of $(\varepsilon', \varepsilon_0')'$. To do so, let $\mathcal{B}$ be the set of all predictors of β with finite second moment under $\mathcal{P}_{n+m}(0, \Omega)$, and redefine

$$\mathcal{C}_1(X_0) = \{\hat{y}_0 = C(e)y \in \mathcal{B} \mid C(\cdot) \text{ is an } m \times n \text{ matrix-valued measurable}$$

$$\text{function satisfying } C(\cdot)X = X_0\}.$$

3.3 A Nonlinear Version of the Gauss–Markov Theorem in Prediction

This section is devoted to establishing a nonlinear version of the Gauss–Markov theorem in prediction. An application to an AR(1) error model is also given.

Decomposition of a GLSP. To state the theorems, we first make a decomposition of a GLSP in the model

$$\begin{pmatrix} y \\ y_0 \end{pmatrix} = \begin{pmatrix} X \\ X_0 \end{pmatrix} \beta + \begin{pmatrix} I_n & 0 \\ \Delta \Sigma^{-1} & I_m \end{pmatrix} \begin{pmatrix} \varepsilon \\ \varepsilon_0 \end{pmatrix}$$

with

$$P \equiv \mathcal{L}\left(\begin{pmatrix} \varepsilon \\ \varepsilon_0 \end{pmatrix}\right) \in \mathcal{P}_{n+m}(0, \Omega), \tag{3.30}$$

where Ω is of the form

$$\Omega = \sigma^2 \begin{pmatrix} \Sigma & 0 \\ 0 & \Sigma_0 - \Delta \Sigma^{-1} \Delta' \end{pmatrix}, \tag{3.31}$$

and $X : n \times k$ is assumed to be of full rank.

Let Z be an $n \times (n - k)$ matrix satisfying

$$N = ZZ' \quad \text{and} \quad Z'Z = I_{n-k}, \tag{3.32}$$

where $N = I_n - X(X'X)^{-1}X'$. The OLS residual vector e is clearly expressed as

$$e = ZZ'y = ZZ'\varepsilon. \tag{3.33}$$

In the sequel, the following matrix identity is often used (see Problem 3.3.1):

$$I_n = X(X'\Sigma^{-1}X)^{-1}X'\Sigma^{-1} + \Sigma Z(Z'\Sigma Z)^{-1}Z', \tag{3.34}$$

where $X(X'\Sigma^{-1}X)^{-1}X'\Sigma^{-1}$ is the projection matrix onto $L(X)$ whose null space is $L(\Sigma Z)$. Here, $L(\cdot)$ denotes the linear subspace spanned by the column vectors of matrix $\cdot$. By using the identity (3.34), the vector ε is decomposed as

$$\begin{aligned} \varepsilon &= X(X'\Sigma^{-1}X)^{-1}X'\Sigma^{-1}\varepsilon + \Sigma Z(Z'\Sigma Z)^{-1}Z'\varepsilon \\ &= XA^{-1}u_1 + \Sigma B^{-1}u_2, \end{aligned} \tag{3.35}$$

where

$$A = X'\Sigma^{-1}X \in \mathcal{S}(k), \quad B = Z'\Sigma Z \in \mathcal{S}(n-k) \tag{3.36}$$

and

$$u_1 = X'\Sigma^{-1}\varepsilon : k \times 1, \quad u_2 = Z'\varepsilon : (n-k) \times 1. \tag{3.37}$$

Since the distribution of $(\varepsilon', \varepsilon_0')'$ is given by (3.30), and since

$$\begin{pmatrix} u_1 \\ u_2 \\ \varepsilon_0 \end{pmatrix} = \begin{pmatrix} X'\Sigma^{-1} & 0 \\ Z' & 0 \\ 0 & I_m \end{pmatrix} \begin{pmatrix} \varepsilon \\ \varepsilon_0 \end{pmatrix},$$

the distribution of u_1, u_2 and ε_0 is described as

$$\mathcal{L}\left(\begin{pmatrix} u_1 \\ u_2 \\ \varepsilon_0 \end{pmatrix}\right) \in \mathcal{P}_{n+m}\left(\begin{pmatrix} 0 \\ 0 \\ 0 \end{pmatrix}, \sigma^2 \begin{pmatrix} A & 0 & 0 \\ 0 & B & 0 \\ 0 & 0 & \Sigma_0 - \Delta\Sigma^{-1}\Delta' \end{pmatrix}\right), \quad (3.38)$$

and hence u_1, u_2 and ε_0 are mutually uncorrelated. Note here that u_1 and u_2 are in one-to-one correspondence with $b(\Sigma) - \beta$ and e respectively, since

$$b(\Sigma) - \beta = A^{-1}u_1 \quad (3.39)$$

and

$$e = Zu_2, \quad \text{or equivalently} \quad u_2 = Z'e. \quad (3.40)$$

Lemma 3.4 *A GLSP* $\hat{y}_0 = C(e)y \in \mathcal{C}_1(X_0)$ *is decomposed as*

$$\hat{y}_0 - y_0 = \left[\tilde{y}_0(\Sigma, \Delta) - y_0\right] + \left[\hat{y}_0 - \tilde{y}_0(\Sigma, \Delta)\right]$$
$$= \{(X_0 - \Delta\Sigma^{-1}X)A^{-1}u_1 - \varepsilon_0\}$$
$$+ \{[C(Zu_2) - \Delta\Sigma^{-1}]\Sigma ZB^{-1}u_2\}$$
$$\equiv Y_1 + Y_2 \quad \text{(say)}. \quad (3.41)$$

Here the two vectors Y_1 and Y_2 depend only on (u_1, ε_0) and u_2 respectively. It is essential for the analysis below that Y_1 is a linear function of (u_1, ε_0).

A nonlinear version of the Gauss–Markov theorem in prediction. Now let us establish a nonlinear version of the Gauss–Markov theorem in prediction. Let

$$V_{ij} = E(Y_i Y_j') \ (i, j = 1, 2). \quad (3.42)$$

Then, from Lemma 3.4, it is easy to see that

$$R(\hat{y}_0, y_0) = V_{11} + V_{12} + V_{21} + V_{22}, \quad (3.43)$$

where

$$V_{11} = E(Y_1 Y_1')$$
$$= E[(\tilde{y}_0(\Sigma, \Delta) - y_0)(\tilde{y}_0(\Sigma, \Delta) - y_0)'] \quad (3.44)$$
$$V_{22} = E(Y_2 Y_2')$$
$$= E[(\hat{y}_0 - \tilde{y}_0(\Sigma, \Delta))(\hat{y}_0 - \tilde{y}_0(\Sigma, \Delta))'] \quad (3.45)$$

and

$$V_{12} = E(Y_1 Y_2')$$
$$= E[(\tilde{y}_0(\Sigma, \Delta) - y_0)(\hat{y}_0 - \tilde{y}_0(\Sigma, \Delta))']$$
$$= V_{21}'. \quad (3.46)$$

If $V_{12} = 0$, then it follows from (3.43) that

$$R(\hat{y}_0, y_0) = V_{11} + V_{22} \geq V_{11} = R(\tilde{y}_0(\Sigma, \Delta), y_0). \tag{3.47}$$

A sufficient condition for $V_{12} = 0$ is the following:

$$E_P(u_1|u_2) = 0 \quad \text{and} \quad E_P(\varepsilon_0|u_2) = 0 \text{ a.s.} \tag{3.48}$$

In fact, when the condition (3.48) holds, we have

$$
\begin{aligned}
V_{12} &= E[E(Y_1|u_2) \, Y_2'] \\
&= E[(X_0 - \Delta\Sigma^{-1}X)A^{-1} \, E(u_1|u_2) \, Y_2' - E(\varepsilon_0|u_2) \, Y_2'] \\
&= 0.
\end{aligned}
\tag{3.49}
$$

Thus, by defining

$$\mathcal{Q}_{n+m}(0, \Omega) = \{P \in \mathcal{P}_{n+m}(0, \Omega) \mid (3.48) \text{ holds}\}, \tag{3.50}$$

the following theorem holds, which was originally established by Kariya and Toyooka (1985) and Eaton (1985).

Theorem 3.5 *Suppose that $\mathcal{L}\left((\varepsilon', \varepsilon_0')'\right) \in \mathcal{Q}_{n+m}(0, \Omega)$. Then, for any GLSP $\hat{y}_0 \in \mathcal{C}_1(X_0)$, its conditional risk matrix is decomposed into two parts:*

$$
\begin{aligned}
R(\hat{y}_0, \ y_0|u_2) &= R(\tilde{y}_0(\Sigma, \Delta), \ y_0) \\
&\quad + (\hat{y}_0 - \tilde{y}_0(\Sigma, \Delta))(\hat{y}_0 - \tilde{y}_0(\Sigma, \Delta))',
\end{aligned}
\tag{3.51}
$$

where

$$R(\hat{y}_0, \ y_0|u_2) = E[(\hat{y}_0 - y_0)(\hat{y}_0 - y_0)'|u_2]. \tag{3.52}$$

Noting that the second term of (3.51) is nonnegative definite, we obtain a nonlinear version of the Gauss–Markov theorem in prediction.

Corollary 3.6 *For any $\hat{y}_0 \in \mathcal{C}_1(X_0)$, the conditional risk matrix is bounded below by the risk (covariance) matrix of the GMP:*

$$R(\tilde{y}_0(\Sigma, \Delta), \ y_0) \leq R(\hat{y}_0, \ y_0|u_2), \tag{3.53}$$

and thus unconditionally

$$R(\tilde{y}_0(\Sigma, \Delta), \ y_0) \leq R(\hat{y}_0, \ y_0). \tag{3.54}$$

The class $\mathcal{Q}_{n+m}(0, \Omega)$ is a broad one and contains several interesting classes of distributions. A typical example is a class of elliptically symmetric distributions, which has been introduced in Section 1.3 of Chapter 1 and includes the normal distribution. To see this, suppose

$$\mathcal{L}\left(\begin{pmatrix} \varepsilon \\ \varepsilon_0 \end{pmatrix}\right) \in \mathcal{E}_{n+m}(0, \Omega).$$

Then we have from Proposition 1.17 of Chapter 1 that

$$\mathcal{L}\left(\begin{pmatrix} u_1 \\ u_2 \\ \varepsilon_0 \end{pmatrix}\right) \in \mathcal{E}_{n+m}\left(\begin{pmatrix} 0 \\ 0 \\ 0 \end{pmatrix}, \sigma^2 \begin{pmatrix} A & 0 & 0 \\ 0 & B & 0 \\ 0 & 0 & \Sigma_0 - \Delta\Sigma^{-1}\Delta' \end{pmatrix}\right), \qquad (3.55)$$

since $(u_1', u_2', \varepsilon_0')'$ is given by a linear transformation of $(\varepsilon', \varepsilon_0')'$. Hence, from Proposition 1.19, the condition (3.48) holds, or equivalently,

$$\mathcal{E}_{n+m}(0, \Omega) \subset \mathcal{Q}_{n+m}(0, \Omega). \qquad (3.56)$$

Thus, Theorem 3.5 and its corollary are valid under the class $\mathcal{E}_{n+m}(0, \Omega)$.

Further decomposition of a GLSP under elliptical symmetry. The quantity

$$Y_1 = \tilde{y}_0(\Sigma, \Delta) - y_0$$
$$= (X_0 - \Delta\Sigma^{-1}X)A^{-1}u_1 - \varepsilon_0$$

can be decomposed into u_1-part and ε_0-part. This suggests a further decomposition. Decompose Y_1 as

$$Y_1 = Y_{11} + Y_{12}$$

with

$$Y_{11} = (X_0 - \Delta\Sigma^{-1}X)A^{-1}u_1, \qquad (3.57)$$

$$Y_{12} = -\varepsilon_0. \qquad (3.58)$$

Here, of course, the following equalities hold:

$$R_{11} \equiv E(Y_{11}Y_{11}')$$
$$= \sigma^2(X_0 - \Delta\Sigma^{-1}X)A^{-1}(X_0 - \Delta\Sigma^{-1}X)', \qquad (3.59)$$

and

$$R_{12} \equiv E(Y_{12}Y_{12}')$$
$$= \sigma^2(\Sigma_0 - \Delta\Sigma^{-1}\Delta'). \qquad (3.60)$$

Note here that under the condition $\mathcal{L}\left((\varepsilon', \varepsilon_0')'\right) \in \mathcal{E}_{n+m}(0, \Omega)$, it holds that

$$E(u_1\varepsilon_0'|u_2) = 0 \text{ a.s.} \qquad (3.61)$$

in addition to (3.48). In fact, by Proposition 1.19,

$$\mathcal{L}\left(\left(\begin{array}{c} u_1 \\ \varepsilon_0 \end{array}\right) \Big| u_2 \right) \in \mathcal{E}_{k+m}\left(\left(\begin{array}{c} 0 \\ 0 \end{array}\right), \ \sigma^2 \left(\begin{array}{cc} A & 0 \\ 0 & \Sigma_0 - \Delta\Sigma^{-1}\Delta' \end{array}\right)\right),$$

where $\mathcal{L}(a|b)$ denotes the conditional distribution of a given b. From this, Theorem 3.5 can be refined as

Theorem 3.7 *Suppose that $\mathcal{L}\left((\varepsilon', \varepsilon_0')'\right) \in \mathcal{E}_{n+m}(0, \ \Omega)$. Then, for any $\hat{y}_0 \in C_1(X_0)$, its conditional risk matrix, given u_2, is decomposed into three parts:*

$$R(\hat{y}_0, \ y_0|u_2) = R_{11} + R_{12} + (\hat{y}_0 - \tilde{y}_0(\Sigma, \Delta))(\hat{y}_0 - \tilde{y}_0(\Sigma, \Delta))'. \quad (3.62)$$

Consequently, by this theorem, the unconditional risk matrix of $\hat{y}_0$ is decomposed as

$$R(\hat{y}_0, \ y_0) = R_{11} + R_{12} + R_2 \quad (3.63)$$

with

$$R_2 = E(Y_2 Y_2') = R(\hat{y}_0, \ \tilde{y}_0(\Sigma, \Delta)).$$

Implications of Theorem 3.7. Let us consider the implications of this risk decomposition.

The first term R_{11} in (3.63) is the expected loss of the prediction efficiency associated with estimating β by the GME $b(\Sigma)$. In fact,

$$Y_{11} = \{X_0 b(\Sigma) + \Delta\Sigma^{-1}[y - X b(\Sigma)]\} - \{X_0 \beta + \Delta\Sigma^{-1}[y - X\beta]\}.$$

However, from (3.59), the closer $X_0 - \Delta\Sigma^{-1}X$ is to zero, the smaller the first term is. Here, the matrix $\Delta\Sigma^{-1}$ is interpreted as a regression coefficient of the model in which X_0 is regressed on X. If $X_0 - \Delta\Sigma^{-1}X = 0$, then the risk matrix is zero and the u_1-part of (3.57) vanishes. On the other hand, in the model with $\Delta = 0$ (or equivalently, $\Delta\Sigma^{-1} = 0$), the first term attains its maximum as a function of Δ. In other words, when y and y_0 are uncorrelated, no reduction of risk is obtained from the R_{11}-part. Consequently,

$$0 \leq R_{11} \leq \sigma^2 X_0 A^{-1} X_0'. \quad (3.64)$$

Models with $\Delta = 0$ include a heteroscedastic model and an SUR model (see Chapter 2). We will treat a serial correlation model as an example of the model with $\Delta \neq 0$.

The second term R_{12} of (3.63) depends on the correlation between ζ and ζ_0. The vector Y_{12} in question is expressed as

$$Y_{12} = \{X_0 \beta + \Delta\Sigma^{-1}[y - X\beta]\} - y_0 = -(\zeta - \Delta\Sigma^{-1}\zeta_0).$$

The third term R_2 of (3.63) is of the most importance, since it reflects the loss of the prediction efficiency caused by estimating unknown Δ and Σ:

$$Y_2 = \tilde{y}_0(\hat{\Sigma}, \hat{\Delta}) - \tilde{y}_0(\Sigma, \Delta).$$

The evaluation of this quantity will be treated in the subsequent chapters in estimation setup where it is supposed that $\Delta = 0$.

Example 3.1 (AR(1) error model)   An example of the model in which $\Delta \neq 0$ holds is a linear regression model with AR(1) error. This model is obtained by letting

$$\zeta = (\zeta_1, \dots, \zeta_n)' \quad \text{and} \quad \zeta_0 = (\zeta_{n+1}, \dots, \zeta_{n+m})'$$

in the model (3.2), and assuming that ζ_j's are generated by

$$\zeta_j = \theta \zeta_{j-1} + \xi_j \quad \text{with} \quad |\theta| < 1.$$

Here, ξ_j's are uncorrelated with mean 0 and variance σ^2. Then the matrix $\Phi \in \mathcal{S}(n + m)$ in (3.3) is expressed as

$$\Phi = \Phi(\sigma^2, \theta) = \frac{\sigma^2}{1 - \theta^2} (\theta^{|i-j|}) : (n + m) \times (n + m) \tag{3.65}$$

(see Example 2.1).

For simplicity, let us consider the case of a one-period-ahead prediction, that is, $m = 1$. In this case,

$$\Phi = \sigma^2 \begin{pmatrix} \Sigma & \Delta' \\ \Delta & \Sigma_0 \end{pmatrix},$$

where

$$\Sigma = \Sigma(\theta) = \frac{1}{1 - \theta^2} (\theta^{|i-j|}) \in \mathcal{S}(n),$$

$$\Delta = \Delta(\theta) = \frac{1}{1 - \theta^2} (\theta^n, \dots, \theta) : 1 \times n \quad \text{and} \tag{3.66}$$

$$\Sigma_0 = \frac{1}{1 - \theta^2} > 0$$

hold. Hence, we have

$$\Delta \Sigma^{-1} = \frac{1}{1 - \theta^2} (\theta^n, \dots, \theta) \begin{pmatrix} 1 & -\theta & & & & 0 \\ -\theta & 1+\theta^2 & -\theta & & & \\ & & \ddots & \ddots & \ddots & \\ & & & \ddots & \ddots & \ddots \\ & & & & \ddots & 1+\theta^2 & -\theta \\ 0 & & & & & -\theta & 1 \end{pmatrix}$$

$$= (0, \dots, 0, \theta) : 1 \times n.$$

Let $X_0 = x_0' : 1 \times k$ and $x_j' : 1 \times k$ be the jth row vector of X. Then the matrices R_{11} and R_{12} are expressed as

$$R_{11} = \sigma^2 (x_0 - \theta x_n)' A^{-1} (x_0 - \theta x_n)$$

and

$$R_{12} = \sigma^2.$$

To define a GLSP, suppose that $P(e = 0) = 0$ and $\mathcal{L}\left((\varepsilon', \varepsilon_0)'\right) \in \mathcal{Q}_{n+1}(0, \Phi)$ (see (3.51)). A typical estimator of θ is

$$\hat{\theta} = \hat{\theta}(e) = \frac{\sum_{j=2}^{n} e_j e_{j-1}}{\sum_{j=1}^{n} e_j^2}. \tag{3.67}$$

Since $\hat{\theta}$ satisfies $\hat{\theta}(ae) = \hat{\theta}(e)$ for any $a > 0$, the GLSP

$$\begin{aligned}
\tilde{y}_0(\hat{\Sigma}, \hat{\Delta}) &= X_0 b(\hat{\Sigma}) + \hat{\Delta} \hat{\Sigma}^{-1} [y - X b(\hat{\Sigma})] \\
&\quad (= x_0' b(\hat{\Sigma}) + \hat{\theta}[y_n - x_n' b(\hat{\Sigma})]) \\
&= C(e) y
\end{aligned} \tag{3.68}$$

satisfies $C(ae) = C(e)$ for any $a > 0$, and hence it belongs to $\mathcal{C}_1(X_0)$, where $\hat{\Sigma} = \Sigma(\hat{\theta})$, $\hat{\Delta} = \Delta(\hat{\theta})$ and

$$C(e) = X_0 (X' \hat{\Sigma}^{-1} X)^{-1} X' \hat{\Sigma}^{-1} + \hat{\Delta} \hat{\Sigma}^{-1} [I_n - X(X' \hat{\Sigma}^{-1} X)^{-1} X' \hat{\Sigma}^{-1}].$$

Thus, by Corollary 3.6, we obtain

$$R(\tilde{y}_0(\hat{\Sigma}, \hat{\Delta}), y_0) \geq R(\tilde{y}_0(\Sigma, \Delta), y_0). \tag{3.69}$$

Furthermore, the GLSP is unbiased as long as $\mathcal{L}(\varepsilon) = \mathcal{L}(-\varepsilon)$, since $\hat{\theta}$ is an even function of e.

For applications to other linear models, see, for example, Harville and Jeske (1992) in which a mixed-effects linear model is considered. See also the references given there.

An identity between two estimators in a missing data problem. Next, we consider the problem of estimating β when the vector y_0 is missing in the model (3.2), where $\Phi \in \mathcal{S}(n + m)$ is assumed. While a GLSE

$$b(\hat{\Sigma}) = (X' \hat{\Sigma}^{-1} X)^{-1} X' \hat{\Sigma}^{-1} y \tag{3.70}$$

is a natural estimator of β in this setup, it is often suggested to use the following two-stage estimator of β. First, by estimating the missing observation y_0 by the GLSP:

$$\tilde{y}_0(\hat{\Sigma}, \hat{\Delta}) = X_0 b(\hat{\Sigma}) + \hat{\Delta} \hat{\Sigma}^{-1} [y - X b(\hat{\Sigma})], \tag{3.71}$$

then substituting $\tilde{y}_0(\hat{\Sigma}, \hat{\Delta})$ into the model (3.2), and thirdly forming a " GLSE " from the estimated model with y_0 replaced by $\tilde{y}_0(\hat{\Sigma}, \hat{\Delta})$:

$$\hat{\beta}(\hat{\Phi}) = (X'\hat{\Phi}^{-1}X)^{-1}X'\hat{\Phi}^{-1}y, \tag{3.72}$$

with

$$\hat{\Phi} = \begin{pmatrix} \hat{\Sigma} & \hat{\Delta}' \\ \hat{\Delta} & \hat{\Sigma}_0 \end{pmatrix}, \quad y = \begin{pmatrix} y \\ \tilde{y}_0(\hat{\Sigma}, \hat{\Delta}) \end{pmatrix} \quad \text{and} \quad X = \begin{pmatrix} X \\ X_0 \end{pmatrix}. \tag{3.73}$$

The following theorem due to Kariya (1988) states that the two-stage estimator $\hat{\beta}(\hat{\Phi})$ thus defined makes no change on the one-stage estimator $b(\hat{\Sigma})$, implying that no efficiency is obtained from the two-stage procedure. More specifically,

Theorem 3.8 *The equality $\hat{\beta}(\hat{\Phi}) = b(\hat{\Sigma})$ holds for any $y \in R^n$.*

Proof. From the definition of the $(n + m) \times 1$ vector y in (3.73), the estimator $\hat{\beta}(\hat{\Phi})$ is rewritten as

$$\hat{\beta}(\hat{\Phi}) = (X'\hat{\Phi}^{-1}X)^{-1}X'\hat{\Phi}^{-1}\begin{pmatrix} I_n \\ H \end{pmatrix} y, \tag{3.74}$$

where

$$H = X_0(X'\hat{\Sigma}^{-1}X)^{-1}X'\hat{\Sigma}^{-1}$$
$$+\hat{\Delta}\hat{\Sigma}^{-1}(I_n - X(X'\hat{\Sigma}^{-1}X)^{-1}X'\hat{\Sigma}^{-1}). \tag{3.75}$$

Replacing I_n in (3.74) and (3.75) by the right-hand side of the following matrix identity

$$I_n = X(X'\hat{\Sigma}^{-1}X)^{-1}X'\hat{\Sigma}^{-1} + \hat{\Sigma}Z(Z'\hat{\Sigma}Z)^{-1}Z' \tag{3.76}$$

yields

$$\begin{pmatrix} I_n \\ H \end{pmatrix} = \begin{pmatrix} X(X'\hat{\Sigma}^{-1}X)^{-1}X'\hat{\Sigma}^{-1} + \hat{\Sigma}Z(Z'\hat{\Sigma}Z)^{-1}Z' \\ X_0(X'\hat{\Sigma}^{-1}X)^{-1}X'\hat{\Sigma}^{-1} + \hat{\Delta}Z(Z'\hat{\Sigma}Z)^{-1}Z' \end{pmatrix}$$
$$= X(X'\hat{\Sigma}^{-1}X)^{-1}X'\hat{\Sigma}^{-1} + \begin{pmatrix} \hat{\Sigma} \\ \hat{\Delta} \end{pmatrix} Z(Z'\hat{\Sigma}Z)^{-1}Z'. \tag{3.77}$$

Now, substituting this into (3.74), we obtain

$$\hat{\beta}(\hat{\Phi}) = b(\hat{\Sigma}) + (X'\hat{\Phi}^{-1}X)^{-1}X'\hat{\Phi}^{-1}\begin{pmatrix} \hat{\Sigma} \\ \hat{\Delta} \end{pmatrix} Z(Z'\hat{\Sigma}Z)^{-1}Z'. \tag{3.78}$$

Here using $\begin{pmatrix} \hat{\Sigma} \\ \hat{\Delta} \end{pmatrix} = \hat{\Phi}\begin{pmatrix} I_n \\ 0 \end{pmatrix}$ and $X'Z = 0$, the second term of the above equality vanishes. This completes the proof.

3.4 A Nonlinear Version of the Gauss–Markov Theorem in Estimation

In this section, the results obtained in the previous sections are applied to the estimation problem of β, which leads to a nonlinear version of the Gauss–Markov theorem. Some relevant topics, including a characterization of elliptical symmetry, are also discussed.

Main theorem with examples. To begin with, we reproduce the general linear regression model considered in this section:

$$y = X\beta + \varepsilon \quad \text{with} \quad \mathcal{L}(\varepsilon) \in \mathcal{P}_n(0, \sigma^2 \Sigma), \tag{3.79}$$

where y is an $n \times 1$ vector, X is an $n \times k$ known matrix of full rank, and $\mathcal{P}_n(0, \sigma^2 \Sigma)$ denotes the class of distributions on R^n with mean 0 and covariance matrix $\sigma^2 \Sigma$ as in (3.7). This model is obtained by letting

$$m = k, \quad X_0 = I_k, \quad \Delta = 0 \quad \text{and} \quad \Sigma_0 = 0 \tag{3.80}$$

in the model (3.2). The class $\mathcal{C}_0(X_0) = \mathcal{C}_0(I_k)$ of linear unbiased predictors of y_0 considered in the previous section clearly reduces to the class $\mathcal{C}_0$ of linear unbiased estimators of β in the model (3.79), where

$$\mathcal{C}_0 = \{\hat{\beta} = Cy \mid C \text{ is a } k \times n \text{ matrix such that } CX = I_k\}. \tag{3.81}$$

Similarly, the class $\mathcal{C}_1(X_0) = \mathcal{C}_1(I_k)$ of GLSPs becomes the class $\mathcal{C}_1$ of GLSEs, where

$$\mathcal{C}_1 = \{\hat{\beta} = C(e)y \in \mathcal{B} \mid C(\cdot) \text{ is a } k \times n \text{ matrix-valued measurable}$$
$$\text{function satisfying } C(\cdot)X = I_k\}. \tag{3.82}$$

Here, $\mathcal{B}$ is the set of all estimators of β with a finite second moment under $\mathcal{P}_n(0, \sigma^2 \Sigma)$, and e is the OLS residual vector given by

$$e = Ny \quad \text{with} \quad N = I_n - X(X'X)^{-1}X'. \tag{3.83}$$

A GLSE $\hat{\beta} = C(e)y \in \mathcal{C}_1$ can be decomposed as

$$\hat{\beta} - \beta = [b(\Sigma) - \beta] + [\hat{\beta} - b(\Sigma)]$$
$$= (X'\Sigma^{-1}X)^{-1}u_1 + C(e)\Sigma Z(Z'\Sigma Z)^{-1}u_2, \tag{3.84}$$

where

$$b(\Sigma) = (X'\Sigma^{-1}X)^{-1}X'\Sigma^{-1}y$$

is the GME of β and

$$u_1 = X'\Sigma^{-1}\varepsilon : k \times 1, \quad u_2 = Z'\varepsilon : (n-k) \times 1.$$

The two terms in (3.84) are uncorrelated under the assumption that

$$\mathcal{L}(\varepsilon) \in \mathcal{Q}_n(0, \sigma^2\Sigma), \tag{3.85}$$

where

$$\mathcal{Q}_n(0, \sigma^2\Sigma) = \{P \in \mathcal{P}_n(0, \sigma^2\Sigma) | E_P(u_1|u_2) = 0 \text{ a.s.}\}, \tag{3.86}$$

the class of distributions in $\mathcal{P}_n(0, \sigma^2\Sigma)$ with conditional mean of u_1 given u_2 being zero. In fact, by using $e = Zu_2$,

$$E[(X'\Sigma^{-1}X)^{-1}u_1 u_2'(Z'\Sigma Z)^{-1}Z'\Sigma C(e)']$$

$$= E[(X'\Sigma^{-1}X)^{-1}u_1 u_2'(Z'\Sigma Z)^{-1}Z'\Sigma C(Zu_2)']$$

$$= E[(X'\Sigma^{-1}X)^{-1} E(u_1|u_2) u_2'(Z'\Sigma Z)^{-1}Z'\Sigma C(Zu_2)']$$

$$= 0.$$

Thus, the following theorem obtains

Theorem 3.9 *Suppose that* $\mathcal{L}(\varepsilon) \in \mathcal{Q}_n(0, \sigma^2\Sigma)$. *Then, for any GLSE* $\hat{\beta} = C(e)y \in \mathcal{C}_1$, *the conditional risk matrix is decomposed into two parts:*

$$R(\hat{\beta}, \beta|u_2) = R(b(\Sigma), \beta) + \left(\hat{\beta} - b(\Sigma)\right)\left(\hat{\beta} - b(\Sigma)\right)'$$

$$= \sigma^2(X'\Sigma^{-1}X)^{-1}$$

$$+ C(Zu_2)\Sigma Z(Z'\Sigma Z)^{-1}u_2 u_2'(Z'\Sigma Z)^{-1}Z'C(Zu_2)'. \tag{3.87}$$

Consequently, the risk matrix of a GLSE in $\mathcal{C}_1$ *is bounded below by that of the GME, that is,*

$$R(\hat{\beta}, \beta) \geq R(b(\Sigma), \beta) = \text{Cov}(b(\Sigma)) = \sigma^2(X'\Sigma^{-1}X)^{-1}. \tag{3.88}$$

This result is due to Kariya (1985a), Kariya and Toyooka (1985) and Eaton (1985). This theorem demonstrates one direction of extension of the original (linear) Gauss–Markov theorem by enlarging the class of estimators to $\mathcal{C}_1$ from $\mathcal{C}_0$ on which the original Gauss–Markov theorem is based. A further extension of this theorem will be given in Chapter 7. On the other hand, some other extension from different standpoints may be possible. For example, one may adopt concentration probability as a criterion for comparison of estimators. This problem has been investigated by, for example, Hwang (1985), Kuritsyn (1986), Andrews and Phillips (1987), Eaton (1986, 1988), Berk and Hwang (1989), Ali and Ponnapalli (1990) and Jensen (1996), some of which will be reviewed in Section 8.2 of Chapter 8.

As applications of Theorem 3.9, we present two examples.

Example 3.2 (Heteroscedastic model)   Suppose that the model (3.79) is of the heteroscedastic structure with p distinct variances, that is,

$$y = \begin{pmatrix} y_1 \\ \vdots \\ y_p \end{pmatrix} : n \times 1, \quad X = \begin{pmatrix} X_1 \\ \vdots \\ X_p \end{pmatrix} : n \times k,$$

$$\varepsilon = \begin{pmatrix} \varepsilon_1 \\ \vdots \\ \varepsilon_p \end{pmatrix} : n \times 1, \tag{3.89}$$

$$\Omega(\theta) = \begin{pmatrix} \theta_1 I_{n_1} & & 0 \\ & \ddots & \\ 0 & & \theta_p I_{n_p} \end{pmatrix} \in \mathcal{S}(n) \tag{3.90}$$

and

$$\theta = (\theta_1, \ldots, \theta_p)' : p \times 1, \tag{3.91}$$

where $y_j : n_j \times 1$, $X_j : n_j \times k$, $\varepsilon_j : n_j \times 1$ and $n = \sum_{j=1}^{p} n_j$ (see Example 2.3). In this model, the GME is given by

$$b(\Omega) = (X'\Omega^{-1}X)^{-1}X'\Omega^{-1}y \quad \text{with} \quad \Omega = \Omega(\theta), \tag{3.92}$$

which is not feasible in the case in which θ is unknown.

Theorem 3.9 guarantees that when $\mathcal{L}(\varepsilon) \in \mathcal{Q}_n(0, \Omega(\theta))$, any GLSE $\hat{\beta} = C(e)y \in \mathcal{C}_1$ satisfies

$$R(\hat{\beta}, \beta) \geq R(b(\Omega), \beta) = \mathrm{Cov}(b(\Omega)) = (X'\Omega^{-1}X)^{-1}, \tag{3.93}$$

where $e = [I_n - X(X'X)^{-1}X']y$. Typical examples are the GLSEs of the form

$$b(\hat{\Omega}) = C(e)y \tag{3.94}$$

with

$$C(e) = (X'\hat{\Omega}^{-1}X)^{-1}X'\hat{\Omega}^{-1} \quad \text{and} \quad \hat{\Omega} = \Omega(\hat{\theta}(e)),$$

where $\hat{\theta}(e) = (\hat{\theta}_1(e), \ldots, \hat{\theta}_p(e))' : p \times 1$ is an estimator of θ such that $\Omega(\hat{\theta}(e)) \in \mathcal{S}(n)$ a.s. and

$$\hat{\theta}_j(ae) = a^2 \hat{\theta}_j(e) \quad \text{for any} \quad a > 0 \quad (j = 1, \ldots, p). \tag{3.95}$$

Such GLSEs clearly include the restricted GLSE and the unrestricted GLSE discussed in Example 2.6.

Example 3.3 (SUR model) Let us consider the p-equation SUR model

$$y = \begin{pmatrix} y_1 \\ \vdots \\ y_p \end{pmatrix} : n \times 1, \quad X = \begin{pmatrix} X_1 & & 0 \\ & \ddots & \\ 0 & & X_p \end{pmatrix} : n \times k,$$

$$\beta = \begin{pmatrix} \beta_1 \\ \vdots \\ \beta_p \end{pmatrix} : k \times 1, \quad \varepsilon = \begin{pmatrix} \varepsilon_1 \\ \vdots \\ \varepsilon_p \end{pmatrix} : n \times 1 \tag{3.96}$$

with covariance structure

$$\Omega = \Sigma \otimes I_m \quad \text{and} \quad \Sigma = (\sigma_{ij}) \in \mathcal{S}(p), \tag{3.97}$$

where

$$y_j : m \times 1, \quad X_j : m \times k_j, \quad \varepsilon_j : m \times 1, \quad n = pm \quad \text{and} \quad k = \sum_{j=1}^{p} k_j.$$

The GME for this model is given by

$$b(\Sigma \otimes I_m) = (X'(\Sigma^{-1} \otimes I_m)X)^{-1} X'(\Sigma^{-1} \otimes I_m)y. \tag{3.98}$$

As an application of Theorem 3.9, we see that the risk matrix of a GLSE $\hat{\beta} = C(e)y \in \mathcal{C}_1$ is bounded below by the covariance matrix of the GME $b(\Sigma \otimes I_m)$:

$$\begin{aligned} R(\hat{\beta}, \beta) &\geq R(b(\Sigma \otimes I_m), \beta) \\ &= \text{Cov}(b(\Sigma \otimes I_m)) \\ &= (X'(\Sigma^{-1} \otimes I_m)X)^{-1}, \end{aligned} \tag{3.99}$$

where e is the OLS residual vector. Typical examples are the GLSEs of the form

$$b(\hat{\Sigma} \otimes I_m) = C(e)y \tag{3.100}$$

with

$$C(e) = (X'(\hat{\Sigma}^{-1} \otimes I_m)X)^{-1} X'(\hat{\Sigma}^{-1} \otimes I_m),$$

where $\hat{\Sigma} = \hat{\Sigma}(e)$ is an estimator of Σ satisfying $\hat{\Sigma}(e) \in \mathcal{S}(p)$ a.s. and

$$\hat{\Sigma}(ae) = a^2 \hat{\Sigma}(e) \quad \text{for any} \quad a > 0. \tag{3.101}$$

Such GLSEs include the restricted Zellner estimator (RZE) and the unrestricted Zellner estimator (UZE). See Example 2.7.

A characterization of elliptical symmetry. Next, we return to the general model (3.79) and discuss the condition

$$E(u_1|u_2) = 0 \text{ a.s.,} \tag{3.102}$$

which defines the class $\mathcal{Q}_n(0, \sigma^2\Sigma)$. While the class $\mathcal{Q}_n(0, \sigma^2\Sigma)$ is much larger than the class $\mathcal{E}_n(0, \sigma^2\Sigma)$ of elliptically symmetric distributions, it depends on the regressor matrix X through u_1 and u_2. The fact that the class $\mathcal{Q}_n(0, \sigma^2\Sigma)$ depends on X is not consistent with our setting in which X is known (and hence X is independent of the error term ε).

This leads to the problem of characterizing the class of distributions that satisfy the following condition:

$$E(u_1|u_2) = 0 \text{ a.s. holds for any } X : n \times k \text{ of full rank.} \tag{3.103}$$

Interestingly, the class of distributions satisfying (3.103) is shown to be a class of elliptically symmetric distributions. This fact follows from Eaton's (1986) theorem in which a characterization of spherically symmetric distribution is given. To state his theorem, it is convenient to enlarge the classes $\mathcal{P}_n(0, \sigma^2\Sigma)$, $\mathcal{Q}_n(0, \sigma^2\Sigma)$ and $\mathcal{E}_n(0, \sigma^2\Sigma)$ to

$$\tilde{\mathcal{P}}_n(0, \Sigma) = \bigcup_{\sigma^2 > 0} \mathcal{P}_n(0, \sigma^2\Sigma),$$

$$\tilde{\mathcal{Q}}_n(0, \Sigma) = \bigcup_{\sigma^2 > 0} \mathcal{Q}_n(0, \sigma^2\Sigma),$$

$$\tilde{\mathcal{E}}_n(0, \Sigma) = \bigcup_{\sigma^2 > 0} \mathcal{E}_n(0, \sigma^2\Sigma) \tag{3.104}$$

respectively, and rewrite the condition (3.102) as

$$E(P_\Sigma \varepsilon | Q_\Sigma \varepsilon) = 0 \text{ a.s.,} \tag{3.105}$$

where

$$P_\Sigma = X(X'\Sigma^{-1}X)^{-1}X'\Sigma^{-1}$$

and

$$Q_\Sigma = I_n - P_\Sigma = \Sigma Z(Z'\Sigma Z)^{-1}Z'.$$

Here, P_Σ is the projection matrix onto $L(X)$ whose null space is $L(\Sigma Z)$. Equivalence between the conditions (3.102) and (3.105) follows since u_2 and $Q_\Sigma \varepsilon$ are in one-to-one correspondence.

For notational simplicity, let us further rewrite the condition (3.105) in terms of

$$\tilde{\varepsilon} = \Sigma^{-1/2}\varepsilon. \tag{3.106}$$

This clearly satisfies

$$\mathcal{L}(\tilde{\varepsilon}) \in \mathcal{E}_n(0, \sigma^2 I_n) \subset \tilde{\mathcal{E}}_n(0, I_n). \tag{3.107}$$

The condition (3.105) is equivalent to

$$E(M_\Sigma \tilde{\varepsilon} | N_\Sigma \tilde{\varepsilon}) = 0 \text{ a.s.,} \tag{3.108}$$

where the matrix

$$M_\Sigma = \Sigma^{-1/2} P_\Sigma \Sigma^{1/2} = \Sigma^{-1/2} X (X' \Sigma^{-1} X)^{-1} X' \Sigma^{-1/2}$$

is the orthogonal projection matrix onto $L(\Sigma^{-1/2}X)$, and

$$N_\Sigma = I_n - M_\Sigma.$$

Equivalence between (3.105) and (3.108) is shown by the one-to-one correspondence between $Q_\Sigma \varepsilon$ and $N_\Sigma \tilde{\varepsilon}$.

Eaton's characterization theorem is given below.

Theorem 3.10 *Suppose that an n-dimensional random vector v has a mean vector. Then it satisfies*

$$\mathcal{L}(\Gamma v) = \mathcal{L}(v) \quad \text{for any } \Gamma \in \mathcal{O}(n), \tag{3.109}$$

if and only if

$$E(x'v | z'v) = 0 \text{ a.s.} \tag{3.110}$$

for any $x, z \in R^n - \{0\}$ such that $x'z = 0$.

Proof. Suppose that v satisfies (3.109). For each $x, z \in R^n - \{0\}$ such that $x'z = 0$, there exists an $n \times n$ orthogonal matrix $\tilde{\Gamma}$ of the form

$$\tilde{\Gamma} = \begin{pmatrix} \gamma_1' \\ \gamma_2' \\ \Gamma_3' \end{pmatrix}$$

with

$$\gamma_1 = x/\|x\|, \quad \gamma_2 = z/\|z\| \quad \text{and} \quad \Gamma_3 : n \times (n-2).$$

Let

$$\tilde{v} = \tilde{\Gamma} v = \begin{pmatrix} \gamma_1' v \\ \gamma_2' v \\ \Gamma_3' v \end{pmatrix} \equiv \begin{pmatrix} \tilde{v}_1 \\ \tilde{v}_2 \\ \tilde{v}_3 \end{pmatrix}.$$

Then it holds that

$$\mathcal{L}(v) = \mathcal{L}(\tilde{v}).$$

Hence, by Proposition 1.17, the marginal distribution of a subvector $(\tilde{v}_1, \tilde{v}_2)' : 2 \times 1$ of $\tilde{v}$ is also spherically symmetric. Therefore, by Proposition 1.19, it holds that the conditional distribution of $\tilde{v}_1$ given $\tilde{v}_2$ is also spherically symmetric. From this,

$$E(\tilde{v}_1 | \tilde{v}_2) = 0 \quad \text{a.s.}$$

is obtained. The above equality is equivalent to (3.110), since correspondence between $\tilde{v}_2$ and $z'v$ is one-to-one.

Conversely, suppose that (3.110) holds. Let $\xi(t)$ be the characteristic function of v:

$$\xi(t) = E[\exp(it'v)] \quad (t = (t_1, \dots, t_n)' \in R^n).$$

It is sufficient to show that

$$\xi(t) = \xi(\Gamma t) \quad \text{for any } \Gamma \in \mathcal{O}(n). \tag{3.111}$$

By using (3.110), observe that for any $x, z \in R^n - \{0\}$ such that $x'z = 0$,

$$\begin{aligned}
x'E[v \exp(iz'v)] &= E[x'v \exp(iz'v)] \\
&= E[E(x'v | z'v) \exp(iz'v)] \\
&= 0.
\end{aligned}$$

This can be restated in terms of $\xi(t)$ as

$$x'\nabla \xi(z) = 0. \tag{3.112}$$

Here,

$$\nabla \xi(t) \equiv \left(\frac{\partial}{\partial t_1} \xi(t), \dots, \frac{\partial}{\partial t_n} \xi(t) \right)' : n \times 1$$

denotes the gradient of $\xi(t)$, which is rewritten by

$$\nabla \xi(t) = iE[v \exp(it'v)].$$

Fix $t \in R^n$ and $\Gamma \in \mathcal{O}(n)$. Then the vector Γt lies in the sphere

$$\mathcal{U}(n; \|t\|) = \{u \in R^n \mid \|u\| = \|t\|\}.$$

Hence, there exists a differentiable function

$$c : (0, 1) \to \mathcal{U}(u; \|t\|)$$

such that $c(\alpha_1) = t$ and $c(\alpha_2) = \Gamma t$ for some $\alpha_1, \alpha_2 \in (0, 1)$. Since $c(\alpha)'c(\alpha) = \|t\|^2$ for any $\alpha \in (0, 1)$, differentiating both sides with respect to α yields

$$\dot{c}(\alpha)'c(\alpha) = 0 \quad \text{for any } \alpha \in (0, 1),$$

where $\dot{c}(\alpha) = \frac{d}{d\alpha}c(\alpha) : n \times 1$. Hence, by letting $x = \dot{c}(\alpha)$ and $z = c(\alpha)$ in (3.112), we have

$$\dot{c}(\alpha)' \, \nabla\xi(c(\alpha)) = 0 \quad \text{for any } \alpha \in (0, 1). \tag{3.113}$$

The left-hand side of this equality is equal to $\frac{d}{d\alpha}\xi(C(\alpha))$. Thus, $\xi(c(\alpha))$ is constant as a function of α on $(0, 1)$. Hence,

$$\xi(\Gamma t) = \xi(C(\alpha_2)) = \xi(C(\alpha_1)) = \xi(t).$$

This completes the proof.

The above theorem is an extension of Cambanis, Huang and Simons (1981). Some related results are found in Fang, Kotz and Ng (1990). Let us apply this theorem to our setting.

Theorem 3.11 *Fix $\Sigma \in \mathcal{S}(n)$ and suppose that ε satisfies $\mathcal{L}(\varepsilon) \in \tilde{\mathcal{P}}_n(0, \Sigma)$. Then the condition (3.103) holds if and only if $\mathcal{L}(\varepsilon) \in \tilde{\mathcal{E}}_n(0, \Sigma)$.*

Proof. The condition (3.103) is equivalent to

$$E(M_\Sigma \tilde{\varepsilon} | N_\Sigma \tilde{\varepsilon}) = 0 \tag{3.114}$$

for any $X : n \times k$ such that $\text{rank} X = k$. Here, the matrices M_Σ and N_Σ are viewed as functions of X. Let $\mathcal{M}$ be the set of all $n \times n$ orthogonal projection matrices of rank k:

$$\mathcal{M} = \{M : n \times n \mid M = M' = M^2, \ \text{rank} M = k\}. \tag{3.115}$$

Then the matrix M_Σ clearly belongs to the set $\mathcal{M}$. Moreover, the set $\mathcal{M}$ is shown to be equivalent to

$$\mathcal{M} = \{\Sigma^{-1/2} X (X'\Sigma^{-1}X)^{-1} X'\Sigma^{-1/2} \mid X : n \times k, \ \text{rank} X = k\}. \tag{3.116}$$

Therefore, we can see that the condition (3.114) is equivalent to

$$E(M\tilde{\varepsilon} | N\tilde{\varepsilon}) = 0 \quad \text{a.s. for any } M \in \mathcal{M}, \tag{3.117}$$

where $N = I_n - M$. It is clear that this condition holds if $\mathcal{L}(\tilde{\varepsilon}) \in \tilde{\mathcal{E}}_n(0, I_n)$, or equivalently $\mathcal{L}(\varepsilon) \in \tilde{\mathcal{E}}_n(0, \Sigma)$. Thus the "if" part of the theorem is proved.

Next, we establish the "only if" part. Since for any $x, z \in R^n - \{0\}$ such that $x'z = 0$ there exists an orthogonal projection matrix $M \in \mathcal{M}$ satisfying $Mx = x$ and $Nz = z$,

$$E(x'\tilde{\varepsilon} | z'\tilde{\varepsilon}) = E(x'M\tilde{\varepsilon} | z'M\tilde{\varepsilon})$$

$$= E[x'E(M\tilde{\varepsilon} | N\tilde{\varepsilon}) | z'N\tilde{\varepsilon}]$$

$$= 0, \tag{3.118}$$

where the second equality follows since $z'N\tilde{\varepsilon}$ is a function of $N\tilde{\varepsilon}$. Thus, from Theorem 3.10, we obtain $\mathcal{L}(\tilde{\varepsilon}) \in \tilde{\mathcal{E}}_n(0, I_n)$, which is equivalent to $\mathcal{L}(\varepsilon) \in \tilde{\mathcal{E}}_n(0, \Sigma)$. This completes the proof.

In addition to the papers introduced above, Berk (1986) and Bischoff, Cremers and Fieger (1991) are important. Berk (1986) obtained a characterization of normality assumption in the spherically symmetric distributions. Bischoff, Cremers and Fieger (1991) also discussed this problem and clarified the relation between an optimality of the ordinary least squares estimator (OLSE) and normality assumption. See also Bischoff (2000).

3.5   An Application to GLSEs with Iterated Residuals

In some models, one may use iterated residuals in forming a GLSE. Typical examples are the iterated versions of the Cochrane–Orcutt estimator and the Prais–Winsten estimator in an AR(1) error model (see e.g., Judge et al., 1985). The MLE for this model is also obtained from the iterated procedure (see Beach and MacKinnon, 1978). In this section, we derive the lower bound for the risk matrices of such iterated estimators by applying Theorem 3.9. The main results are due to Toyooka (1987). Furthermore, as a similar result to Theorem 3.8, we also discuss the case in which the use of iterated residuals makes no change on the one with no iteration.

GLSE with iterated residuals. Let us consider the model

$$y = X\beta + \varepsilon \quad \text{with} \quad \mathcal{L}(\varepsilon) \in \mathcal{P}_n(0, \sigma^2 \Sigma), \tag{3.119}$$

where y is an $n \times 1$ vector and X is an $n \times k$ known matrix of full rank. Suppose that the matrix Σ is given by a function of an unknown vector θ:

$$\Sigma = \Sigma(\theta), \tag{3.120}$$

where its functional form is supposed to be known.

To define a GLSE with iterated residuals, or simply, an iterated GLSE, let

$$e_{[1]} = y - Xb(I_n) = Ny \quad \text{with} \quad N = I_n - X(X'X)^{-1}X' \tag{3.121}$$

be the first-step OLS residual vector. A first-step GLSE is defined as

$$\hat{\beta}_{[1]} = b(\hat{\Sigma}_{[1]}) = (X'\hat{\Sigma}_{[1]}^{-1}X)^{-1}X'\hat{\Sigma}_{[1]}^{-1}y \tag{3.122}$$

with

$$\hat{\Sigma}_{[1]} = \Sigma(\hat{\theta}(e_{[1]})),$$

where $\hat{\theta} = \hat{\theta}(e_{[1]})$ is an estimator of θ. The first-step GLSE is nothing but the usual GLSE (without iteration) that we have considered so far. Next, we define the second-step residual vector as

$$e_{[2]} = y - X\hat{\beta}_{[1]} = [I_n - X(X'\hat{\Sigma}_{[1]}^{-1}X)^{-1}X'\hat{\Sigma}_{[1]}^{-1}]y \tag{3.123}$$

and construct the second-step estimator of θ by $\hat{\theta}(e_{[2]})$. Here we note that the functional form of $\hat{\theta}$ is fixed throughout the iteration procedure. This leads to the second-step iterated GLSE

$$\hat{\beta}_{[2]} = b(\hat{\Sigma}_{[2]}) = (X'\hat{\Sigma}_{[2]}^{-1}X)^{-1}X'\hat{\Sigma}_{[2]}^{-1}y \tag{3.124}$$

with

$$\hat{\Sigma}_{[2]} = \Sigma(\hat{\theta}(e_{[2]})).$$

Similarly, we can formulate the ith step iterated GLSE as

$$\hat{\beta}_{[i]} = b(\hat{\Sigma}_{[i]}) = (X'\hat{\Sigma}_{[i]}^{-1}X)^{-1}X'\hat{\Sigma}_{[i]}^{-1}y \tag{3.125}$$

with

$$\hat{\Sigma}_{[i]} = \Sigma(\hat{\theta}(e_{[i]})),$$

where

$$e_{[i]} = y - X\hat{\beta}_{[i-1]} = [I_n - X(X'\hat{\Sigma}_{[i-1]}^{-1}X)^{-1}X'\hat{\Sigma}_{[i-1]}^{-1}]y \tag{3.126}$$

is the ith step residual vector.

Here let us confirm that what we called the "iterated GLSE" above is in fact a GLSE in our sense, or equivalently, that is, we must prove that the iterated GLSE $\hat{\beta}_{[i]}$ is rewritten as

$$\hat{\beta}_{[i]} = C_i(e)y$$

for some function $C_i(e)$ such that $C_i(e)X = I_k$, where $e = e_{[1]}$. To do so, it is sufficient to show that the ith step residual vector $e_{[i]}$ depends only on the first-step residual vector $e_{[1]}$. Let

$$Q_{[i]} = \hat{\Sigma}_{[i]}Z(Z'\hat{\Sigma}_{[i]}Z)^{-1}Z' \quad (i \geq 1) \quad \text{and} \quad Q_{[0]} = N, \tag{3.127}$$

where Z is an $n \times (n-k)$ matrix satisfying

$$N = ZZ' \quad \text{and} \quad Z'Z = I_{n-k}.$$

Then, from the matrix identify

$$I_n - X(X'\hat{\Sigma}^{-1}X)^{-1}X'\hat{\Sigma}^{-1} = \hat{\Sigma}Z(Z'\hat{\Sigma}Z)^{-1}Z',$$

we can see that for $i \geq 1$,

$$\begin{aligned} e_{[i]} &= \hat{\Sigma}_{[i-1]}Z(Z'\hat{\Sigma}_{[i-1]}Z)^{-1}Z'y \\ &= \hat{\Sigma}_{[i-1]}Z(Z'\hat{\Sigma}_{[i-1]}Z)^{-1}Z'ZZ'y \\ &= \hat{\Sigma}_{[i-1]}Z(Z'\hat{\Sigma}_{[i-1]}Z)^{-1}Z'e_{[1]} \\ &= Q_{[i-1]}e_{[1]}, \end{aligned} \tag{3.128}$$

where the second equality follows from $Z'Z = I_{n-k}$ and the third from $e_{[1]} = Z'y$. This equality shows that $e_{[i]}$ depends only on $e_{[1]}$. In fact, the matrix $Q_{[i-1]}$ in (3.128) is a function of $e_{[i-1]}$. Substituting $i = 2$ into (3.128) shows that $e_{[2]}$ depends only on $e_{[1]}$. And successive substitution yields that the ith step residual vector $e_{[i]}$ is a function of $e_{[1]}$.

Thus, we have the following result due to Toyooka (1987).

Proposition 3.12 *The ith step iterated GLSE $\hat{\beta}_{[i]}$ in (3.125) can be rewritten as $\hat{\beta}_{[i]} = C_i(e_{[1]})y$ for some function C_i such that $C_i(e_{[1]})X = I_k$.*

Hence, all the results that have been proved for class $\mathcal{C}_1$ are applicable. For example, if an iterated GLSE $\hat{\beta} = C(e_{[1]})y$ satisfies

$$C(ae_{[1]}) = C(e_{[1]}) \quad \text{for any} \ \ a > 0, \tag{3.129}$$

and if the function C is continuous, then the iterated GLSE has a finite second moment. If $\mathcal{L}(\varepsilon) = \mathcal{L}(-\varepsilon)$ and the GLSE satisfies

$$C(e_{[1]}) = C(-e_{[1]}), \tag{3.130}$$

then it is unbiased as long as the first moment is finite.

Applying Theorem 3.9 with

$$u_1 = X'\Sigma^{-1}\varepsilon : k \times 1 \quad \text{and} \quad u_2 = Z'\varepsilon : (n-k) \times 1$$

to the above situation yields

Theorem 3.13 *Suppose that $\mathcal{L}(\varepsilon) \in \mathcal{Q}_n(0, \sigma^2\Sigma)$. Then the conditional risk matrix of the ith step iterated GLSE $\hat{\beta}_{[i]} = C_i(e_{[1]})y$ is decomposed into two parts:*

$$R(\hat{\beta}_{[i]}, \beta | u_2) = R(b(\Sigma), \beta) + (\hat{\beta}_{[i]} - b(\Sigma))(\hat{\beta}_{[i]} - b(\Sigma))' \tag{3.131}$$
$$= \sigma^2(X'\Sigma^{-1}X)^{-1}$$
$$+ C_i(Zu_2)\Sigma Z(Z'\Sigma Z)^{-1}u_2 u_2'(Z'\Sigma Z)^{-1}Z'\Sigma C_i(Zu_2)',$$

where $b(\Sigma) = (X'\Sigma^{-1}X)^{-1}X'\Sigma^{-1}y$ is the GME of β. Thus, the risk matrix is bounded below by that of the GME:

$$R(\hat{\beta}_{[i]}, \beta) \geq R(b(\Sigma), \beta) = \text{Cov}(b(\Sigma))$$
$$= \sigma^2(X'\Sigma^{-1}X)^{-1}. \tag{3.132}$$

The MLE derived under normal distribution is given as a solution of the estimating equations, and it is usually approximated by an iterated GLSE because it is regarded as the limit of iteration.

An identity between iterated GLSEs. It is not trivial whether an iterated GLSE is superior to a GLSE with no iteration. In fact, in an AR(1) error model, Magee, Ullah and Srivastava (1987), Magee (1985), Kobayashi (1985) and Toyooka (1987) obtained asymptotic expansions of the risk matrices of a typical GLSE and its iterated version, and pointed out that there is no difference in asymptotic efficiency among them up to the second order. In an SUR model, Srivastava and Giles (1987) reported that higher steps iterated GLSEs are generally inferior to the first or second step GLSE. In this section, we discuss the case in which the iterated GLSE makes no change on the one with no iteration. For notational simplicity, we write the first step residual $e_{[1]}$ as e without suffix.

Proposition 3.14 *If an estimator $\hat{\theta} = \hat{\theta}(e)$ satisfies the following invariance property*

$$\hat{\theta}(e + Xg) = \hat{\theta}(e) \quad for \ any \ g \in R^k, \tag{3.133}$$

then, for any iterated residuals $e_{[i]}$ $(i \geq 2)$, the following identity holds:

$$\hat{\theta}(e_{[i]}) = \hat{\theta}(e). \tag{3.134}$$

Thus, for any $i \geq 2$, the ith step iterated GLSE is the same as the one with no iteration.

Proof. For any $i \geq 2$, there exists a function $v_i(e)$ satisfying

$$e_{[i]} = e + Xv_i(e), \tag{3.135}$$

because from (3.128), we obtain

$$Z'e_{[i]} = Z'e. \tag{3.136}$$

Hence, we have $\hat{\theta}(e_{[i]}) = \hat{\theta}(e + Xv_i(e)) = \hat{\theta}(e)$. This completes the proof.

Examples of the estimator $\hat{\theta}$ satisfying (3.133) are the unrestricted GLSE in a heteroscedastic model and the UZE in an SUR model.

Example 3.4 (Heteroscedastic model)   In the heteroscedastic model in Example 3.2, decompose the OLS residual vector e as

$$e = y - Xb(I_n) = (e_1', \ldots, e_p')' \quad with \quad e_j : n_j \times 1. \tag{3.137}$$

Then, for any $g \in R^k$, the vector $e + Xg$ is written as

$$e + Xg = \begin{pmatrix} e_1 + X_1 g \\ \vdots \\ e_p + X_p g \end{pmatrix}. \tag{3.138}$$

The unrestricted GLSE is a GLSE $b(\Omega(\hat{\theta}))$ with

$$\hat{\theta} = \hat{\theta}(e) = \begin{pmatrix} \hat{\theta}_1 \\ \vdots \\ \hat{\theta}_p \end{pmatrix} : p \times 1,$$

where

$$\hat{\theta}_j = \hat{\theta}_j(e) = e'_j N_j e_j / (n_j - r_j), \tag{3.139}$$
$$r_j = \text{rank} X_j,$$
$$N_j = I_{n_j} - X_j (X'_j X_j)^+ X'_j.$$

Since $N_j X_j = 0$ holds for $j = 1, \ldots, p$, we obtain from (3.138) that

$$\hat{\theta}_j(e + Xg) = (e_j + X_j g)' N_j (e_j + X_j g)/(n_j - r_j)$$
$$= e'_j N_j e_j / (n_j - r_j)$$
$$= \hat{\theta}_j(e), \tag{3.140}$$

and hence $\hat{\theta}(e + Xg) = \hat{\theta}(e)$ for any $g \in R^k$. Thus, from Proposition 3.14, the iterated versions of the unrestricted GLSE are the same as the one with no iteration.

Example 3.5 (SUR model)   In the p-equation SUR model in Example 3.3, the OLS residual vector is given by

$$e = y - Xb(I_p \otimes I_m) = (e'_1, \ldots, e'_p)' \quad \text{with} \quad e_j : m \times 1. \tag{3.141}$$

For any $g = (g'_1, \ldots, g'_p)' \in R^k$ with $g_j : k_j \times 1$, the vector $e + Xg$ is written as

$$e + Xg = \begin{pmatrix} e_1 + X_1 g_1 \\ \vdots \\ e_p + X_p g_p \end{pmatrix}. \tag{3.142}$$

The UZE is a GLSE $b(\hat{\Sigma} \otimes I_m)$ with

$$\hat{\Sigma} = S(e) = (e'_i N_* e_j), \tag{3.143}$$

where $N_* = I_m - X_* (X'_* X_*)^+ X'_*$ and $X_* = (X_1, \ldots, X_p) : m \times k$. Since $N_* X_j = 0$ holds for $j = 1, \ldots, p$, we have from (3.142) that

$$S(e + Xg) = ((e_i + X_i g_i)' N_* (e_j + X_j g_j)) = (e'_i N_* e_j) = S(e). \tag{3.144}$$

Hence, from Proposition 3.14, the iterated versions of the UZE are the same as the one with no iteration.

3.6 Problems

3.1.1 Verify that Φ in (3.3) is equal to

$$\begin{pmatrix} I_n & 0 \\ \Delta\Sigma^{-1} & I_m \end{pmatrix} \Omega \begin{pmatrix} I_n & 0 \\ \Delta\Sigma^{-1} & I_m \end{pmatrix}'$$

with Ω in (3.6).

3.2.1 Show that the unbiasedness of a linear predictor $\hat{y}_0 = Cy$ is equivalent to $CX = X_0$.

3.2.2 Complete the proof of Theorem 3.1 by showing (3.16), (3.17) and

$$V_{12} = 0.$$

3.2.3 Establish Proposition 3.3.

3.3.1 Establish the following four matrix identities:

(1) $\Sigma^{-1/2}X(X'\Sigma^{-1}X)^{-1}X'\Sigma^{-1/2} + \Sigma^{1/2}Z(Z'\Sigma Z)^{-1}Z'\Sigma^{1/2} = I_n$;

(2) $X(X'\Sigma^{-1}X)^{-1}X'\Sigma^{-1} + \Sigma Z(Z'\Sigma Z)^{-1}Z' = I_n$;

(3) $\Sigma^{-1}X(X'\Sigma^{-1}X)^{-1}X'\Sigma^{-1} + Z(Z'\Sigma Z)^{-1}Z' = \Sigma^{-1}$;

(4) $X(X'\Sigma^{-1}X)^{-1}X' + \Sigma Z(Z'\Sigma Z)^{-1}Z'\Sigma = \Sigma$.

3.3.2 Verify Lemma 3.4.

3.4.1 Modify Eaton's characterization theorem (Theorem 3.10) to the case in which v may not have a finite first moment. The answer will be found in Eaton (1986).

3.4.2 By modifying Lemma 3.1 (and its proof) of Bischoff, Cremers and Fieger (1991) to the setup treated in Section 3.4, prove the following statement, which can be understood as the converse of the nonlinear Gauss–Markov theorem in Section 3.4: Let $P \equiv \mathcal{L}(\varepsilon) \in \mathcal{P}_n(0, \sigma^2\Sigma)$. If the GME $b(\Sigma)$ satisfies

$$\mathrm{Cov}_P(b(\Sigma)) \leq R_P(\hat{\beta}, \beta)$$

for any $\hat{\beta} \in \mathcal{C}_1$, then P belongs to the class $\mathcal{Q}_n(0, \sigma^2\Sigma)$.

3.4.3 In the two-equation heteroscedastic model (which is obtained by letting $p = 2$ in Example 3.2), suppose that the error term ε is normally distributed: $\mathcal{L}(\varepsilon) = N_n(0, \Omega)$.

(1) Show that $\hat{\beta}_j = (X'_j X_j)^{-1}X'_j y_j$ is in $\mathcal{C}_1$ ($j = 1, 2$), that is, the OLSE calculated from each homoscedastic model is a GLSE.

(2) Find a sufficient condition for which the restricted and unrestricted GLSEs $b(\hat{\Omega})$ in Example 2.6 satisfies

$$\text{Cov}(b(\hat{\Omega})) \leq \text{Cov}(\hat{\beta}_j).$$

(3) Evaluate the exact covariance matrix of the unrestricted GLSE.

See Khatri and Shah (1974) in which various results on the efficiency of the GLSEs are derived. See also Taylor (1977, 1978), Swamy and Mehta (1979) and Kubokawa (1998). In Kubokawa (1998), several GLSEs that improve the unrestricted GLSE are obtained.

3.4.4 Consider the two-equation SUR model (which is obtained by letting $p = 2$ in Example 3.3).

(1) Show that when the regressor matrix X_2 of the second equation satisfies $X_2 = (X_1, G)$ for some $G : m \times (k_2 - k_1)$, the GME

$$b(\Sigma \otimes I_m) = \begin{pmatrix} b_1(\Sigma \otimes I_m) \\ b_2(\Sigma \otimes I_m) \end{pmatrix} \quad \text{with } b_j(\Sigma \otimes I_m) : k_j \times 1$$

satisfies

$$b_1(\Sigma \otimes I_m) = b_1(I_2 \otimes I_m) \quad (= (X_1'X_1)^{-1}X_1'y_1).$$

(2) Extend the result above to the case in which $L(X_1) \subset L(X_2)$.

(3) Show that

$$b(\Sigma \otimes I_m) = b(I_2 \otimes I_m)$$

holds if and only if $L(X_1) = L(X_2)$.

See Revankar (1974, 1976), Kariya (1981b) and Srivastava and Giles (1987).

3.4.5 In the two-equation SUR model, suppose that the error term ε is normally distributed: $\mathcal{L}(\varepsilon) = N_n(0, \Sigma \otimes I_m)$.

(1) Derive the exact covariance matrix of the UZE under the assumption $X_1'X_2 = 0$.

(2) Derive the exact covariance matrix of the UZE for general X.

For (2), see Kunitomo (1977) and Mehta and Swamy (1976).

4

Efficiency of GLSE with Applications to SUR and Heteroscedastic Models

4.1 Overview

In Chapter 3, it was shown that the risk matrix of a generalized least squares estimator (GLSE) is bounded below by the risk (covariance) matrix of the Gauss–Markov estimator (GME) under a certain condition. The difference between the two risk matrices reflects the loss of efficiency caused by estimating the unknown parameters in the covariance matrix of the error term. In this chapter, we consider the problem of evaluating the efficiency of a GLSE by deriving an effective upper bound for its risk matrix relative to that of the GME. The upper bound thus derived serves as a measure of the efficiency of a GLSE.

To describe the problem precisely, let the general linear regression model be

$$y = X\beta + \varepsilon, \tag{4.1}$$

where y is an $n \times 1$ vector and X is an $n \times k$ known matrix of full rank. Let $\mathcal{P}_n(0, \Omega)$ be the set of distributions on R^n with mean 0 and covariance matrix Ω, and assume that the distribution of the error term ε is in $\mathcal{P}_n(0, \Omega)$, that is,

$$\mathcal{L}(\varepsilon) \in \mathcal{P}_n(0, \Omega) \quad \text{with} \quad \Omega \in \mathcal{S}(n), \tag{4.2}$$

where as before, $\mathcal{L}(\cdot)$ denotes the distribution of $\cdot$ and $\mathcal{S}(n)$ the set of $n \times n$ positive definite matrices. In our context, the covariance matrix Ω is supposed to

Generalized Least Squares Takeaki Kariya and Hiroshi Kurata
© 2004 John Wiley & Sons, Ltd ISBN: 0-470-86697-7 (PPC)

be a function of an unknown but estimable parameter vector θ, say

$$\Omega = \Omega(\theta), \tag{4.3}$$

where the functional form of $\Omega(\cdot)$ is known.

In this model, the GME of β is given by

$$b(\Omega) = (X'\Omega^{-1}X)^{-1}X'\Omega^{-1}y, \tag{4.4}$$

which is not feasible in application. A feasible estimator of β is a GLSE of the form

$$b(\hat{\Omega}) = (X'\hat{\Omega}^{-1}X)^{-1}X'\hat{\Omega}^{-1}y, \tag{4.5}$$

where $\hat{\Omega}$ is defined by

$$\hat{\Omega} = \Omega(\hat{\theta}) \tag{4.6}$$

and $\hat{\theta} = \hat{\theta}(e)$ is an estimator of θ based on the ordinary least squares (OLS) residual vector

$$e = Ny \quad \text{with} \quad N = I_n - X(X'X)^{-1}X'. \tag{4.7}$$

Then clearly, $b(\hat{\Omega})$ belongs to the class

$$\mathcal{C}_1 = \{\hat{\beta} = C(e)y \mid C(\cdot) \text{ is a } k \times n \text{ matrix-valued function}$$
$$\text{on } R^n \text{ satisfying } C(\cdot)X = I_k\}, \tag{4.8}$$

since $b(\hat{\Omega})$ is written as $b(\hat{\Omega}) = C(e)y$ by letting $C(e) = (X'\hat{\Omega}^{-1}X)^{-1}X'\hat{\Omega}^{-1}$. The risk matrix of a GLSE $b(\hat{\Omega})$ is given by

$$R(b(\hat{\Omega}), \beta) = E\{(b(\hat{\Omega}) - \beta)(b(\hat{\Omega}) - \beta)'\}.$$

Let Z be any $n \times (n - k)$ matrix satisfying

$$Z'X = 0, \quad Z'Z = I_{n-k} \quad \text{and} \quad ZZ' = N \tag{4.9}$$

and will be fixed for a given X throughout this chapter. When $\mathcal{L}(\varepsilon)$ belongs to the class $\mathcal{Q}_n(0, \Omega)$, where

$$\mathcal{Q}_n(0, \Omega) = \{P \in \mathcal{P}_n(0, \Omega) \mid E_P(u_1|u_2) = 0 \text{ a.s.}\}$$

with

$$u_1 = X'\Omega^{-1}\varepsilon : k \times 1 \quad \text{and} \quad u_2 = Z'\varepsilon : (n - k) \times 1,$$

the risk matrix of a GLSE $b(\hat{\Omega})$ is bounded below by the covariance matrix of the GME:

$$R(b(\hat{\Omega}), \beta) \geq \text{Cov}(b(\Omega)) = (X'\Omega^{-1}X)^{-1}. \tag{4.10}$$

(See Theorem 3.9.) Recall that the class $\mathcal{Q}_n(0, \Omega)$ contains the class $\mathcal{E}_n(0, \Omega)$ of elliptically symmetric distribution, which in turn contains the normal distribution $N_n(0, \Omega)$.

By this inequality, finite sample efficiency of a GLSE can be measured by such quantities as

$$\delta_1 = R(b(\hat{\Omega}), \beta) - \text{Cov}(b(\Omega)), \tag{4.11}$$

$$\delta_2 = |R(b(\hat{\Omega}), \beta)| / |\text{Cov}(b(\Omega))|, \tag{4.12}$$

$$\delta_3 = \frac{1}{k} tr \left\{ R(b(\hat{\Omega}), \beta) \, [\text{Cov}(b(\Omega))]^{-1} \right\}, \tag{4.13}$$

where $|A|$ denotes the determinant of matrix A. However, since the GLSE $b(\hat{\Omega})$ is in general nonlinear in y, it is difficult to evaluate these δ's in an explicit form. In fact, the risk matrices of the GLSEs introduced in the previous chapters have not been analytically derived except for several simple cases. Furthermore, even in such simple cases, the quantities δ's are quite complicated functions of θ and X in general, and hence, these δ's cannot be tractable measures for the efficiency of a GLSE.

To overcome this difficulty, we formulate the problem as that of finding an effective upper bound for the risk matrix $R(b(\hat{\Omega}), \beta)$ of the form

$$R(b(\hat{\Omega}), \beta) \leq \alpha(b(\hat{\Omega})) \, \text{Cov}(b(\Omega)), \tag{4.14}$$

and adopt the quantity $\alpha(b(\hat{\Omega}))$ as an alternative measure of the efficiency of $b(\hat{\Omega})$ instead of the above δ's. Here $\alpha \equiv \alpha(b(\hat{\Omega}))$ is a nonrandom real-valued function associated with a GLSE $b(\hat{\Omega})$. This function usually takes a much simpler form than δ's take, and is expected to be a tractable measure of the efficiency. Since the inequality (4.14) is based on the matrix ordering, it holds that for any $a \in R^k$,

$$\text{Var}(a'b(\Omega)) \leq R(a'b(\hat{\Omega}), a'\beta) \leq \alpha(b(\hat{\Omega})) \, \text{Var}(a'b(\Omega))$$

with

$$R(a'b(\hat{\Omega}), a'\beta) = E\{[a'b(\hat{\Omega}) - a'\beta]^2\} = a'R(b(\hat{\Omega}), \beta)a.$$

In particular, for each element $b_j(\hat{\Omega})$ of $b(\hat{\Omega}) = (b_1(\hat{\Omega}), \dots, b_k(\hat{\Omega}))'$, it holds that

$$\text{Var}(b_j(\Omega)) \leq R(b_j(\hat{\Omega}), \beta_j) \leq \alpha(b(\hat{\Omega})) \, \text{Var}(b_j(\Omega)).$$

To get a clearer idea of our concept of efficiency and of its usefulness, we demonstrate without proofs an example of the inequality (4.14) for the unrestricted Zellner estimator (UZE) in a two-equation seemingly unrelated regression (SUR) model. A detailed discussion will be given in subsequent sections.

Example 4.1 (Two-equation SUR model)   The two-equation SUR model is given by

$$y = \begin{pmatrix} y_1 \\ y_2 \end{pmatrix} : n \times 1, \quad X = \begin{pmatrix} X_1 & 0 \\ 0 & X_2 \end{pmatrix} : n \times k, \quad (4.15)$$

$$\beta = \begin{pmatrix} \beta_1 \\ \beta_2 \end{pmatrix} : k \times 1, \quad \varepsilon = \begin{pmatrix} \varepsilon_1 \\ \varepsilon_2 \end{pmatrix} : n \times 1,$$

where

$$y_j : m \times 1, \quad X_j : m \times k_j, \quad n = 2m \text{ and } k = k_1 + k_2.$$

We assume that the error term is distributed as the normal distribution $N_n(0, \Omega)$ with $\Omega = \Sigma \otimes I_m$ and $\Sigma = (\sigma_{ij}) \in \mathcal{S}(2)$.

As is discussed in Examples 2.7 and 3.3, the UZE for this model is a GLSE $b(\hat{\Sigma} \otimes I_m)$ with $\hat{\Sigma} = S$:

$$b(S \otimes I_m) = (X'(S^{-1} \otimes I_m)X)^{-1}X'(S^{-1} \otimes I_m)y,$$

where

$$S = Y'[I_m - X_*(X_*'X_*)^+ X_*']Y, \quad (4.16)$$

$$X_* = (X_1, X_2) : m \times k,$$

$$Y = (y_1, y_2) : m \times 2.$$

The UZE is an unbiased GLSE with a finite second moment, and the matrix S is distributed as $W_2(\Sigma, q)$, the Wishart distribution with mean $q\Sigma$ and degrees of freedom q, where

$$q = m - \operatorname{rank} X_*.$$

In Section 4.3, it is shown that the UZE satisfies

$$(X'(\Sigma^{-1} \otimes I_m)X)^{-1} \leq \operatorname{Cov}(b(S \otimes I_m))$$

$$\leq \alpha(b(S \otimes I_m)) (X'(\Sigma^{-1} \otimes I_m)X)^{-1} \quad (4.17)$$

with

$$\alpha(b(S \otimes I_m)) = E\left[\frac{(w_1 + w_2)^2}{4w_1 w_2}\right] = 1 + \frac{2}{q - 3}, \quad (4.18)$$

where $w_1 \leq w_2$ denotes the latent roots of the matrix $W = \Sigma^{-1/2} S \Sigma^{-1/2}$ and the matrix $(X'(\Sigma^{-1} \otimes I_m)X)^{-1}$ is the covariance matrix of the GME $b(\Sigma \otimes I_m)$. In view of this inequality, it is reasonable to adopt the upper bound $\alpha = \alpha_q = \alpha(b(S \otimes I_m))$ in (4.18) as a measure of the efficiency of the UZE $b(S \otimes I_m)$ relative to the GME. In fact, we can see that whatever X may be, the covariance matrix of the UZE is not greater than $[1 + 2/(q - 3)] (X'(\Sigma^{-1} \otimes I_m)X)^{-1}$. This clearly shows that the UZE can be almost as efficient as the GME when the sample size is large, since the upper bound is a monotonically decreasing function of q and satisfies

$$\lim_{q \to \infty} \alpha_q = 1.$$

Furthermore, as for δ's in (4.11), (4.12) and (4.13), we can obtain the following:

(1) Bound for δ_1:

$$0 \le \delta_1 \le \frac{2}{q-3} (X'(\Sigma^{-1} \otimes I_m)X)^{-1}; \tag{4.19}$$

(2) Bound for δ_2:

$$1 \le \delta_2 \le \left[1 + \frac{2}{q-3}\right]^k; \tag{4.20}$$

(3) Bound for δ_3:

$$1 \le \delta_3 \le 1 + \frac{2}{q-3}. \tag{4.21}$$

Also, it follows immediately from (4.17) that for each subvector $b_j(S \otimes I_m)$: $k_j \times 1$ $(j = 1, 2)$ of the UZE, the following inequality holds:

$$\text{Cov}(b_j(\Sigma \otimes I_m)) \le \text{Cov}(b_j(S \otimes I_m))$$

$$\le \left\{1 + \frac{2}{q-3}\right\} \text{Cov}(b_j(\Sigma \otimes I_m)). \tag{4.22}$$

The results of this example will be established in the p-equation model in Section 4.4. Example 4.1 ends here.

In this book, the problem of deriving an upper bound of the form (4.14) will be called *an upper bound problem*.

Chapters 4 and 5 will be devoted to the description of various aspects of this problem. In these two chapters, the problem is first formulated in a general framework and an effective upper bound is derived when the model satisfies some appropriate assumptions. More precisely, for a GLSE $b(\hat{\Omega})$ in (4.5), the upper bound $\alpha(b(\hat{\Omega}))$ in (4.14) will be derived when either of the following two conditions is satisfied:

(1) The conditional mean and covariance matrix of $b(\hat{\Omega})$ given $\hat{\Omega}$ are of the following form:

$$E[b(\hat{\Omega})|\hat{\Omega}] = \beta \quad \text{and} \quad \text{Cov}(b(\hat{\Omega})|\hat{\Omega}) = H(\hat{\Omega}, \Omega), \tag{4.23}$$

where the function $H : \mathcal{S}(n) \times \mathcal{S}(n) \to \mathcal{S}(k)$ is defined by

$$H = H(\hat{\Omega}, \Omega)$$
$$= (X'\hat{\Omega}^{-1}X)^{-1}X'\hat{\Omega}^{-1}\Omega\hat{\Omega}^{-1}X(X'\hat{\Omega}^{-1}X)^{-1}. \tag{4.24}$$

We call the structure of H *simple covariance structure* in the sequel;

(2) The covariance matrix $\Omega = \Omega(\theta_1, \theta_2)$ takes the form

$$\Omega = \theta_1 \Sigma(\theta_2) \tag{4.25}$$

with

$$\Sigma(\theta_2)^{-1} = I_n + \lambda_n(\theta_2) \, C \quad \text{and } \theta_2 \in \Theta,$$

where C is an $n \times n$ known symmetric matrix, Θ is a set in R^1 and λ_n is a continuous real-valued function on Θ.

Although the condition (4.23) with (4.24) may seem to be quite restrictive, as will be seen soon, some important GLSEs satisfy this condition when X and Ω have certain structures. Typical examples are the UZE in an SUR model and the unrestricted GLSE in a heteroscedastic model. On the other hand, the covariance structure in (4.25) has already been introduced in the previous chapters. Typically, the Anderson model, an equi-correlated model and a two-equation heteroscedastic model satisfy (4.25), which will be treated in Chapter 5.

With this overview, this chapter develops a theory on the basis of the following sections:

4.2 GLSEs with a Simple Covariance Structure

4.3 Upper Bound for the Covariance Matrix of a GLSE

4.4 Upper Bound Problem for the UZE in an SUR Model

4.5 Upper Bound Problem for a GLSE in a Heteroscedastic Model

4.6 Empirical Example: CO_2 Emission Data.

In Section 4.2, some fundamental properties of the GLSEs will be studied. In Section 4.3, an upper bound of the form (4.14) for the covariance matrix of a GLSE will be obtained under a general setting. The results are applied to typical GLSEs in the SUR model and the heteroscedastic model in Sections 4.4 and 4.5 respectively. In Section 4.6, an empirical example on CO_2 emission data is given.

4.2   GLSEs with a Simple Covariance Structure

In this section, we define a class of GLSEs that have simple covariance structure and investigate its fundamental properties.

A nonlinear version of the Gauss–Markov theorem.   The following general linear regression model

$$y = X\beta + \varepsilon \tag{4.26}$$

with

$$\mathcal{L}(\varepsilon) \in \mathcal{P}_n(0, \Omega) \quad \text{and} \quad \Omega = \Omega(\theta) \in \mathcal{S}(n)$$

is studied, where $y : n \times 1$ and $X : n \times k$ is a known matrix of full rank. Let C_3 be a class of GLSEs of the form (4.5) having the simple covariance structure in (4.23) with (4.24), that is,

$$C_3 = \{b(\hat{\Omega}) \in C_1 | \ E[b(\hat{\Omega})|\hat{\Omega}] = \beta \ \text{ and } \ \text{Cov}(b(\hat{\Omega})|\hat{\Omega}) = H(\hat{\Omega}, \Omega)\}. \quad (4.27)$$

A GLSE $b(\hat{\Omega})$ in C_3 is unbiased, since taking the expectations of both sides of $E(b(\hat{\Omega})|\hat{\Omega}) = \beta$ clearly implies

$$E[b(\hat{\Omega})] = \beta.$$

Hence, the conditional covariance matrix $H(\hat{\Omega}, \Omega)$ is written as

$$H(\hat{\Omega}, \Omega) = E\{(b(\hat{\Omega}) - \beta)(b(\hat{\Omega}) - \beta)'|\hat{\Omega}\},$$

which in turn implies that the (unconditional) covariance matrix is given by the expected value of $H(\hat{\Omega}, \Omega)$:

$$\text{Cov}(b(\hat{\Omega})) = E[H(\hat{\Omega}, \Omega)]. \quad (4.28)$$

The reason the conditional covariance matrix $H(\hat{\Omega}, \Omega)$ is called simple lies in the fact that this property is shared by all the linear unbiased estimators, that is, $C_0 \subset C_3$ holds, where

$$C_0 = \{\hat{\beta} = Cy \mid C \text{ is a } k \times n \text{ matrix such that } CX = I_k\} \quad (4.29)$$

is the class of linear unbiased estimators. To see this, note first that a linear unbiased estimator $\hat{\beta} = Cy$ is a GLSE $b(\hat{\Omega}_0)$ with $\hat{\Omega}_0 = [C'C + N]^{-1}$ (see Proposition 2.3). Since the matrix $\hat{\Omega}_0$ is nonrandom, the conditional distribution of $b(\hat{\Omega})$ given $\hat{\Omega}_0$ is the same as the unconditional one. Hence, we have

$$E[b(\hat{\Omega}_0)|\hat{\Omega}_0] = \beta \quad \text{ and } \quad \text{Cov}(b(\hat{\Omega}_0)|\hat{\Omega}_0) = H(\hat{\Omega}_0, \Omega), \quad (4.30)$$

implying $C_0 \subset C_3$. Note that this holds without distinction of the structure of X and Ω. Typical examples are the GME $b(\Omega)$ with covariance matrix

$$H(\Omega, \Omega) = (X'\Omega^{-1}X)^{-1}, \quad (4.31)$$

and the OLSE $b(I_n) = (X'X)^{-1}X'y$ with covariance matrix

$$H(I_n, \Omega) = (X'X)^{-1}X'\Omega X(X'X)^{-1}. \quad (4.32)$$

The covariance matrix (4.28) of a GLSE in the class C_3 is bounded below by that of the GME $b(\Omega)$:

Theorem 4.1 *For any GLSE $b(\hat{\Omega}) \in C_3$, the following inequality holds:*

$$H(\Omega, \Omega) \leq H(\hat{\Omega}, \Omega) \quad \text{a.s.} \quad (4.33)$$

And thus,

$$\text{Cov}(b(\Omega)) \leq \text{Cov}(b(\hat{\Omega})). \quad (4.34)$$

Proof. Using the matrix identity

$$I_n = X(X'\Omega^{-1}X)^{-1}X'\Omega^{-1} + \Omega Z(Z'\Omega Z)^{-1}Z' \qquad (4.35)$$

yields

$$
\begin{aligned}
H(\hat{\Omega}, \Omega) &= (X'\hat{\Omega}^{-1}X)^{-1}X'\hat{\Omega}^{-1}\Omega\hat{\Omega}^{-1}X(X'\hat{\Omega}^{-1}X)^{-1} \\
&= (X'\hat{\Omega}^{-1}X)^{-1}X'\hat{\Omega}^{-1}[X(X'\Omega^{-1}X)^{-1}X'\Omega^{-1} + \Omega Z(Z'\Omega Z)^{-1}Z'] \\
&\qquad \times \Omega\hat{\Omega}^{-1}X(X'\hat{\Omega}^{-1}X)^{-1} \\
&= (X'\Omega^{-1}X)^{-1} \\
&\quad + (X'\hat{\Omega}^{-1}X)^{-1}X'\hat{\Omega}^{-1}\Omega Z(Z'\Omega Z)^{-1}Z'\Omega\hat{\Omega}^{-1}X(X'\hat{\Omega}^{-1}X)^{-1} \quad (4.36) \\
&\geq (X'\Omega^{-1}X)^{-1} \\
&= H(\Omega, \Omega),
\end{aligned}
$$

where the inequality follows since the second term of (4.36) is nonnegative definite. This completes the proof.

This theorem is due to Kariya (1981a).

Examples of GLSEs. The class $\mathcal{C}_3$ contains several typical GLSEs including the unrestricted GLSE in a heteroscedastic model and the UZE in the SUR model.

Example 4.2 (Heteroscedastic model)   In this example, it is shown that the unrestricted GLSE in a heteroscedastic model belongs to the class $\mathcal{C}_3$, when the distribution of the error term is normal. The heteroscedastic model considered here is given by

$$
y = \begin{pmatrix} y_1 \\ \vdots \\ y_p \end{pmatrix} : n \times 1, \quad X = \begin{pmatrix} X_1 \\ \vdots \\ X_p \end{pmatrix} : n \times k,
$$

$$
\varepsilon = \begin{pmatrix} \varepsilon_1 \\ \vdots \\ \varepsilon_p \end{pmatrix} : n \times 1,
$$

$$
\Omega = \Omega(\theta) = \begin{pmatrix} \theta_1 I_{n_1} & & 0 \\ & \ddots & \\ 0 & & \theta_p I_{n_p} \end{pmatrix} \in \mathcal{S}(n), \qquad (4.37)
$$

where $n = \sum_{j=1}^{p} n_j$, $y_j : n_j \times 1$, $X_j : n_j \times k$, $\varepsilon_j : n_j \times 1$ and

$$
\theta = \begin{pmatrix} \theta_1 \\ \vdots \\ \theta_p \end{pmatrix} : p \times 1.
$$

We assume that the error term ε is normally distributed:

$$\mathcal{L}(\varepsilon) = N_n(0, \Omega).$$

The unrestricted GLSE is a GLSE $b(\hat{\Omega})$ with $\hat{\Omega} = \Omega(s)$, where

$$s = (s_1^2, \ldots, s_p^2)' : p \times 1, \tag{4.38}$$

and

$$
\begin{aligned}
s_j^2 &= y_j'[I_{n_j} - X_j(X_j'X_j)^+ X_j']y_j/q_j \\
&= \varepsilon_j'[I_{n_j} - X_j(X_j'X_j)^+ X_j']\varepsilon_j/q_j, \\
q_j &= n_j - rank\, X_j.
\end{aligned}
\tag{4.39}
$$

The conditional mean of $b(\hat{\Omega})$ given $\hat{\Omega}$ is calculated as follows:

$$
\begin{aligned}
E(b(\hat{\Omega})|\hat{\Omega}) &= \beta + (X'\hat{\Omega}^{-1}X)^{-1}E(X'\hat{\Omega}^{-1}\varepsilon|\hat{\Omega}) \\
&= \beta + (X'\hat{\Omega}^{-1}X)^{-1}E(X'\hat{\Omega}^{-1}\varepsilon|s),
\end{aligned}
\tag{4.40}
$$

where the last equality is due to one-to-one correspondence between $\hat{\Omega}$ and s. By using the structure of the matrix $\hat{\Omega}$ and the independence between $X_i'\varepsilon_i$'s and s_j^2's (Problem 4.2.1), we obtain

$$
\begin{aligned}
E(X'\hat{\Omega}^{-1}\varepsilon|s) &= \sum_{j=1}^{p} s_j^{-2}E(X_j'\varepsilon_j|s) \\
&= \sum_{j=1}^{p} s_j^{-2}E(X_j'\varepsilon_j) \\
&= 0,
\end{aligned}
$$

proving

$$E[b(\hat{\Omega})|\hat{\Omega}] = \beta. \tag{4.41}$$

Similarly, it can be easily proved that

$$E(X'\hat{\Omega}^{-1}\varepsilon\varepsilon'\hat{\Omega}^{-1}X|\hat{\Omega}) = X'\hat{\Omega}^{-1}\Omega\hat{\Omega}^{-1}X, \tag{4.42}$$

which implies that

$$\mathrm{Cov}(b(\hat{\Omega})|\hat{\Omega}) = H(\hat{\Omega}, \Omega). \tag{4.43}$$

Thus, from (4.41) and (4.43), we see that the unrestricted GLSE $b(\hat{\Omega})$ belongs to C_3, where

$$C_3 = \{b(\hat{\Omega}) \in C_1 \mid E(b(\hat{\Omega})|\hat{\Omega}) = \beta,\ \mathrm{Cov}(b(\hat{\Omega})|\hat{\Omega}) = H(\hat{\Omega}, \Omega)\}. \tag{4.44}$$

Hence, by Theorem 4.1, the following two inequalities are obtained:

$$\text{Cov}(b(\Omega)) = (X'\Omega^{-1}X)^{-1} = H(\Omega, \Omega)$$
$$\leq H(\hat{\Omega}, \Omega) = \text{Cov}(b(\hat{\Omega})|\hat{\Omega}) \qquad (4.45)$$

and

$$\text{Cov}(b(\Omega)) \leq \text{Cov}(b(\hat{\Omega})), \qquad (4.46)$$

where the latter follows by taking the expectations of both sides of (4.45).

Finally, it is earmarked for later discussion that the conditional distribution of the unrestricted GLSE $b(\hat{\Omega})$ is normal, that is,

$$\mathcal{L}(b(\hat{\Omega})|\hat{\Omega}) = N_k(\beta, \ H(\hat{\Omega}, \Omega)). \qquad (4.47)$$

Example 4.3 (SUR model)   This example shows that the UZE in an SUR model belongs to the class $\mathcal{C}_3$ under the assumption that the error term is normally distributed. The SUR model treated here is given by

$$y = \begin{pmatrix} y_1 \\ \vdots \\ y_p \end{pmatrix} : n \times 1, \quad X = \begin{pmatrix} X_1 & & 0 \\ & \ddots & \\ 0 & & X_p \end{pmatrix} : n \times k,$$

$$\beta = \begin{pmatrix} \beta_1 \\ \vdots \\ \beta_p \end{pmatrix} : k \times 1, \quad \varepsilon = \begin{pmatrix} \varepsilon_1 \\ \vdots \\ \varepsilon_p \end{pmatrix} : n \times 1,$$

$$\Omega = \Sigma \otimes I_m \quad \text{and} \quad \Sigma = (\sigma_{ij}) \in \mathcal{S}(p), \qquad (4.48)$$

where

$$y_j : m \times 1, \ X_j : m \times k_j, \ n = pm \ \text{and} \ k = \sum_{j=1}^{p} k_j.$$

It is assumed that the error term is normally distributed:

$$\mathcal{L}(\varepsilon) = N_n(0, \Sigma \otimes I_m).$$

The UZE is a GLSE $b(\hat{\Omega})$ with $\hat{\Omega} = \hat{\Sigma} \otimes I_m$ and $\hat{\Sigma} = S$, where the matrix S is defined as

$$S = Y'N_*Y = U'N_*U. \qquad (4.49)$$

Here,

$$Y = (y_1, \dots, y_p) : m \times p,$$
$$U = (\varepsilon_1, \dots, \varepsilon_p) : m \times p,$$
$$N_* = I_m - X_*(X_*'X_*)^+X_*' : m \times m,$$
$$X_* = (X_1, \dots, X_p) : m \times k. \qquad (4.50)$$

Recall that

$$\mathcal{L}(S) = W_p(\Sigma, q) \quad \text{with} \quad q = m - rank X_*, \tag{4.51}$$

where $W_p(\Sigma, q)$ denotes the Wishart distribution with mean $q\Sigma$ and degrees of freedom q.

To calculate the conditional mean of the UZE, note firstly that

$$X_j' N_* = 0 \quad \text{for any } j = 1, \ldots, p \tag{4.52}$$

holds and therefore the statistic S is independent of $X_i' \varepsilon_j$'s (Problem 4.2.2). Hence, in a similar way as in Example 4.2, we obtain

$$
\begin{aligned}
E(X' \hat{\Omega}^{-1} \varepsilon | \hat{\Omega}) &= E[X'(S^{-1} \otimes I_m) \varepsilon | S] \\
&= \begin{pmatrix} \sum_{j=1}^{p} s^{1j} E(X_1' \varepsilon_j | S) \\ \vdots \\ \sum_{j=1}^{p} s^{pj} E(X_p' \varepsilon_j | S) \end{pmatrix} \quad \text{with} \quad S^{-1} = (s^{ij}) \\
&= \begin{pmatrix} \sum_{j=1}^{p} s^{1j} E(X_1' \varepsilon_j) \\ \vdots \\ \sum_{j=1}^{p} s^{pj} E(X_p' \varepsilon_j) \end{pmatrix} \\
&= 0, \tag{4.53}
\end{aligned}
$$

establishing the conditional unbiasedness of the UZE, where the first equality follows from the one-to-one correspondence between $\hat{\Omega}$ and S and the third from the independence between $X_i' \varepsilon_j$'s and S.

Similarly, it is easy to show that

$$E[X'(S^{-1} \otimes I_m) \varepsilon \varepsilon'(S^{-1} \otimes I_m) X | S] = X'(S^{-1} \Sigma S^{-1} \otimes I_m) X. \tag{4.54}$$

This implies that

$$\text{Cov}(b(S \otimes I_m) | S) = H(S \otimes I_m, \Sigma \otimes I_m). \tag{4.55}$$

Thus, the UZE is also in the class $\mathcal{C}_3$, where

$$\mathcal{C}_3 = \{ b(\hat{\Omega}) \in \mathcal{C}_1 \mid E(b(\hat{\Omega}) | \hat{\Omega}) = \beta, \ \text{Cov}(b(\hat{\Omega}) | \hat{\Omega}) = H(\hat{\Omega}, \Omega) \}. \tag{4.56}$$

The two inequalities corresponding to (4.45) and (4.46) are given respectively as

$$
\begin{aligned}
\text{Cov}(b(\Sigma \otimes I_m)) &= (X'(\Sigma^{-1} \otimes I_m) X)^{-1} = H(\Sigma \otimes I_m, \Sigma \otimes I_m) \\
&\leq H(S \otimes I_m, \Sigma \otimes I_m) = \text{Cov}(b(S \otimes I_m) | S) \tag{4.57}
\end{aligned}
$$

and

$$\text{Cov}(b(\Sigma \otimes I_m)) \leq \text{Cov}(b(S \otimes I_m)). \tag{4.58}$$

Again, observe that the conditional distribution of the UZE is normal, that is,

$$\mathcal{L}(b(S \otimes I_m) | S) = N_k(\beta, \ H(S \otimes I_m, \Sigma \otimes I_m)). \tag{4.59}$$

4.3   Upper Bound for the Covariance Matrix of a GLSE

This section is devoted to deriving an effective upper bound $\alpha(b(\hat{\Omega}))$ of the form

$$\text{Cov}(b(\hat{\Omega})) \le \alpha(b(\hat{\Omega})) \, \text{Cov}(b(\Omega)) \tag{4.60}$$

for a GLSE $b(\hat{\Omega})$ in the class $\mathcal{C}_3$ in a general setup.

To begin with, we fix the model

$$y = X\beta + \varepsilon \tag{4.61}$$

with

$$\mathcal{L}(\varepsilon) \in \mathcal{P}_n(0, \Omega) \quad \text{and} \quad \Omega = \Omega(\theta) \in \mathcal{S}(n),$$

where $y : n \times 1$ and $X : n \times k$ is of full rank. For a GLSE $b(\hat{\Omega}) \in \mathcal{C}_3$, one way of obtaining an upper bound in (4.60) is deriving an upper bound for the conditional covariance matrix

$$\text{Cov}(b(\hat{\Omega})|\hat{\Omega}) = H(\hat{\Omega}, \Omega)$$

relative to the covariance matrix of the GME $b(\Omega)$

$$\text{Cov}(b(\Omega)) = H(\Omega, \Omega) = (X'\Omega^{-1}X)^{-1}.$$

In fact, suppose that a real-valued random function

$$L(\hat{\Omega}, \Omega) : \mathcal{S}(n) \times \mathcal{S}(n) \to [0, \infty)$$

of $\hat{\Omega}$ and Ω satisfies the following inequality:

$$H(\hat{\Omega}, \Omega) \le L(\hat{\Omega}, \Omega) \, H(\Omega, \Omega) \text{ a.s.} \tag{4.62}$$

For any L satisfying (4.62), the expected value of $L(\hat{\Omega}, \Omega)$ is one of the upper bounds in the sense of (4.60), since taking the expectations of both sides of (4.62) yields

$$\text{Cov}(b(\hat{\Omega})) \le \alpha(b(\hat{\Omega})) \, \text{Cov}(b(\Omega)) \tag{4.63}$$

with

$$\alpha(b(\hat{\Omega})) = E\{L(\hat{\Omega}, \Omega)\}.$$

Clearly, $\alpha(b(\hat{\Omega}))$ thus derived is a nonrandom positive function associated with a GLSE $b(\hat{\Omega})$. Thus, the problem is to find an effective one among the upper bounds $L(\hat{\Omega}, \Omega)$s that satisfy (4.62).

Such a function L always exists. A possible choice of L is the largest latent root of the matrix

$$H(\Omega, \Omega)^{-1/2} \, H(\hat{\Omega}, \Omega) \, H(\Omega, \Omega)^{-1/2}, \tag{4.64}$$

because the inequality (4.62) is equivalent to

$$H(\Omega, \Omega)^{-1/2} \, H(\hat{\Omega}, \Omega) \, H(\Omega, \Omega)^{-1/2} \le L(\hat{\Omega}, \Omega) \, I_k. \tag{4.65}$$

Although this is a natural choice of L, we do not use this since it generally depends on the regressor matrix X in a complicated way and hence it cannot be a tractable measure of the efficiency of a GLSE.

Reduction of variables. To avoid this difficulty, we first reduce the variables as follows:

$$A = X'\Omega^{-1}X \in \mathcal{S}(k),$$
$$\overline{X} = \Omega^{-1/2}XA^{-1/2} : n \times k,$$
$$P = P(\hat{\Omega}, \Omega) = \Omega^{-1/2}\hat{\Omega}\Omega^{-1/2} \in \mathcal{S}(n). \tag{4.66}$$

Here, the regressor matrix X is transformed to the $n \times k$ matrix $\overline{X}$, which takes a value in the compact subset $\mathcal{F}_{n,k}$ of R^{nk}, where

$$\mathcal{F}_{n,k} = \{U : n \times k \mid U'U = I_k\}. \tag{4.67}$$

The matrices $H(\Omega, \Omega)$ and $H(\hat{\Omega}, \Omega)$ in question are rewritten in terms of $\overline{X}$, A and P as

$$H(\Omega, \Omega) = A^{-1}, \tag{4.68}$$
$$H(\hat{\Omega}, \Omega) = A^{-1/2}(\overline{X}'P^{-1}\overline{X})^{-1}\overline{X}'P^{-2}\overline{X}(\overline{X}'P^{-1}\overline{X})^{-1}A^{-1/2}, \tag{4.69}$$

respectively. Thus, the left-hand side of the inequality (4.65) becomes

$$\Phi(\overline{X}, P) = (\overline{X}'P^{-1}\overline{X})^{-1}\overline{X}'P^{-2}\overline{X}(\overline{X}'P^{-1}\overline{X})^{-1}. \tag{4.70}$$

As a tractable upper bound, we aim to derive a function

$$l_0(P) : \mathcal{S}(n) \to [0, \infty)$$

that depends only on P and satisfies

$$\Phi(\overline{X}, P) \le l_0(P)I_k \quad \text{a.s. for any } \overline{X} \in \mathcal{F}_{n,k}. \tag{4.71}$$

Since $l_0(P)$ is free from $\overline{X}$ as well as X, it is expected to take a simple form. For convenience, we often write the function $l_0(P)$ as a function of $\hat{\Omega}$ and Ω, that is,

$$l_0(P) = L_0(\hat{\Omega}, \Omega). \tag{4.72}$$

The inequality (4.71) is rewritten as

$$H(\hat{\Omega}, \Omega) \leq L_0(\hat{\Omega}, \Omega) \, H(\Omega, \Omega) \text{ a.s.} \tag{4.73}$$

and this holds for any $X : n \times k$ of full rank. From this, we obtain

$$\text{Cov}(b(\hat{\Omega})) \leq \alpha_0(b(\hat{\Omega})) \, \text{Cov}(b(\Omega)) \tag{4.74}$$

with

$$\alpha_0(b(\hat{\Omega})) = E[L_0(\hat{\Omega}, \Omega)] = E[l_0(P)]. \tag{4.75}$$

Thus, the problem is formulated as one of obtaining $L_0(\hat{\Omega}, \Omega) = l_0(P)$, as described in (4.72).

Derivation of $l_0(P)$. The following inequality, known as the *Kantorovich inequality* is useful for our purpose.

Lemma 4.2 *For any $n \times 1$ vector u with $u'u = 1$ and any $\Lambda \in \mathcal{S}(n)$,*

$$1 \leq u'\Lambda u \, u'\Lambda^{-1}u \leq \frac{(\lambda_1 + \lambda_n)^2}{4\lambda_1\lambda_n} \tag{4.76}$$

holds, where $0 < \lambda_1 \leq \cdots \leq \lambda_n$ are the latent roots of Λ.

Proof. The left-hand side of (4.76) is an easy consequence of Cauchy–Schwarz inequality in (2.70).

To show the right-hand side, note that the matrix Λ can be set as a diagonal matrix. In fact, let $\Gamma \in \mathcal{O}(n)$ be an orthogonal matrix that diagonalizes Λ:

$$\Lambda = \Gamma \tilde{\Lambda} \Gamma' \text{ with } \tilde{\Lambda} = \begin{pmatrix} \lambda_1 & & 0 \\ & \ddots & \\ 0 & & \lambda_n \end{pmatrix},$$

where $\mathcal{O}(n)$ is the group of $n \times n$ orthogonal matrices. And let $\mathcal{U}(n) = \{u \in R^n \mid \|u\| = 1\}$. Then

$$\sup_{u'u=1} u'\Lambda u u'\Lambda^{-1}u = \sup_{u\in\mathcal{U}(n)} u'\Gamma\tilde{\Lambda}\Gamma'uu'\Gamma\tilde{\Lambda}^{-1}\Gamma'u$$

$$= \sup_{u\in\mathcal{U}(n)} u'\tilde{\Lambda}uu'\tilde{\Lambda}^{-1}u,$$

where the second equality follows since the group $\mathcal{O}(n)$ acts transitively on $\mathcal{U}(n)$ via the group action $u \to \Gamma'u$ (see Example 1.2). For each $u \equiv (u_1, \ldots, u_n)' \in \mathcal{U}(n)$, let $W \equiv W(u)$ be a discrete random variable such that

$$P(W = \lambda_j) = u_j^2 \quad (j = 1, \ldots, n).$$

Then we can write

$$u'\tilde{\Lambda}uu'\tilde{\Lambda}^{-1}u = E(W)E(W^{-1}).$$

On the other hand, for any $\lambda \in [\lambda_1, \lambda_n]$, the following inequality holds:

$$0 \le (\lambda - \lambda_1)(\lambda_n - \lambda)$$
$$= (\lambda_1 + \lambda_n - \lambda)\lambda - \lambda_1\lambda_n,$$

from which

$$\frac{1}{\lambda} \le \frac{\lambda_1 + \lambda_n - \lambda}{\lambda_1\lambda_n}$$

follows. Since the random variable W satisfies $P(W \in [\lambda_1, \lambda_n]) = 1$,

$$E(W^{-1}) \le \frac{\lambda_1 + \lambda_n - E(W)}{\lambda_1\lambda_n}$$

holds, and hence

$$E(W)E(W^{-1}) \le \frac{\{\lambda_1 + \lambda_n - E(W)\}E(W)}{\lambda_1\lambda_n}$$
$$= -\frac{1}{\lambda_1\lambda_n}\left\{E(W) - \frac{\lambda_1 + \lambda_n}{2}\right\}^2 + \frac{(\lambda_1 + \lambda_n)^2}{4\lambda_1\lambda_n}$$
$$\le \frac{(\lambda_1 + \lambda_n)^2}{4\lambda_1\lambda_n}.$$

Here, the extreme right-hand side of the above inequality is free from u. This completes the proof.

Note that the upper bound in (4.76) is attainable. In fact, let η_j be the latent vector corresponding to λ_j. Then by letting

$$u = \frac{1}{\sqrt{2}}(\eta_1 + \eta_n),$$

we can see that $u'\Lambda uu'\Lambda^{-1}u = (\lambda_1 + \lambda_n)/4\lambda_1\lambda_n$.

The Kantorovich inequality has been used and expanded as an important tool for evaluating the efficiency of the OLSE relative to the GME. In Chapter 14 of Rao and Rao (1998), several extensions of this inequality are summarized.

The following lemma derives $l_0(P)$ described above. The proof here is from Kurata and Kariya (1996). The result can also be obtained from Wang and Shao (1992) in which a matrix version of the Kantorovich inequality is derived. Wang and Ip (1999) established a matrix version of the Wielandt inequality, which is closely related to the Kantorovich inequality.

Lemma 4.3 *Let*

$$\pi_1 \leq \cdots \leq \pi_n$$

be the latent roots of $P = P(\hat{\Omega}, \Omega)$. Then the following inequality holds:

$$\Phi(\overline{X}, P) \leq l_0(P)\, I_k \quad \text{for any } \overline{X} \in \mathcal{F}_{n,k},$$

where

$$l_0(P) = \frac{(\pi_1 + \pi_n)^2}{4\pi_1 \pi_n}. \tag{4.77}$$

Proof. Note first that

$$\Phi(\overline{X}\Delta) = \Delta' \Phi(\overline{X})\Delta \tag{4.78}$$

holds for any $k \times k$ orthogonal matrix Δ, which implies that the latent roots of $\Phi(\overline{X})$ are invariant under the transformation $\overline{X} \to \overline{X}\Delta$. For a given $\overline{X}$, choose a $k \times k$ orthogonal matrix Δ_1 such that $\Delta_1' \overline{X}' P^{-1} \overline{X} \Delta_1$ is diagonal and let $V = (v_1, \ldots, v_k) = \overline{X}\Delta_1$. Then V is clearly in $\mathcal{F}_{n,k}$, and $\Phi(\overline{X})$ and $\Phi(V)$ have the same latent roots. The function $\Phi(V)$ is written as

$$\Phi(V) = (V'P^{-1}V)^{-1}V'P^{-2}V(V'P^{-1}V)^{-1}$$

$$= \begin{pmatrix} v_1'P^{-1}v_1 & & 0 \\ & \ddots & \\ 0 & & v_k'P^{-1}v_k \end{pmatrix}^{-1} \begin{pmatrix} v_1'P^{-2}v_1 & \cdots & v_1'P^{-2}v_k \\ \vdots & & \vdots \\ v_k'P^{-2}v_1 & \cdots & v_k'P^{-2}v_k \end{pmatrix}$$

$$\times \begin{pmatrix} v_1'P^{-1}v_1 & & 0 \\ & \ddots & \\ 0 & & v_k'P^{-1}v_k \end{pmatrix}^{-1}$$

$$= \begin{pmatrix} \dfrac{v_1'P^{-2}v_1}{(v_1'P^{-1}v_1)^2} & \cdots & \dfrac{v_1'P^{-2}v_k}{v_1'P^{-1}v_1\, v_k'P^{-1}v_k} \\ \vdots & & \vdots \\ \dfrac{v_k'P^{-2}v_1}{v_k'P^{-1}v_k\, v_1'P^{-1}v_1} & \cdots & \dfrac{v_k'P^{-2}v_k}{(v_k'P^{-1}v_k)^2} \end{pmatrix}$$

$$= (\phi_{ij}) \quad \text{(say)} \tag{4.79}$$

with

$$\phi_{ij} = \phi(v_i, v_j) = \frac{v_i'P^{-2}v_j}{v_i'P^{-1}v_i\, v_j'P^{-1}v_j}. \tag{4.80}$$

Next, choose a $k \times k$ orthogonal matrix such that $\Delta_2'\Phi(V)\Delta_2 = \Phi(V\Delta_2)$ is diagonal. Then the matrix $\Phi(U)$ with $U = (u_1, \ldots, u_k) = V\Delta_2 \in \mathcal{F}_{n,k}$ is a diagonal matrix with diagonal elements $\phi(u_j, u_j)$ $(j = 1, \ldots, k)$.

Therefore, to find the least upper bound $l_0(P)$ that satisfies $\Phi(U) \leq l_0(P)I_k$, it suffices to maximize $\phi(u_j, u_j)$ under $u'_j u_j = 1$. Hence, we have

$$
\begin{aligned}
\sup_{u'_j u_j = 1} \phi(u_j, u_j) &= \sup_{u'_j u_j = 1} \frac{(u'_j P^{-2} u_j)(u'_j u_j)}{(u'_j P^{-1} u_j)^2} \\
&= \sup_{u_j \in R^k - \{0\}} \frac{(u'_j P^{-2} u_j)(u'_j u_j)}{(u'_j P^{-1} u_j)^2} \\
&= \sup_{u_j \in R^k - \{0\}} \frac{(u'_j P^{-1} u_j)(u'_j P u_j)}{(u'_j u_j)^2} \\
&= \sup_{u'_j u_j = 1} (u'_j P^{-1} u_j)(u'_j P u_j) \\
&= \frac{(\pi_1 + \pi_n)^2}{4\pi_1 \pi_n},
\end{aligned}
\tag{4.81}
$$

where the first equality follows from $u'_j u_j = 1$, the second from the invariance under the transformation $u_j \to a u_j$ for $a > 0$, the third from transforming $u_j \to P^{1/2} u_j$ and the last from Lemma 4.2. This completes the proof.

The next theorem is due to Kurata and Kariya (1996).

Theorem 4.4 *For any GLSE $b(\hat{\Omega})$ in C_3, let $P = P(\hat{\Omega}, \Omega) = \Omega^{-1/2} \hat{\Omega} \Omega^{-1/2}$ and $L_0(\hat{\Omega}, \Omega) = l_0(P) = (\pi_1 + \pi_n)^2 / 4\pi_1 \pi_n$, where $\pi_1 \leq \cdots \leq \pi_n$ are the latent roots of P. Then the following inequality holds:*

$$
\mathrm{Cov}(b(\hat{\Omega})) \leq \alpha_0(b(\hat{\Omega})) \, \mathrm{Cov}(b(\Omega))
\tag{4.82}
$$

with

$$
\alpha_0(b(\hat{\Omega})) = E[L_0(\hat{\Omega}, \Omega)] = E[l_0(P)].
\tag{4.83}
$$

Combining it with Theorem 4.1, a bound for the covariance matrix of a GLSE $b(\hat{\Omega})$ is obtained as

$$
\mathrm{Cov}(b(\Omega)) \leq \mathrm{Cov}(b(\hat{\Omega})) \leq \alpha_0(b(\hat{\Omega})) \, \mathrm{Cov}(b(\Omega)),
\tag{4.84}
$$

from which the bounds for the δ's in (4.11) to (4.13) are also obtained.

An interpretation of $l_0(P) = L_0(\hat{\Omega}, \Omega)$. The upper bound $l_0(P) = L_0(\hat{\Omega}, \Omega)$ can be viewed as a loss function for choosing an estimator $\hat{\Omega}$ in a GLSE $b(\hat{\Omega})$. To explain this more precisely, let $x = \pi_n / \pi_1$ (≥ 1) and rewrite $l_0(P)$ as a function of x:

$$
l_0(P) = \frac{(1 + x)^2}{4x}.
\tag{4.85}
$$

Obviously, $l_0(P)$ is a monotonically increasing function of x such that $f(x) \geq 1$, and the minimum is attained when $x = 1$, or equivalently, when

$$P = \gamma I_n \quad \text{for some } \gamma > 0, \tag{4.86}$$

where $P = P(\hat{\Omega}, \Omega) = \Omega^{-1/2} \hat{\Omega} \Omega^{-1/2}$. Therefore, $l_0(P)$ is regarded as a measure of the sphericity of the matrix P: as P becomes more and more spherical, $l_0(P)$ becomes closer and closer to its minimum. Since the condition (4.86) is nothing but

$$\hat{\Omega} = \gamma \Omega \quad \text{for some } \gamma > 0, \tag{4.87}$$

the smaller the upper bound $L_0(\hat{\Omega}, \Omega)$ is, the better the estimation efficiency of $\hat{\Omega}$ for Ω becomes. Note that the GLSE $b(\hat{\Omega})$ that satisfies (4.87) is the GME $b(\Omega)$, since

$$b(\gamma \Omega) = b(\Omega) \quad \text{for any } \gamma > 0.$$

In this sense, the upper bound $l_0(P) = L_0(\hat{\Omega}, \Omega)$ can be viewed as a loss function, and the problem of finding an efficient GLSE $b(\hat{\Omega})$ via $l_0(P)$ is equivalent to the problem of deriving an optimal estimator $\hat{\Omega}$ for the covariance matrix Ω with respect to the loss function $L_0(\hat{\Omega}, \Omega)$.

Invariance properties of $L_0(\hat{\Omega}, \Omega)$ as a loss function. As a loss function for estimating the covariance matrix Ω, the upper bound $L_0(\hat{\Omega}, \Omega)$ has the following two invariance properties (see Problem 4.3.1):

(i) Invariance under the transformation $(\hat{\Omega}, \Omega) \to (\hat{\Omega}^{-1}, \Omega^{-1})$:

$$L_0(\hat{\Omega}, \Omega) = L_0(\hat{\Omega}^{-1}, \Omega^{-1}); \tag{4.88}$$

(ii) Invariance under the transformation $(\hat{\Omega}, \Omega) \to (a\hat{\Omega}, b\Omega)$ for $a, b > 0$:

$$L_0(a\hat{\Omega}, b\Omega) = L_0(\hat{\Omega}, \Omega) \quad \text{for any } a, b > 0. \tag{4.89}$$

We make a brief comment on the implication of the invariance property (i), which is called the *symmetric inverse* property by Bilodeau (1990) in the context of the two-equation heteroscedastic model. This means that the loss function L_0 equally penalizes the overestimates and the underestimates of Ω. Thus, in constructing a GLSE $b(\hat{\Omega})$, choosing an optimal estimator $\hat{\Omega}$ of Ω with respect to the loss function L_0 is equivalent to choosing an optimal one $\hat{\Psi}$ of $\Psi \equiv \Omega^{-1}$. Further investigation of this topic will be given in Problem 4.3.2, Sections 4.4, 4.5 and Chapter 5. In Problem 4.3.2, we consider this property in a simple setting in which the variance of the normal distribution is estimated.

Upper bound for the OLSE. Next, let us apply Theorem 4.4 to the OLSE $b(I_n)$.

Corollary 4.5 *Let*

$$\omega_1 \leq \cdots \leq \omega_n$$

be the latent roots of Ω. Then the following inequality holds:

$$\text{Cov}(b(\Omega)) \leq \text{Cov}(b(I_n)) \leq \alpha_0(b(I_n)) \, \text{Cov}(b(\Omega)) \tag{4.90}$$

with

$$\alpha_0(b(I_n)) = \frac{(\omega_1 + \omega_n)^2}{4\omega_1\omega_n}.$$

Proof. Since $P = P(I_n, \Omega) = \Omega^{-1}$, the upper bound $\alpha_0(b(I_n))$ is obtained as

$$\alpha_0(b(I_n)) = l_0(\Omega^{-1}) = L_0(I_n, \Omega)$$
$$= L_0(I_n, \Omega^{-1}) = l_0(\Omega), \tag{4.91}$$

where the third equality is due to (4.88).

The upper bound $\alpha_0(b(I_n))$ is available when the ratio ω_n/ω_1 of the latent roots of Ω is known. Hillier and King (1987) treated this situation and obtained bounds for linear combinations of the GME and the OLSE $b(I_n)$.

A greater but simpler upper bound. Although the general expression (4.83) of the upper bound $\alpha_0(b(\hat{\Omega}))$ is quite simple, it is not easy to evaluate it explicitly except for some simple cases as in Corollary 4.5. The difficulty lies in the fact that the distribution of the extreme latent roots of a random matrix usually takes a very complicated form. Hence, we derive an alternative upper bound. While the upper bound derived below is greater than α_0, it is quite simple and is still effective as well as useful.

Lemma 4.6 *For any positive definite matrix $P \in \mathcal{S}(n)$ with latent roots $\pi_1 \leq \cdots \leq \pi_n$, the following inequality holds:*

$$l_0(P) \leq l_1(P), \tag{4.92}$$

where

$$l_1(P) = \left[\frac{1}{n} \sum_{j=1}^{n} \pi_j \Big/ \prod_{j=1}^{n} \pi_j^{1/n} \right]^n = \frac{(tr\,P)^n}{n^n \, |P|}. \tag{4.93}$$

The equality holds if and only if

$$\pi_2 = \cdots = \pi_{n-1} = (\pi_1 + \pi_n)/2.$$

Proof. Fix $\pi_1 \leq \pi_n$ and let

$$
f(\pi_2, \ldots, \pi_{n-1}) = \left[\frac{1}{n} \sum_{j=1}^{n} \pi_j \Big/ \prod_{j=1}^{n} \pi_j^{1/n} \right]^n = \left[\frac{1}{n} \sum_{j=1}^{n} \pi_j \right]^n \Big/ \prod_{j=1}^{n} \pi_j
$$

as a function of $\pi_2, \ldots, \pi_{n-1}$. Differentiating f with respect to π_i ($i = 2, \ldots, n - 1$) yields

$$
\frac{\partial}{\partial \pi_i} f(\pi_2, \ldots, \pi_{n-1}) = \frac{\left[\frac{1}{n} \sum_{j=1}^{n} \pi_j \right]^{n-1}}{\prod_{j=1}^{n} \pi_j} \frac{\pi_i - \frac{1}{n} \sum_{j=1}^{n} \pi_j}{\pi_i}.
$$

Here, $\frac{\partial}{\partial \pi_i} f = 0$ ($i = 2, \ldots, n - 1$) implies $\pi_2 = \cdots = \pi_{n-1}$. Hence, to find the minimum of f, we can set

$$
\pi_2 = \cdots = \pi_{n-1} = c
$$

without loss of generality. Let

$$
g(c) = f(c, \ldots, c)
$$

$$
= \frac{\left\{ \frac{1}{n} [\pi_1 + \pi_n + (n-2)c] \right\}^n}{\pi_1 \pi_n c^{n-2}}.
$$

By differentiating g with respect to c, we can easily see that $g(c)$ takes its (unique) minimum at $c = (\pi_1 + \pi_n)/2$ and

$$
g\left(\frac{\pi_1 + \pi_n}{2} \right) = \frac{(\pi_1 + \pi_n)^2}{4\pi_1 \pi_n}.
$$

This completes the proof.

Setting $P = P(\hat{\Omega}, \Omega) = \Omega^{-1/2} \hat{\Omega} \Omega^{-1/2}$ in $l_1(P)$ yields

$$
l_1(P) = \frac{[tr(\hat{\Omega} \Omega^{-1})]^n}{n^n |\hat{\Omega} \Omega^{-1}|} \equiv L_1(\hat{\Omega}, \Omega), \tag{4.94}
$$

which implies the following result due to Kurata and Kariya (1996).

Theorem 4.7 *For any GLSE* $b(\hat{\Omega}) \in C_3$,

$$
\mathrm{Cov}(b(\hat{\Omega})) \leq \alpha_0(b(\hat{\Omega})) \, \mathrm{Cov}(b(\Omega))
$$

$$
\leq \alpha_1(b(\hat{\Omega})) \, \mathrm{Cov}(b(\Omega)) \tag{4.95}
$$

with

$$
\alpha_1(b(\hat{\Omega})) = E\{L_1(\hat{\Omega}, \Omega)\} = E\{l_1(P)\}. \tag{4.96}
$$

The upper bound $\alpha_1 = E(L_1)$ is much more easily evaluated than α_0, as will be shown in an SUR model and a heteroscedastic model.

As well as L_0, the function $L_1(\hat{\Omega}, \Omega) = l_1(P)$ also reflects the efficiency of estimating Ω by an estimator $\hat{\Omega}$, and therefore, it can be viewed as a loss function for choosing an estimator $\hat{\Omega}$ in $b(\hat{\Omega})$. In application, it is often the case that Ω is structured and then the distinct latent roots of P are reduced to the case of a lower dimension. For example, in the two-equation SUR model, $P = \Sigma^{-1/2}\hat{\Sigma}\Sigma^{-1/2} \otimes I_m$ where $\Sigma \in \mathcal{S}(2)$. In such a case, the following proposition will be useful.

Proposition 4.8 *For any* $\Psi \in \mathcal{S}(2)$,

$$l_1(\Psi) = l_0(\Psi) \tag{4.97}$$

holds.

4.4 Upper Bound Problem for the UZE in an SUR Model

In this section, we treat the SUR model and evaluate the upper bounds for the covariance matrices of various typical GLSEs including the UZE and the OLSE. The following two problems of practical importance are studied:

(1) To clarify the condition under which the UZE is more efficient than the OLSE;

(2) To derive a GLSE whose upper bound is smaller than that of the UZE.

Preliminaries. The SUR model considered in this section is the one given in Example 4.3, where it was shown that the UZE

$$b(S \otimes I_m) = (X'(S^{-1} \otimes I_m)X)^{-1}X'(S^{-1} \otimes I_m)y$$

defined in (4.49) belongs to the class $\mathcal{C}_3$ in (4.56). To treat the two problems described in the above paragraph, we first note the following result:

Proposition 4.9 *For any GLSE of the form*

$$b(\hat{\Sigma} \otimes I_m) = (X'(\hat{\Sigma}^{-1} \otimes I_m)X)^{-1}X'(\hat{\Sigma}^{-1} \otimes I_m)y,$$

if $\hat{\Sigma}$ *is a measurable function of the matrix S, say*

$$\hat{\Sigma} = \hat{\Sigma}(S), \tag{4.98}$$

then the GLSE $b(\hat{\Sigma} \otimes I_m)$ *belongs to the class* $\mathcal{C}_3$, *that is,*

$$E[b(\hat{\Sigma} \otimes I_m)|\hat{\Sigma}] = \beta,$$

$$\mathrm{Cov}(b(\hat{\Sigma} \otimes I_m)|\hat{\Sigma}) = H(\hat{\Sigma} \otimes I_m, \Sigma \otimes I_m). \tag{4.99}$$

Proof. See Problem 4.4.1.

The UZE and the OLSE are particular cases obtained by letting $\hat{\Sigma}(S) = S$ and $\hat{\Sigma}(S) = I_p$ respectively, where the latter is a constant function. Thus, the results in the previous two sections are applicable to the GLSEs $b(\hat{\Sigma} \otimes I_m)$.

Derivation of upper bound. For a GLSE $b(\hat{\Sigma} \otimes I_m)$ in C_3, let

$$P = P(\hat{\Sigma} \otimes I_m, \Sigma \otimes I_m) = \Sigma^{-1/2} \hat{\Sigma} \Sigma^{-1/2} \otimes I_m \qquad (4.100)$$

and let $\pi_1 \leq \cdots \leq \pi_p$ be the latent roots of the matrix $\Sigma^{-1/2} \hat{\Sigma} \Sigma^{-1/2}$. Then the latent roots of P are given by π_1 (with multiplicity m), π_2 (with multiplicity m), $\ldots, \pi_p$ (with multiplicity m). Hence, the upper bound $\alpha_0(b(\hat{\Sigma} \otimes I_m))$ is obtained by Theorem 4.4 as

$$\alpha_0(b(\hat{\Sigma} \otimes I_m)) = E\{L_0(\hat{\Sigma}, \Sigma)\} \qquad (4.101)$$

with

$$L_0(\hat{\Sigma}, \Sigma) = l_0(\Sigma^{-1/2} \hat{\Sigma} \Sigma^{-1/2}) = \frac{(\pi_1 + \pi_p)^2}{4\pi_1 \pi_p}. \qquad (4.102)$$

Here we used the following equality:

$$L_0(\hat{\Sigma} \otimes I_m, \Sigma \otimes I_m) = l_0(\Sigma^{-1/2} \hat{\Sigma} \Sigma^{-1/2} \otimes I_m) = \frac{(\pi_1 + \pi_p)^2}{4\pi_1 \pi_p}$$

$$= l_0(\Sigma^{-1/2} \hat{\Sigma} \Sigma^{-1/2}) = L_0(\hat{\Sigma}, \Sigma) \qquad (4.103)$$

Furthermore, applying Lemma 4.6 and Theorem 4.7 to $L_0(\hat{\Sigma}, \Sigma)$ yields the greater upper bound

$$\alpha_1(b(\hat{\Sigma} \otimes I_m)) = E\{L_1(\hat{\Sigma}, \Sigma)\}, \qquad (4.104)$$

where

$$L_1(\hat{\Sigma}, \Sigma) = l_1(P) = \frac{(tr\,P)^p}{p^p |P|} = \frac{[tr(\hat{\Sigma} \Sigma^{-1})]^p}{p^p |\hat{\Sigma} \Sigma^{-1}|}. \qquad (4.105)$$

The results given above are summarized as follows: for any $b(\hat{\Sigma} \otimes I_m) \in C_3$,

$$\mathrm{Cov}(b(\Sigma \otimes I_m)) \leq \mathrm{Cov}(b(\hat{\Sigma} \otimes I_m))$$

$$\leq \alpha_0(b(\hat{\Sigma} \otimes I_m))\, \mathrm{Cov}(b(\Sigma \otimes I_m))$$

$$\leq \alpha_1(b(\hat{\Sigma} \otimes I_m))\, \mathrm{Cov}(b(\Sigma \otimes I_m)). \qquad (4.106)$$

Note that the function $L_0(\hat{\Sigma}, \Sigma)$ as well as $L_1(\hat{\Sigma}, \Sigma)$ can be understood as a loss function for choosing an estimator $\hat{\Sigma}$ of Σ in the GLSE $b(\hat{\Sigma} \otimes I_m) \in C_3$.

When the model consists of two equations, that is, when $p = 2$, Proposition 4.8 guarantees that

$$L_0(\hat{\Sigma}, \Sigma) = L_1(\hat{\Sigma}, \Sigma). \qquad (4.107)$$

The upper bounds α_0 and α_1 for the covariance matrices of the UZE $b(S \otimes I_m)$ and the OLSE $b(I_p \otimes I_m)$ are obtained by letting $\hat{\Sigma} = S$ and $\hat{\Sigma} = I_p$ respectively. As for the UZE, it is clear that

$$\alpha_0(b(S \otimes I_m)) = E[l_0(W)] \quad \text{with} \quad l_0(W) = \frac{(w_1 + w_p)^2}{4w_1 w_p}, \qquad (4.108)$$

and

$$\alpha_1(b(S \otimes I_m)) = E[l_1(W)] \quad \text{with} \quad l_1(W) = \frac{[tr(W)]^p}{p^p |W|}, \qquad (4.109)$$

where

$$W = P(S, \Sigma) = \Sigma^{-1/2} S \Sigma^{-1/2}, \qquad (4.110)$$

and

$$w_1 \leq \cdots \leq w_p$$

are the latent roots of the matrix W. The matrix W is distributed as the Wishart distribution $W_p(I_p, q)$ with mean $q I_p$ and degrees of freedom

$$q = m - rank\, X_*,$$

where $X_* = (X_1, \ldots, X_p) : m \times k$.

On the other hand, in the case of the OLSE $b(I_p \otimes I_m)$, it is readily seen that

$$\alpha_0(b(I_p \otimes I_m)) = l_0(\Sigma) \quad \text{with} \quad l_0(\Sigma) = \frac{(\lambda_1 + \lambda_p)^2}{4\lambda_1 \lambda_p}, \qquad (4.111)$$

and

$$\alpha_1(b(I_p \otimes I_m)) = l_1(\Sigma) \quad \text{with} \quad l_1(\Sigma) = \frac{(tr\,\Sigma)^p}{p^p |\Sigma|}, \qquad (4.112)$$

where $\lambda_1 \leq \cdots \leq \lambda_p$ are the latent roots of Σ. We used

$$L_0(I_p, \Sigma) = l_0(\Sigma^{-1}) = l_0(\Sigma) = L_0(I_p, \Sigma^{-1}).$$

The two upper bounds (4.111) and (4.112) can be viewed as measures of the sphericity of the matrix Σ. When $\Sigma = \gamma I_p$ for some $\gamma > 0$, the two bounds are 1.

Efficiency comparison between the UZE and the OLSE. As will be seen below, the upper bound $\alpha_0(b(S \otimes I_m))$ takes quite a complicated form. Hence, we first evaluate the greater upper bound $\alpha_1(b(S \otimes I_m))$.

Theorem 4.10 *Suppose that $q > p + 1$. Then, for the UZE:*

(1) *The greater upper bound α_1 is given by*

$$\alpha_1(b(S \otimes I_m)) = \prod_{j=1}^{p-1} \left(1 + \frac{[(p-2)/p]\,j + 2}{q - j - 2} \right); \tag{4.113}$$

(2) *Moreover,*

$$\lim_{q \to \infty} \alpha_1(b(S \otimes I_m)) = 1; \tag{4.114}$$

(3) *In particular, when $p = 2$ and $q > 3$, it holds that*

$$\alpha_0(b(S \otimes I_m)) = \alpha_1(b(S \otimes I_m)) = 1 + \frac{2}{q - 3}. \tag{4.115}$$

Proof. The greater upper bound is given by

$$\alpha_1(b(S \otimes I_m)) = E_q\{l_1(W)\} \quad \text{with} \quad l_1(W) = \frac{[tr(W)]^p}{p^p |W|}, \tag{4.116}$$

where the suffix q denotes the expectation under the Wishart distribution $W_p(I_p, q)$ with degrees of freedom q. The probability density function (pdf) of W is given by

$$d(q)\,|W|^{(q-p-1)/2} \exp\left(-\frac{tr\,W}{2} \right) \tag{4.117}$$

with

$$d(q) = \frac{1}{2^{pq/2}\Gamma_p(q/2)},$$

which is positive on the set $\mathcal{S}(p)$ of $p \times p$ positive definite matrices. By absorbing the denominator $|W|$ of $l_1(W)$ into the pdf (4.117), we have

$$E_q[l_1(W)] = \frac{1}{p^p} \frac{d(q)}{d(q-2)} \times E_{q-2}[(tr\,W)^p]$$

$$= \frac{1}{p^p} \frac{d(q)}{d(q-2)} \times \frac{2^p \Gamma\left(\frac{p(q-2)}{2} + p \right)}{\Gamma\left(\frac{p(q-2)}{2} \right)}. \tag{4.118}$$

From this, (4.113) is readily obtained, since

$$\Gamma(x + n) = x(x + 1)\dots(x + n - 1)\Gamma(x).$$

On the right-hand side of (4.118), we used the fact that $tr\,W$ is distributed as the χ^2 distribution with degrees of freedom $p(q - 2)$ (see Proposition 1.7 in Chapter 1 and Problem 4.4.2). Thus, (1) is proved, (2) is obvious, and (3) is due to (4.107). This completes the proof.

The above theorem is due to Kurata and Kariya (1996). The paper is an extension of Kariya (1981a) in which the case of $p = 2$ is dealt with. Note that these upper bounds are free from unknown parameters and from X.

Corollary 4.11 *For any* $p \geq 2$,

$$\lim_{q \to \infty} \alpha_0(b(S \otimes I_m)) = 1. \tag{4.119}$$

Proof. It is obvious from (2) of the above theorem that

$$1 \leq \alpha_0(b(S \otimes I_m)) \leq \alpha_1(b(S \otimes I_m)) \to 1 \quad \text{as} \quad q \to \infty.$$

It is interesting to see that the quantity $l_1(W)$ coincides with the reciprocal of the likelihood ratio test statistic for testing the sphericity of the covariance matrix of a multivariate normal distribution. See, for example, Section 8.3 of Muirhead (1982) and Khatri and Srivastava (1971).

By comparing the upper bounds derived above, the upper bound for the covariance matrix of the OLSE can be smaller than that of the UZE when the correlation coefficients $\rho_{ij} = \sigma_{ij}/(\sigma_{ii}\sigma_{jj})^{1/2}$ are close to zero and/or the degrees of freedom q is small. When $p = 2$, the condition $\rho_{12} = 0$ is clearly equivalent (under the normality condition) to the independence of the two submodels $y_j = X_j\beta_j + \varepsilon_j$ $(j = 1, 2)$. For testing the null hypothesis $H : \sigma_{12} = 0$, one can use the locally optimal test derived by Kariya (1981b). See also Kariya, Fujikoshi and Krishnaiah (1984, 1987) and Davis (1989) in which some multivariate extension of Kariya (1981b) is given. When the model consists of $p(\geq 2)$ equations, the Lagrange multiplier test for $H : \sigma_{ij} = 0$ $(i \neq j;\ i, j = 1, \ldots p)$ derived by Breusch and Pagan (1980) is widely used.

Expression of $\alpha_0(b(S \otimes I_m))$ when $p \geq 3$. Kurata and Kariya (1996) gave an explicit evaluation of the upper bound $\alpha_0(b(S \otimes I_m))$ for the covariance matrix of the UZE for general p. However, the expression is quite complicated and it involves the zonal polynomials when $p \geq 4$. Hence, only the result for $p = 3$ is presented here without proof.

Theorem 4.12 *Let* $p = 3$ *and suppose that* $q > 4$, *then the following equality holds:*

$$\alpha_0(b(S \otimes I_m)) = c(q) \sum_{k=0}^{\infty} \frac{\Gamma(3q/2 + k)}{3^k k!} \sum_{j=0}^{2} a_j \frac{\Gamma(k + j + 5)}{\Gamma(q/2 + k + j + 3)}$$

$$\times \sum_{s=0}^{k} \binom{k}{s} \frac{\Gamma(s + 2)}{\Gamma(s + 4)} \sum_{t=0}^{\infty} \frac{(-q/2 + 2)_t\ (k + j + 5)_t\ (s + 2)_t}{(q/2 + k + j + 3)_t\ (s + 4)_t\ t!},$$

$$\tag{4.120}$$

where $(a_0, a_1, a_2) = (4, -4, 1)$ *and*

$$c(q) = \frac{\pi^{9/2}\Gamma(q/2 - 2)}{4 \times 3^{3q/2}\Gamma_3(3/2)\Gamma_3(q/2)}.$$

Derivation of the "optimal" GLSE. In the case in which $p = 2$, we treat the problem of finding an optimal GLSE $b(\hat{\Sigma}_B \otimes I_m)$ in the sense that it satisfies

$$\alpha_0(b(\hat{\Sigma}_B \otimes I_m)) \leq \alpha_0(b(\hat{\Sigma} \otimes I_m)) \tag{4.121}$$

for any GLSE $b(\hat{\Sigma} \otimes I_m)$ that belongs to an appropriate subclass of C_3.

To this end, consideration is limited to a GLSE $b(\hat{\Sigma} \otimes I_m)$ such that $\hat{\Sigma}$ is a function of S, say

$$\hat{\Sigma} = \hat{\Sigma}(S).$$

The matrix S is distributed as the Wishart distribution:

$$\mathcal{L}(S) = W_2(\Sigma, q).$$

Since $b(\hat{\Sigma} \otimes I_m)$ with $\hat{\Sigma} = \hat{\Sigma}(S)$ is in C_3, the upper bound α_0 is described as

$$\alpha_0(b(\hat{\Sigma} \otimes I_m)) = E[L_0(\hat{\Sigma}, \Sigma)], \tag{4.122}$$

where

$$L_0(\hat{\Sigma}, \Sigma) = L_1(\hat{\Sigma}, \Sigma) = \frac{[tr(\hat{\Sigma}\Sigma^{-1})]^2}{4|\hat{\Sigma}\Sigma^{-1}|}.$$

(See Proposition 4.8.) Therefore, the problem of finding an optimal GLSE $b(\hat{\Sigma}_B \otimes I_m)$ is nothing but the problem of deriving the optimal estimator

$$\hat{\Sigma}_B = \hat{\Sigma}_B(S)$$

of Σ with respect to the loss function L_0. The results in the sequel are due to Bilodeau (1990).

We begin with describing a reasonable subclass of C_3. Let $\mathcal{G}$ be the group of 2×2 nonsingular lower-triangular matrices:

$$\mathcal{G} \equiv \mathcal{G}_T(2) = \{L = (l_{ij}) \in \mathcal{G}\ell(2) \mid l_{ij} = 0 \ (i < j)\}.$$

The group $\mathcal{G}$ acts on the space $\mathcal{S}(2)$ of S via the group action

$$S \to LSL', \tag{4.123}$$

where $L \in \mathcal{G}$. For relevant definitions, see Section 1.4 of Chapter 1. Since

$$\mathcal{L}(LSL') = W_2(L\Sigma L', q)$$

by Proposition 1.5, the action (4.123) on the space of S induces the following action on the space $\mathcal{S}(2)$ of Σ:

$$\Sigma \to L\Sigma L'. \tag{4.124}$$

Thus, it is natural to consider the class of GLSEs $b(\hat{\Sigma} \otimes I_m)$ that satisfy

$$\hat{\Sigma}(LSL') = L\hat{\Sigma}(S)L' \quad \text{for any } L \in \mathcal{G}. \tag{4.125}$$

An estimator $\hat{\Sigma}$ with this property is called an equivariant estimator of Σ with respect to the group $\mathcal{G}$.

Let $\mathcal{F}$ denote the class of equivariant estimators with respect to $\mathcal{G}$.

Lemma 4.13 *The class $\mathcal{F}$ is characterized by*

$$\mathcal{F} = \left\{ \hat{\Sigma}(S) = TDT' \,\middle|\, D = \begin{pmatrix} d_1 & 0 \\ 0 & d_2 \end{pmatrix}, \, d_j > 0, \, j = 1, 2 \right\}, \tag{4.126}$$

where the matrix $T = (t_{ij})$ is the unique 2×2 lower-triangular matrix with positive diagonal elements such that $S = TT'$, that is, the Cholesky decomposition of S (see Lemma 1.8 and Proposition 1.9).

Proof. Suppose that (4.125) holds for any $L \in \mathcal{G}$. Letting $S = I_2$ in (4.125) yields

$$\hat{\Sigma}(LL') = L\hat{\Sigma}(I_2)L'. \tag{4.127}$$

Since this equality holds for all diagonal matrices L with diagonal elements ± 1, and since $LL' = I_2$, we have

$$\hat{\Sigma}(I_2) = L\hat{\Sigma}(I_2)L' \quad \text{for any } L = \begin{pmatrix} \pm 1 & 0 \\ 0 & \pm 1 \end{pmatrix}. \tag{4.128}$$

This shows that the matrix $\hat{\Sigma}(I_2)$ is diagonal, say

$$\hat{\Sigma}(I_2) = D = \begin{pmatrix} d_1 & 0 \\ 0 & d_2 \end{pmatrix},$$

where d_j's are nonrandom (see Problem 4.4.3). Further, d_j's are positive, since the matrix $\hat{\Sigma}(I_2)$ is positive definite. Here put

$$L = T$$

in (4.127). This yields

$$\hat{\Sigma}(S) = TDT'. \tag{4.129}$$

Conversely, an estimator $\hat{\Sigma}(S)$ of the form (4.129) clearly satisfies (4.125). Thus, (4.126) follows. This completes the proof.

This leads to a reasonable subclass $\mathcal{C}_3^F$ of $\mathcal{C}_3$, where

$$\mathcal{C}_3^F = \{ b(\hat{\Sigma}) \in \mathcal{C}_3 \mid \hat{\Sigma} \in \mathcal{F} \}. \tag{4.130}$$

While the UZE $b(S \otimes I_m)$ is in $\mathcal{C}_3^F$, the OLSE $b(I_2 \otimes I_m)$ is not.

In the definition of $\mathcal{F}$, we can set $d_1 = 1$ without loss of generality, because the loss function $L_0(\hat{\Sigma}, \Sigma)$ has the following invariance property:

$$L_0(a\hat{\Sigma}, \Sigma) = L_0(\hat{\Sigma}, \Sigma) \quad \text{for any } a > 0, \tag{4.131}$$

(see (4.89)). This fact corresponds to the scale-invariance property of GLSEs: $b(a\Psi \otimes I_m) = b(\Psi \otimes I_m)$ (see (2.80)).

It is easily observed that the loss function $L_0(\hat{\Sigma}, \Sigma)$ is also invariant under $\mathcal{G}$:

$$L_0(L\hat{\Sigma}L', L\Sigma L') = L_0(\hat{\Sigma}, \Sigma) \quad \text{for any } L \in \mathcal{G}. \tag{4.132}$$

Theorem 4.14 *For any $\hat{\Sigma}(S) = TDT'$ in $\mathcal{F}$ with*

$$D = \begin{pmatrix} 1 & 0 \\ 0 & d \end{pmatrix},$$

the risk function

$$R(\hat{\Sigma}, \Sigma) = E[L_0(\hat{\Sigma}, \Sigma)]$$

is evaluated as

$$R(\hat{\Sigma}, \Sigma) = \frac{1}{4}\left[\frac{(q-1)(q+1)}{d(q-3)(q-2)} + \frac{2(q-1)}{q-2} + \frac{d(q-1)}{q-2}\right], \tag{4.133}$$

and the minimum risk equivariant estimator $\hat{\Sigma}_B$ of Σ is given by

$$\hat{\Sigma}_B(S) = TD_BT' \quad \text{with} \quad D_B = \begin{pmatrix} 1 & 0 \\ 0 & \sqrt{\frac{q+1}{q-3}} \end{pmatrix}. \tag{4.134}$$

Proof. We first show that the risk function $R(\hat{\Sigma}, \Sigma)$ in question is constant as a function of Σ. (This fact is valid under a much more general setup where the group acts transitively (Section 1.4 of Chapter 1) on parameter space. See Chapter 3 of Lehmann (1983) for general results.) Let L be the lower-triangular matrix with positive diagonal elements such that $\Sigma = LL'$. Then transforming $S = LWL'$ and noting that

$$\mathcal{L}(W) = W_2(I_2, q)$$

yields

$$R(\hat{\Sigma}, \Sigma) = E_\Sigma\{L_0(\hat{\Sigma}(S), \Sigma)\}$$
$$= E_{I_2}\{L_0(\hat{\Sigma}(LWL'), LL')\}, \tag{4.135}$$

where E_Σ and E_{I_2} denote the expectation under $W_2(\Sigma, q)$ and $W_2(I_2, q)$ respectively. By using (4.125) and (4.132), the right-hand side of (4.135) is further evaluated as

$$E_{I_2}\{L_0(L\hat{\Sigma}(W)L', LL')\} = E_{I_2}\{L_0(\hat{\Sigma}(W), I_2)\}. \tag{4.136}$$

Thus, the risk function does not depend on Σ and hence, we can assume $\Sigma = I_2$ without loss of generality.

Next, to evaluate the risk function, let $T = (t_{ij})$ be the lower-triangular matrix with positive diagonal elements such that $W = TT'$. Then, by Proposition 1.9 of Chapter 1, all the nonzero elements of T are independent,

$$\mathcal{L}(t_{ii}^2) = \chi_{q-i+1}^2 \ (i = 1, 2) \quad \text{and} \quad \mathcal{L}(t_{21}^2) = \chi_1^2.$$

Therefore, we have for any $\hat{\Sigma}(W) = TDT'$ in $\mathcal{F}$,

$$E\left[L_0(TDT', I_2)\right] = E\left\{ \frac{[tr(TDT')]^2}{4|TDT'|} \right\}$$

$$= E\left\{ \frac{\left[t_{11}^2 + t_{21}^2 + dt_{22}^2\right]^2}{4d\, t_{11}^2 t_{22}^2} \right\}$$

$$= \frac{1}{4d}\, E\left\{ \frac{t_{11}^2}{t_{22}^2} + \frac{t_{21}^4}{t_{11}^2 t_{22}^2} + d^2\frac{t_{22}^2}{t_{11}^2} + 2\frac{t_{21}^2}{t_{22}^2} + 2d + 2d\frac{t_{21}^2}{t_{11}^2} \right\}$$

$$= \frac{1}{4d}\left\{ \frac{q}{q-3} + \frac{3}{(q-2)(q-3)} \right.$$

$$\left. + \frac{d^2(q-1)}{q-2} + \frac{2}{q-3} + 2d + \frac{2d}{q-2} \right\},$$

proving (4.133). Here, we used the following formulas: let W be a random variable such that $\mathcal{L}(W) = \chi_m^2$, then

$$E(W^2) = m(m+2),$$

$$E(W) = m,$$

$$E(W^{-1}) = \frac{1}{m-2}.$$

See Problem 4.4.2. Thus, for example, the term $E[t_{21}^4/(t_{11}^2 t_{22}^2)]$ is evaluated as follows: let $W_{ij} = t_{ij}^2$.

$$E\left[\frac{t_{21}^4}{t_{11}^2 t_{22}^2} \right] = E(W_{21}^2)E(W_{11}^{-1})E(W_{22}^{-1})$$

$$= 3 \times \frac{1}{q-2} \times \frac{1}{[(q-1)-2]}$$

$$= \frac{3}{(q-2)(q-3)}.$$

Minimizing with respect to d yields the result. This completes the proof.

Consequently, the GLSE $b(\hat{\Sigma}_B \otimes I_m)$ is optimum in the class $\mathcal{C}_3^F$, because it satisfies

$$\alpha_0(b(\hat{\Sigma}_B \otimes I_m)) \leq \alpha_0(b(\hat{\Sigma} \otimes I_m)) \tag{4.137}$$

for any $b(\hat{\Sigma} \otimes I_m) \in \mathcal{C}_3^F$. The optimal GLSE $b(\hat{\Sigma}_B)$ has a smaller upper bound than the UZE has, because the UZE belongs to the class $\mathcal{C}_3^E$.

Finally, we note that the function $L_0(\hat{\Sigma}, \Sigma)$ satisfies

$$L_0(\hat{\Sigma}^{-1}, \Sigma^{-1}) = L_0(\hat{\Sigma}, \Sigma). \tag{4.138}$$

This means that as a loss function for choosing an estimator $\hat{\Sigma}$, the function L_0 symmetrically penalizes the underestimate and the overestimate.

4.5 Upper Bound Problems for a GLSE in a Heteroscedastic Model

In this section, we will treat a heteroscedastic model and evaluate the upper bounds for the covariance matrices of several typical GLSEs including the unrestricted GLSE and the OLSE. Since the implications of the results derived here are quite similar to those of Section 4.4, we often omit the details.

General results. The model considered here is the p-equation heteroscedastic model treated in Example 4.2, where we have observed that the unrestricted GLSE defined in (4.38) belongs to the class $\mathcal{C}_3$.

To state the results in a general setup, consider a GLSE of the form

$$b(\hat{\Omega}) = (X'\hat{\Omega}^{-1}X)^{-1}X'\hat{\Omega}^{-1}y \quad \text{with} \quad \hat{\Omega} = \Omega(\hat{\theta}), \tag{4.139}$$

where $\hat{\theta}$ is an estimator of $\theta = (\theta_1, \ldots, \theta_p)'$, and suppose that $\hat{\theta}$ depends only on the statistic

$$s = \begin{pmatrix} s_1^2 \\ \vdots \\ s_p^2 \end{pmatrix} : p \times 1$$

in (4.38), say

$$\hat{\theta} = \begin{pmatrix} \hat{\theta}_1 \\ \vdots \\ \hat{\theta}_p \end{pmatrix} = \begin{pmatrix} \hat{\theta}_1(s) \\ \vdots \\ \hat{\theta}_p(s) \end{pmatrix} = \hat{\theta}(s) : p \times 1. \tag{4.140}$$

Then, it is easily seen that a GLSE $b(\hat{\Omega})$ with $\hat{\Omega} = \Omega(\hat{\theta}(s))$ belongs to the class $\mathcal{C}_3$, that is,

$$E[b(\hat{\Omega})|\hat{\Omega}] = \beta,$$

$$\text{Cov}(b(\hat{\Omega})|\hat{\Omega}) = H(\hat{\Omega}, \Omega), \qquad (4.141)$$

where the function H is defined by

$$H(\hat{\Omega}, \Omega) = (X'\hat{\Omega}^{-1}X)^{-1}X'\hat{\Omega}^{-1}\Omega\hat{\Omega}^{-1}X(X'\hat{\Omega}^{-1}X)^{-1}. \qquad (4.142)$$

The results in Section 4.3 yield the upper bound $\alpha_0(b(\hat{\Omega}))$ and $\alpha_1(b(\hat{\Omega}))$ for the covariance matrix of a GLSE $b(\hat{\Omega}) \in \mathcal{C}_3$. In fact, the latent roots of the matrix $\Omega^{-1/2}\hat{\Omega}\Omega^{-1/2}$ are given by π_1 (with multiplicity n_1), π_2 (with multiplicity n_2), $\dots, \pi_p$ (with multiplicity n_p), where

$$\pi_j = \hat{\theta}_j/\theta_j \ (j = 1, \dots, p).$$

Let $\pi_{(j)}$ be the j-th smallest one:

$$\pi_{(1)} \leq \cdots \leq \pi_{(p)}. \qquad (4.143)$$

Then $\alpha_0(b(\hat{\Omega}))$ is obtained from Theorem 4.4 as

$$\text{Cov}(b(\Omega)) \leq \text{Cov}(b(\hat{\Omega})) \leq \alpha_0(b(\hat{\Omega})) \, \text{Cov}(b(\Omega)) \qquad (4.144)$$

with

$$\alpha_0(b(\hat{\Omega})) = E\left\{ \frac{(\pi_{(1)} + \pi_{(p)})^2}{4\pi_{(1)}\pi_{(p)}} \right\}. \qquad (4.145)$$

Here in (4.144),

$$b(\Omega) = (X'\Omega^{-1}X)^{-1}X'\Omega^{-1}y$$

is the GME and hence

$$\text{Cov}(b(\Omega)) = (X'\Omega^{-1}X)^{-1}.$$

Further, applying Lemma 4.6 and Theorem 4.7 yields the greater upper bound $\alpha_1(b(\hat{\Omega}))$:

$$\alpha_0(b(\hat{\Omega})) \leq \alpha_1(b(\hat{\Omega}))$$

$$= E\left\{ \left(\frac{\frac{1}{p}\sum_{j=1}^{p} \pi_{(j)}}{\prod_{j=1}^{p} \pi_{(j)}^{1/p}} \right)^p \right\}$$

$$= E\left\{ \left(\frac{\frac{1}{p}\sum_{j=1}^{p} \pi_j}{\prod_{j=1}^{p} \pi_j^{1/p}} \right)^p \right\}, \qquad (4.146)$$

where the last equality follows from the permutation invariance. Thus, it follows that for any $b(\hat{\Omega}) \in \mathcal{C}_3$,

$$\mathrm{Cov}(b(\Omega)) \le \mathrm{Cov}(b(\hat{\Omega}))$$

$$\le \alpha_0(b(\hat{\Omega})) \, \mathrm{Cov}(b(\Omega))$$

$$\le \alpha_1(b(\hat{\Omega})) \, \mathrm{Cov}(b(\Omega)). \tag{4.147}$$

When $p = 2$, that is, when the model consists of two homoscedastic equations, Proposition 4.8 guarantees that

$$\alpha_0(b(\hat{\Omega})) = \alpha_1(b(\hat{\Omega})).$$

Both the unrestricted GLSE and the OLSE $b(I_n) = (X'X)^{-1}X'y$ are special cases of (4.139). The unrestricted GLSE is a GLSE

$$b(\hat{\Omega}_0) = (X'\hat{\Omega}_0^{-1}X)^{-1}X'\hat{\Omega}_0^{-1}y$$

with

$$\hat{\Omega}_0 = \Omega(s) = \begin{pmatrix} s_1^2 I_{n_1} & & 0 \\ & \ddots & \\ 0 & & s_p^2 I_{n_p} \end{pmatrix},$$

which is obtained by letting $\hat{\theta}(s) = s$ in (4.140). On the other hand, the OLSE is given by

$$b(I_n) = (X'X)^{-1}X'y,$$

which is a GLSE $b(\hat{\Omega})$ with

$$\hat{\theta}(s) \equiv 1_p \equiv \begin{pmatrix} 1 \\ \vdots \\ 1 \end{pmatrix} : p \times 1$$

and hence

$$\hat{\Omega} = \Omega(1_p) = \begin{pmatrix} I_{n_1} & & 0 \\ & \ddots & \\ 0 & & I_{n_p} \end{pmatrix} = I_n.$$

Efficiency comparison between the unrestricted GLSE and the OLSE. By letting $\hat{\theta}(s) = s$, the two upper bounds for the covariance matrix of the unrestricted GLSE $b(\hat{\Omega}_0)$ are obtained as

$$\alpha_0(b(\hat{\Omega}_0)) = E\left\{ \frac{(v_{(1)} + v_{(p)})^2}{4 v_{(1)} v_{(p)}} \right\} \tag{4.148}$$

and

$$\alpha_1(b(\hat{\Omega}_0)) = E\left\{\left(\frac{\frac{1}{p}\sum_{j=1}^{p} v_j}{\prod_{j=1}^{p} v_j^{1/p}}\right)^p\right\}, \tag{4.149}$$

where

$$v_{(1)} \leq \cdots \leq v_{(p)}$$

are the ordered values of $v_j = s_j^2/\theta_j$'s $(j = 1, \ldots, p)$.

As for the OLSE $b(I_p \otimes I_m)$, we have the following two equalities:

$$\alpha_0(b(I_n)) = \frac{(\theta_{(1)} + \theta_{(p)})^2}{4\theta_{(1)}\theta_{(p)}} \tag{4.150}$$

and

$$\alpha_1(b(I_n)) = \left[\frac{\frac{1}{p}\sum_{j=1}^{p} \theta_j}{\prod_{j=1}^{p} \theta_j^{1/p}}\right]^p, \tag{4.151}$$

where

$$\theta_{(1)} \leq \cdots \leq \theta_{(p)}$$

are the ordered values of $\theta_1, \ldots, \theta_p$. When $p = 2$, the two upper bounds reduce to

$$\alpha_0(b(I_n)) = \alpha_1(b(I_n)) = \frac{(\theta_1 + \theta_2)^2}{4\theta_1\theta_2} = \frac{(1 + \eta)^2}{4\eta}, \tag{4.152}$$

where

$$\eta = \theta_1/\theta_2.$$

The upper bounds (4.150) and (4.151) can be viewed as measures of the sphericity of the matrix Ω. This suggests that the OLSE can be more efficient than the unrestricted GLSE in the sense of the upper bounds, if the matrix Ω is close to the identity matrix up to a multiplicative constant. If not, the unrestricted GLSE is preferable. To see this more precisely, we first evaluate the upper bound $\alpha_0(b(\hat{\Omega}_0))$ in (4.148) for the case in which $p = 2$.

The following theorem is due to Kariya (1981a).

Theorem 4.15 *Let $p = 2$ and suppose that $q_j > 2$ $(j = 1, 2)$. Then the upper bound for the covariance matrix of the unrestricted GLSE $b(\hat{\Omega}_0)$ is given by*

$$\alpha_0(b(\hat{\Omega}_0)) = \alpha_1(b(\hat{\Omega}_0)) = 1 + \frac{1}{2(q_1 - 2)} + \frac{1}{2(q_2 - 2)}, \tag{4.153}$$

that is,

$$\text{Cov}(b(\Omega)) \leq \text{Cov}(b(\hat{\Omega}_0))$$

$$\leq \left[1 + \frac{1}{2(q_1 - 2)} + \frac{1}{2(q_2 - 2)}\right] \text{Cov}(b(\Omega)). \qquad (4.154)$$

Proof. This result will be proved as a special case of Theorem 4.16.

The upper bound $\alpha_0(b(\hat{\Omega}_0))$ is a monotonically decreasing function of the degrees of freedom q_1 and q_2. Furthermore, as q_j's go to infinity, it converges to 1:

$$\lim_{q_1, q_2 \to \infty} \alpha_0(b(\hat{\Omega}_0)) = 1. \qquad (4.155)$$

Comparing $\alpha_0(b(\hat{\Omega}_0))$ with $\alpha_0(b(I_n))$, we see that $\alpha_0(b(I_n)) \leq \alpha_0(b(\hat{\Omega}_0))$ holds if and only if the variance ratio $\eta = \theta_1/\theta_2$ satisfies

$$(1 + 2c) - 2\sqrt{c(1 + c)} \leq \eta \leq (1 + 2c) + 2\sqrt{c(1 + c)}, \qquad (4.156)$$

where $c = \alpha_0(b(\hat{\Omega}_0)) - 1$. This interval always contains 1 and

$$(1 + 2c) \pm 2\sqrt{c(1 + c)} \to 1 \quad \text{as} \quad q_1, q_2 \to \infty.$$

Thus, we see that the OLSE can be more efficient in terms of the upper bound than the unrestricted GLSE in the context of the upper bounds when the ratio η is close to 1 and the sample size is small.

Derivation of the optimal GLSE. Next, we derive a GLSE $b(\hat{\Omega}_B)$ that is optimal in the sense that

$$\alpha_0(b(\hat{\Omega}_B)) \leq \alpha_0(b(\hat{\Omega})) \qquad (4.157)$$

holds for any GLSE $b(\hat{\Omega})$ in an appropriate subclass of $\mathcal{C}_3$.

To do so, let $p = 2$ and consider the GLSE

$$b(\hat{\Omega}) = (X'\hat{\Omega}^{-1}X)^{-1}X'\hat{\Omega}^{-1}y$$

with $\hat{\Omega} = \Omega(\hat{\theta})$, where

$$\hat{\theta} = \begin{pmatrix} \hat{\theta}_1 \\ \hat{\theta}_2 \end{pmatrix} = \begin{pmatrix} \hat{\theta}_1(s) \\ \hat{\theta}_2(s) \end{pmatrix} = \hat{\theta}(s) : 2 \times 1 \qquad (4.158)$$

and $s = (s_1^2, s_2^2)'$ is defined in (4.38). It follows from (4.144) that, the GLSE defined above satisfies

$$\text{Cov}(b(\Omega)) \leq \text{Cov}(b(\hat{\Omega})) \leq \alpha_0(b(\hat{\Omega})) \, \text{Cov}(b(\Omega)) \qquad (4.159)$$

with

$$\alpha_0(b(\hat{\Omega})) = E[L_0(\hat{\theta}, \theta)] \quad \text{and}$$

$$L_0(\hat{\theta}, \theta) = \frac{[(\hat{\theta}_1/\theta_1) + (\hat{\theta}_2/\theta_2)]^2}{4(\hat{\theta}_1/\theta_1)(\hat{\theta}_2/\theta_2)}. \tag{4.160}$$

Thus, the problem is reduced to that of finding an optimal estimator

$$\hat{\theta}_B = \hat{\theta}_B(s) : 2 \times 1$$

of $\theta = (\theta_1, \theta_2)'$ with respect to the loss function L_0.

Let $\mathcal{G} = (0, \infty) \times (0, \infty)$. The group $\mathcal{G}$ acts on the space $\mathcal{X} = (0, \infty) \times (0, \infty)$ of $s = (s_1^2, s_2^2)'$ via the group action

$$s = \begin{pmatrix} s_1^2 \\ s_2^2 \end{pmatrix} \rightarrow \begin{pmatrix} g_1 s_1^2 \\ g_2 s_2^2 \end{pmatrix} \equiv gs, \tag{4.161}$$

where $g = (g_1, g_2) \in \mathcal{G}$. Since

$$\mathcal{L}(q_j s_j^2/\theta_j) = \chi_{q_j}^2 \quad (j = 1, 2),$$

this action induces the following action on the space $\Theta = (0, \infty) \times (0, \infty)$ of $\theta = (\theta_1, \theta_2)'$:

$$\theta = \begin{pmatrix} \theta_1 \\ \theta_2 \end{pmatrix} \rightarrow \begin{pmatrix} g_1 \theta_1 \\ g_2 \theta_2 \end{pmatrix} \equiv g\theta. \tag{4.162}$$

Hence, it is natural to restrict our consideration to the estimators $\hat{\theta}(s)$ satisfying

$$\hat{\theta}(gs) = \begin{pmatrix} \hat{\theta}_1(gs) \\ \hat{\theta}_2(gs) \end{pmatrix} = \begin{pmatrix} g_1 \hat{\theta}_1(s) \\ g_2 \hat{\theta}_2(s) \end{pmatrix} \equiv g\hat{\theta}(s) \tag{4.163}$$

for any $g = (g_1, g_2) \in \mathcal{G}$. An estimator $\hat{\theta}$ that satisfies this condition is called an *equivariant estimator* of θ with respect to $\mathcal{G}$. The loss function L_0 in (4.160) is invariant under these transformations:

$$L_0(g\hat{\theta}, g\theta) = L_0(\hat{\theta}, \theta) \quad \text{for any } g \in \mathcal{G}. \tag{4.164}$$

As is easily shown (see Problem 4.5.1), the class, say $\mathcal{F}$, of equivariant estimators is characterized by

$$\mathcal{F} = \left\{ \hat{\theta}(s) = \begin{pmatrix} a_1 s_1^2 \\ a_2 s_2^2 \end{pmatrix} \mid a_j > 0, \ j = 1, 2 \right\}. \tag{4.165}$$

This leads to the class C_3^F of GLSEs, where

$$C_3^F = \{b(\hat{\Omega}) \in C_3 \mid \hat{\Omega} = \Omega(\hat{\theta}), \ \hat{\theta} \in \mathcal{F}\}. \tag{4.166}$$

While the unrestricted GLSE $b(\hat{\Omega}_0)$ is a member of C_3^F, the OLSE is not.

As in the case of the SUR model, we can set $a_2 = 1$ in the definition of the class $\mathcal{F}$ without loss of generality, because the loss function L_0 is scale-invariant in the sense that

$$L_0(a\hat{\theta}, \theta) = L_0(\hat{\theta}, \theta) \quad \text{for any } a > 0. \tag{4.167}$$

This corresponds to the fact that $b(a\hat{\Omega}) = b(\hat{\Omega})$ for any $a > 0$. The following theorem due to Bilodeau (1990) derives the best equivariant estimator $\hat{\theta}_B$ with respect to the loss function L_0.

Theorem 4.16 *Let $p = 2$ and suppose $q_j > 2$ ($j = 1, 2$). Then, for any estimator of the form*

$$\hat{\theta} = \begin{pmatrix} ds_1^2 \\ s_2^2 \end{pmatrix} \in \mathcal{F},$$

the risk function

$$R(\hat{\theta}, \theta) = E[L_0(\hat{\theta}, \theta)]$$

is given by

$$R(\hat{\theta}, \theta) = 1 + \frac{(1 - d)^2}{4d} + \frac{d^{-1}}{2(q_1 - 2)} + \frac{d}{2(q_2 - 2)}, \tag{4.168}$$

and it is minimized when

$$d = \sqrt{\frac{q_1(q_2 - 2)}{(q_1 - 2)q_2}} \equiv d_B.$$

Thus, the GLSE $b(\hat{\Omega}_B)$ with $\hat{\Omega}_B = \Omega(\hat{\theta}_B)$ where

$$\hat{\theta}_B = \begin{pmatrix} d_B s_1^2 \\ s_2^2 \end{pmatrix} \tag{4.169}$$

satisfies (4.157) among the subclass C_3^F.

Proof. For any $\hat{\theta} = (ds_1^2, s_2^2)'$, let

$$\eta = \theta_1/\theta_2 \quad \text{and} \quad \hat{\eta} = \hat{\theta}_1/\hat{\theta}_2.$$

First observe that

$$L_0(\hat{\theta}, \theta) = 1 + \frac{(\hat{\eta} - \eta)^2}{4\hat{\eta}\eta}$$

$$= 1 + \frac{1}{4}\left[\frac{\hat{\eta}}{\eta} + \frac{\eta}{\hat{\eta}} - 2\right]$$

$$\equiv \tilde{L}_0(\hat{\eta}/\eta), \tag{4.170}$$

where

$$\tilde{L}_0(t) = 1 + \frac{1}{4}\left(t + \frac{1}{t} - 2\right).$$

Let $x_j = q_j s_j^2/\theta_j$. Then x_j's are independently distributed as $\chi_{q_j}^2$, and $\hat{\eta}/\eta$ is expressed as

$$\frac{\hat{\eta}}{\eta} = d\,\frac{q_2\,x_1}{q_1\,x_2},$$

from which both

$$E\left(\hat{\eta}/\eta\right) = d\frac{q_2}{q_2 - 2} \quad \text{and} \quad E\left(\eta/\hat{\eta}\right) = \frac{1}{d}\frac{q_1}{q_1 - 2}$$

follow. This yields (4.168). The right-hand side of (4.168) is clearly minimized when $d = d_B$. This completes the proof.

By using

$$b(a\hat{\Omega}) = b(\hat{\Omega}) \quad \text{and} \quad \Omega(a\hat{\theta}) = a\Omega(\hat{\theta}),$$

another expression of $\hat{\theta}_B$ is obtained:

$$\hat{\theta}_B = \begin{pmatrix} \sqrt{\frac{q_1}{q_1-2}}\,s_1^2 \\ \sqrt{\frac{q_2}{q_2-2}}\,s_2^2 \end{pmatrix}.$$

which of course yields the GLSE $b(\hat{\Omega}_B)$. The result of Theorem 4.15 follows by substituting $d = 1$ in the right-hand side of (4.168).

The loss function $\tilde{L}_0(\hat{\eta}/\eta)$ satisfies the following property

$$\tilde{L}_0(t) = \tilde{L}_0(1/t), \tag{4.171}$$

implying that, as a loss function for choosing an estimator $\hat{\eta}$ of η, the function $\tilde{L}_0$ equally penalizes underestimates and overestimates.

The case in which $p \geq 3$. Next, we treat the case in which $p \geq 3$ and consider the efficiency of the unrestricted GLSE $b(\hat{\Omega}_0)$ with $\hat{\Omega}_0 = \Omega(s)$. The upper bound $\alpha_0(b(\hat{\Omega}_0))$ for the covariance matrix of the unrestricted GLSE is given by (4.148). Although this expression is simple, the evaluation is difficult since the upper bound α_0 depends on the extreme values $v_{(1)}$ and $v_{(p)}$. The following result is due to Kurata and Kariya (1996).

Theorem 4.17 *When $q_j > 2$ ($j = 1, \ldots, p$), the upper bound for the covariance matrix of the unrestricted GLSE $b(\hat{\Omega}_0)$ is given by*

$$\alpha_0(b(\hat{\Omega}_0)) \leq \alpha_1(b(\hat{\Omega}_0))$$

$$= \frac{1}{p^p} \sum \frac{p!}{p_1! \cdots p_p!} \prod_{j=1}^{p} \left(\frac{2}{q_j}\right)^{p_j-1} \frac{\Gamma(q_j/2 + n_j - 1)}{\Gamma(q_j/2)},$$

where the sum on the right-hand side carries over $p_j \geq 0$ and $\sum_{j=1}^{p} p_j = p$. Furthermore,

$$1 \leq \alpha_0(b(\hat{\Omega}_0)) \leq \alpha_1(b(\hat{\Omega}_0)) \to 1 \tag{4.172}$$

as $q_1, \ldots, q_p \to \infty$.

Proof. The proof is straightforward and is hence omitted.

The implication of this result is quite similar to that of Section 4.4 and is also omitted.

4.6   Empirical Example: CO_2 Emission Data

In this section, by using the GLSEs in a two-equation SUR model, we give an example of analysis on CO_2 emission data. The discussion here is a continuation of Section 2.5 of Chapter 2.

Two-equation SUR model. The data in Tables 2.1 and 2.2 show the volume of CO_2 emission and GNP in Japan and the USA from 1970 to 1996. The values of GNP are deflated by the 1990 prices.

Let us consider the following SUR model consisting of two simple linear regression equations:

$$y = X\beta + \varepsilon \tag{4.173}$$

with $n = mp$, $m = 27$, $p = 2$, $k = k_1 + k_2$, $k_1 = k_2 = 2$, and

$$y = \begin{pmatrix} y_1 \\ y_2 \end{pmatrix} : n \times 1, \quad X = \begin{pmatrix} X_1 & 0 \\ 0 & X_2 \end{pmatrix} : n \times k,$$

$$\beta = \begin{pmatrix} \beta_1 \\ \beta_2 \end{pmatrix} : k \times 1, \quad \varepsilon = \begin{pmatrix} \varepsilon_1 \\ \varepsilon_2 \end{pmatrix} : n \times 1,$$

where

$$y_j = \begin{pmatrix} \log(CO2_{j1}) \\ \vdots \\ \log(CO2_{jm}) \end{pmatrix} : m \times 1,$$

$$X_j = \begin{pmatrix} 1 & \log(GNP_{j1}) \\ \vdots & \vdots \\ 1 & \log(GNP_{jm}) \end{pmatrix} : m \times k_j,$$

$$\beta_j = \begin{pmatrix} \beta_{j1} \\ \beta_{j2} \end{pmatrix} : k_j \times 1, \quad \varepsilon_j = \begin{pmatrix} \varepsilon_{j1} \\ \vdots \\ \varepsilon_{jm} \end{pmatrix} : m \times 1.$$

Here, the suffix $j = 1$ denotes Japan and $j = 2$ the USA. So β_{12} and β_{22} are understood as the elasticity of CO_2 relative to the GNP in Japan and the USA, respectively. Suppose that the error term ε is distributed as the normal distribution:

$$\mathcal{L}(\varepsilon) = N_n(0, \Sigma \otimes I_m) \text{ with } \Sigma = (\sigma_{ij}) \in \mathcal{S}(2), \quad (4.174)$$

where $\mathcal{S}(2)$ denotes the set of 2×2 positive definite matrices.

We begin by estimating the model using the OLSE:

$$b(I_2 \otimes I_m) = (X'X)^{-1}X'y \quad (4.175)$$

$$= \begin{pmatrix} (X_1'X_1)^{-1}X_1'y_1 \\ (X_2'X_2)^{-1}X_2'y_2 \end{pmatrix} : k \times 1,$$

which of course gives an equivalent result when applying the OLS estimation procedure to the two equations separately. The estimate of $\beta = (\beta_1', \beta_2')'$ obtained by the OLSE is

$$b(I_2 \otimes I_m) = ((4.754, \ 0.364)', \ (6.102, \ 0.275)')'.$$

Thus, the models are estimated as

$$\log(CO2_1) = 4.754 + 0.364 \ \log(GNP_1) \quad : \text{Japan}$$

$$\log(CO2_2) = 6.102 + 0.275 \ \log(GNP_2) \quad : \text{USA}. \quad (4.176)$$

The elasticities of CO_2 in Japan and the USA are estimated at 0.364 and 0.275 respectively.

The estimation procedure above ignores the correlation between the two regression equations. The matrix Σ is estimated by

$$\hat{\Sigma} = (\hat{\sigma}_{ij}) = (e_i'e_j/m) : 2 \times 2 \quad (4.177)$$

$$= \begin{pmatrix} 0.002515 & 0.001456 \\ 0.001456 & 0.001439 \end{pmatrix}, \quad (4.178)$$

where e is the OLS residual vector defined by

$$e = [I_n - X(X'X)^{-1}X']y$$

$$= \begin{pmatrix} [I_m - X_1(X_1'X_1)^{-1}X_1']y_1 \\ [I_m - X_2(X_2'X_2)^{-1}X_2']y_2 \end{pmatrix}$$

$$= \begin{pmatrix} e_1 \\ e_2 \end{pmatrix} : n \times 1. \tag{4.179}$$

The correlation between the two regression equations is large and hence needs to be taken into account in the estimation of the coefficients. In fact, the correlation coefficient

$$\rho = \sigma_{12}/(\sigma_{11}\sigma_{22})^{1/2}$$

is estimated as

$$\hat{\rho} = \frac{\hat{\sigma}_{12}}{(\hat{\sigma}_{11}\hat{\sigma}_{22})^{1/2}} \tag{4.180}$$

$$= \frac{0.001456}{(0.002515 \times 0.001439)^{1/2}}$$

$$= 0.7656. \tag{4.181}$$

Let us use this information by the SUR model to improve the efficiency of the OLSE. In our context, the efficiency may be discussed in terms of the upper bounds. As is shown in Corollary 4.5, the upper bound $\alpha_0(b(I_2 \otimes I_m))$ of the covariance matrix of the OLSE is given by

$$\alpha_0(b(I_2 \otimes I_m)) = \frac{(\lambda_1 + \lambda_2)^2}{4\lambda_1\lambda_2} = \frac{(tr(\Sigma))^2}{4|\Sigma|}, \tag{4.182}$$

where $\lambda_1 \leq \lambda_2$ denotes the latent roots of Σ. By replacing Σ in the right-hand side of the above equality by the estimate $\hat{\Sigma}$ in (4.178), the upper bound $\alpha_0(b(I_2 \otimes I_m))$ is estimated as

$$\alpha_0(b(I_2 \otimes I_m)) = \frac{(tr(\hat{\Sigma}))^2}{4|\hat{\Sigma}|} = 2.6095,$$

which is quite large.

On the other hand, the restricted Zellner estimator (RZE) is given by

$$b(\hat{\Sigma} \otimes I_m) = (X'(\hat{\Sigma}^{-1} \otimes I_m)X)^{-1}X'(\hat{\Sigma}^{-1} \otimes I_m)y$$

$$= ((4.686, \ 0.375)', \ (6.065, \ 0.279)')'. \tag{4.183}$$

Thus, the model is estimated as

$$\log(CO2_1) = 4.686 + 0.375 \ \log(GNP_1) \ : Japan$$

$$\log(CO2_2) = 6.065 + 0.279 \ \log(GNP_2) \ : USA. \tag{4.184}$$

The UZE is an alternative GLSE for this model. The UZE is a GLSE $b(\hat{\Sigma} \otimes I_m)$ with $\hat{\Sigma} = S/q$, where

$$S = Y'_*[I_m - X_*(X'_*X_*)^+ X'_*]Y_*. \tag{4.185}$$

Here, A^+ denotes the Moore–Penrose generalized inverse of A, and

$$Y_* = (y_1, y_2) : m \times 2, \quad X_* = (X_1, X_2) : m \times k,$$

$$q = m - rank X_* = 27 - 3 = 24.$$

The matrix Σ is estimated by $\hat{\Sigma} = S/q$ as

$$S/q = \frac{1}{24} \begin{pmatrix} 0.06013 & 0.02925 \\ 0.02925 & 0.02563 \end{pmatrix}$$

$$= \begin{pmatrix} 0.002505 & 0.001219 \\ 0.001219 & 0.001068 \end{pmatrix}. \tag{4.186}$$

From this, the value of the UZE is given by

$$b(S/q \otimes I_m) = b(S \otimes I_m)$$

$$= (X'(S^{-1} \otimes I_m)X)^{-1}X'(S^{-1} \otimes I_m)y$$

$$= ((4.673, 0.378)', (6.057, 0.280)')'. \tag{4.187}$$

Hence, the model is estimated by

$$\log(CO2_1) = 4.673 + 0.378 \ \log(GNP_1) \ : Japan$$

$$\log(CO2_2) = 6.057 + 0.280 \ \log(GNP_2) \ : USA. \tag{4.188}$$

The elasticity of CO_2 in Japan and the USA is given by 0.378 and 0.280 respectively. The above three estimates commonly indicate that the elasticity of CO_2 in Japan is higher than the elasticity of CO_2 in the USA.

Concerning the efficiency of the UZE, the upper bound $\alpha(b(S \otimes I_m))$ for the covariance matrix $Cov(b(S \otimes I_m))$ of the UZE relative to that of the GME $b(\Sigma \otimes I_m)$ is given by

$$\alpha_0(b(S \otimes I_m)) = 1 + \frac{2}{q-3} = 1 + \frac{2}{24-3} = 1.0952. \tag{4.189}$$

Hence, the result obtained by the RZE and the UZE may be more dependable than that of the OLSE. Consequently, so long as the specification of the model is accepted, it is better to use the GLSEs.

Modification of model. The SUR model considered above requires that the marginal distribution of the error term ε_j of each regression equation is normal with covariance matrix $\sigma_{jj} I_m$:

$$\mathcal{L}(\varepsilon_j) = N_m(0, \sigma_{jj} I_m) \quad (j = 1, 2). \tag{4.190}$$

However, as has been observed in Section 2.5, the values of the Durbin–Watson statistic

$$DW_j = \frac{\sum_{i=2}^{m}(e_{ji} - e_{j,i-1})^2}{\sum_{i=1}^{m} e_{ji}^2} \quad (j = 1, 2)$$

calculated from each equation may suggest the presence of serial correlation among the error terms ε_{ji}'s, where

$$e_j = \begin{pmatrix} e_{j1} \\ \vdots \\ e_{jm} \end{pmatrix} : m \times 1 \quad (j = 1, 2).$$

In fact, we obtain

$$DW_1 = 0.5080 \quad \text{and} \quad DW_2 = 0.4113,$$

which shows that the assumption (4.190) is inappropriate. Hence, we modify the model in order to dissolve the serial correlation.

In Nawata (2001), it is shown that the optimal model for Japanese data in terms of AIC (Akaike Information Criterion) among the following six models is (3):

(1) $\log(CO_2) = \beta_1 + \beta_2 \log(GNP)$,

(2) $\log(CO_2) = \beta_1 + \beta_2 \log(GNP) + \beta_3[\log(GNP)]^2$,

(3) $\log(CO_2) = \beta_1 + \beta_2 \log(GNP) + \beta_3[\log(GNP)]^2 + \beta_4[\log(GNP)]^3$,

(4) $\log(CO_2) = \beta_1 + \beta_2 D + \beta_3 \log(GNP)$,

(5) $\log(CO_2) = \beta_1 + \beta_2 \log(GNP) + \beta_3 \left[D \log(GNP) \right]$,

(6) $\log(CO_2) = \beta_1 + \beta_2 D + \beta_3 \log(GNP) + \beta_4 \left[D \log(GNP) \right]$,

where $D = (d_1, \ldots, d_m)'$ is a dummy variable such that

$$d_i = \begin{cases} 0 & (i = 1, \ldots, 11) \\ 1 & (i = 12, \ldots, 27(= m)) \end{cases}$$

Arguing in the same way, we can see that the model (6) is optimal for USA data. Calculation of AIC and several relevant statistics including t-values, R^2, Durbin–Watson statistics and so on, is left to the readers (see Problem 4.6.1). Note that model (6) is rewritten as

$$\log(CO2_{2i}) = \beta_1 + \beta_3 \log(GNP_{2i}) + \varepsilon_{2i} \quad (i = 1, \ldots, 11)$$

$$\log(CO2_{2i}) = \beta_1^* + \beta_3^* \log(GNP_{2i}) + \varepsilon_{2i} \quad (i = 12, \ldots, 27), \tag{4.191}$$

where $\beta_1^* = \beta_1 + \beta_2$ and $\beta_3^* = \beta_3 + \beta_4$. Here, $\beta_2 = 0$ ($\beta_4 = 0$) in the model (6) is equivalent to $\beta_1 = \beta_1^*$ ($\beta_3 = \beta_3^*$) in (4.191).

The newly adopted model is the model (4.173) with k_1, k_2, k, X_1 and X_2 replaced by $k_1 = 4$, $k_2 = 4$, $k = k_1 + k_2 = 8$,

$$X_1 = \begin{pmatrix} 1 & \log(\text{GNP}_{11}) & \left[\log(\text{GNP}_{11})\right]^2 & \left[\log(\text{GNP}_{11})\right]^3 \\ \vdots & \vdots & \vdots & \vdots \\ 1 & \log(\text{GNP}_{1m}) & \left[\log(\text{GNP}_{1m})\right]^2 & \left[\log(\text{GNP}_{1m})\right]^3 \end{pmatrix}$$

$: m \times k_1 \ :$ Japan

and

$$X_2 = \begin{pmatrix} 1 & d_1 & \log(\text{GNP}_{21}) & d_1 \log(\text{GNP}_{21}) \\ \vdots & \vdots & \vdots & \vdots \\ 1 & d_m & \log(\text{GNP}_{2m}) & d_m \log(\text{GNP}_{2m}) \end{pmatrix}$$

$: m \times k_2 \ :$ USA.

The model is estimated by the OLSE $b(I_2 \otimes I_m)$ in (4.175) as

$$\log(\text{CO2}_1) = -419.30 + 224.85 \ \log(\text{GNP}_1) - 39.549 \ [\log(\text{GNP}_1)]^2$$
$$+ 2.319 \ [\log(\text{GNP}_1)]^3 \ : \text{Japan}$$

$$\log(\text{CO2}_2) = 5.721 - 2.370 \ D + 0.323 \ \log(\text{GNP}_2)$$
$$+ 0.268 \ D \log(\text{GNP}_2) \ : \text{USA}. \tag{4.192}$$

The matrix Σ is estimated by $\hat{\Sigma}$ in (4.177) as

$$\hat{\Sigma} = \begin{pmatrix} 0.0006597 & 0.0001202 \\ 0.0001202 & 0.0002989 \end{pmatrix},$$

from which $\hat{\rho}$ in (4.180) is calculated as

$$\hat{\rho} = 0.2706.$$

This suggests high efficiency of the OLSE. In fact, the upper bound $\alpha_0(b(I_2 \otimes I_m))$ in (4.182) is estimated by

$$\alpha_0(b(I_2 \otimes I_m)) = \frac{(tr(\hat{\Sigma}))^2}{4|\hat{\Sigma}|} = 1.2571,$$

which is much smaller than that of the previous model.

The RZE $b(\hat{\Sigma} \otimes I_m)$ in (4.183) estimates the model by

$$\log(\text{CO2}_1) = -402.77 + 216.15 \ \log(\text{GNP}_1) - 38.027 \ \left[\log(\text{GNP}_1)\right]^2$$
$$+ 2.230 \ \left[\log(\text{GNP}_1)\right]^3 \ : \text{Japan}$$

$$\log(\text{CO2}_2) = 5.910 - 2.456 \ D + 0.300 \ \log(\text{GNP}_2)$$
$$+ 0.280 \ D \log(\text{GNP}_2) \ : \text{USA}. \tag{4.193}$$

On the other hand, the estimate of Σ obtained via the matrix S in (4.185) is given by

$$S/q = \frac{1}{20} \begin{pmatrix} 0.013365 & 0.0042747 \\ 0.0042747 & 0.0073109 \end{pmatrix}$$

$$= \begin{pmatrix} 0.0006683 & 0.0002137 \\ 0.0002137 & 0.0003656 \end{pmatrix},$$

where $q = m - rank X_* = 27 - 7 = 20$. The UZE $b(S \otimes I_m)$ estimates the model by

$$\log(CO2_1) = -388.17 + 208.48 \ \log(GNP_1) - 36.683 \ [\log(GNP_1)]^2$$

$$+ 2.152 \ [\log(GNP_1)]^3 \ : \text{Japan}$$

$$\log(CO2_2) = 6.079 - 2.538 \ D + 0.280 \ \log(GNP_2)$$

$$+ 0.290 \ D \log(GNP_2) \ : \text{USA}. \tag{4.194}$$

The upper bound $\alpha_0(b(S \otimes I_m))$ in (4.189) is given by

$$\alpha_0(b(S \otimes I_m)) = 1 + \frac{2}{20 - 3} = 1.1177.$$

Clearly,

$$\alpha_0(b(S \otimes I_m)) < \alpha_0(b(I_2 \otimes I_m))$$

holds. In this case, the estimate of ρ is not so large and hence the efficiency of the OLSE is expected to be high. However, since the sample size $m(= 27)$ of each equation is moderate, the upper bound $\alpha_0(b(S \otimes I_m))$ becomes smaller than that of the OLSE.

4.7 Problems

4.2.1 In the heteroscedastic model in Example 4.2, show that $X_i' \varepsilon_i$'s and s_j^2's are independent $(i, j = 1, \ldots, p)$.

4.2.2 In the SUR model in Example 4.3, show that S and $X_i' \varepsilon_j$'s are independent $(i, j = 1, \ldots, p)$.

4.3.1 Verify the two invariance properties of $L_0(\hat{\Omega}, \Omega)$: (4.88) and (4.89).

4.3.2 Suppose that the random variables $y_1, \ldots, y_n$ are independent with

$$\mathcal{L}(y_j) = N(0, \sigma^2).$$

As is well known, the minimum variance unbiased estimator (UMVUE) of σ^2 in this setup is given by $S^2 = \frac{1}{n} \sum_{i=1}^{n} y_j^2$. On the other hand, let $\mathcal{E}$ be a class of estimators of the form

$$\hat{\sigma}^2 = \hat{\sigma}^2(c) = cS^2 \quad \text{with } c > 0.$$

(1) Find the optimal estimator with respect to the loss function

$$L_1(\hat{\sigma}^2, \sigma^2) = \left(\frac{\hat{\sigma}^2}{\sigma^2} - 1\right)^2$$

in the class $\mathcal{E}$.

(2) Find the optimal estimator with respect to the loss function

$$L_2(\hat{\sigma}^2, \sigma^2) = \left(\frac{\sigma^2}{\hat{\sigma}^2} - 1\right)^2$$

in the class $\mathcal{E}$.

(3) Find the optimal estimator with respect to the following symmetric inverse loss function

$$L_0\left(\hat{\sigma}^2(c), \sigma^2\right) = \frac{1}{4}\left[\frac{\sigma^2}{\hat{\sigma}^2} + \frac{\hat{\sigma}^2}{\sigma^2} - 2\right]$$

in the class $\mathcal{E}$.

4.3.3 Show the following version of Kantorovich inequality: If a function f satisfies $0 < m \le f(x) \le M$ on $a \le x \le b$, then

$$\int_a^b f(x)dx \int_a^b \frac{1}{f(x)}dx \le \frac{(m+M)^2}{4mM}(b-a)^2.$$

4.3.4 Consider the efficiency measure δ_2 in (4.12). When the efficiency of the OLSE relative to the GME is measured by

$$\delta_2 = |\mathrm{Cov}(b(I_n))|/|\mathrm{Cov}(b(\Omega))| = |X'\Omega X||X'\Omega^{-1}X|/|X'X|^2,$$

find a proof for

$$1 \le \delta_2 \le \prod_{j=1}^k \frac{(\omega_j + \omega_{n-j+1})^2}{4\omega_j\omega_{n-j+1}},$$

where $0 < \omega_1 \le \cdots \le \omega_n$ are the latent roots of Ω, and $n \ge 2k$ (that is, $n - k \ge k$) is assumed. See Bloomfield and Watson (1975) and Knott (1975).

4.4.1 Establish Proposition 4.9.

4.4.2 Let w be a random variable such that $\mathcal{L}(w) = \chi_q^2$. Establish the following equality:

$$E(w^r) = \frac{2^r\Gamma\left(\frac{q}{2} + r\right)}{\Gamma\left(\frac{q}{2}\right)}.$$

4.4.3 Show that the matrix $\hat{\Sigma}(I_2)$ in (4.128) is diagonal.

4.4.4 As is stated in Bilodeau (1990), an unattractive aspect of the optimal GLSE $b(\hat{\Sigma}_B \otimes I_m)$ derived in Theorem 4.14 is that it depends on the coordinate system in which the matrix S is expressed. Bilodeau (1990) proposes a randomized estimator of the form $b(\hat{\Sigma} \otimes I_m)$ with

$$\hat{\Sigma} = \begin{cases} \hat{\Sigma}_{P_1} \equiv P_1 T D_B T' P_1' & \text{with probability } 1/2 \\ \hat{\Sigma}_{P_2} \equiv P_2 T D_B T' P_2' & \text{with probability } 1/2, \end{cases}$$

where P_1 and P_2 are the 2×2 permutation matrices. Discuss the property of this estimator. Further, consider the following nonrandomized version of this:

$$\hat{\Sigma} = \frac{1}{2} \sum_{j=1}^{2} \hat{\Sigma}_{P_j}.$$

See Remarks 2 and 3 of Bilodeau (1990). Some related results in the context of estimation of covariance matrix will be found in Takemura (1984) and Perron (1992).

4.5.1 Verify (4.165).

4.6.1 Complement the analysis in Section 4.6 by calculating several relevant statistics including AIC, t-values, R^2, and so on.

5

Efficiency of GLSEs with Applications to a Serial Correlation Model

5.1 Overview

In Chapter 4, on the basis of our definition, we evaluated the finite sample efficiency of various typical generalized least squares estimators (GLSEs) possessing a simple covariance structure (see (4.24)). However, the structure is not necessarily shared by GLSEs in the models with serially correlated errors. Hence, in this chapter, we treat the problem of deriving an upper bound for the risk matrix of a GLSE in some models with serially correlated errors. We also treat the case of a two-equation heteroscedastic model. To describe the problem, let the general linear regression model be

$$y = X\beta + \varepsilon, \tag{5.1}$$

where

$$y : n \times 1, \ X : n \times k, \ \text{and} \ rank \ X = k.$$

We assume that the distribution $P \equiv \mathcal{L}(\varepsilon)$ of the error term ε satisfies

$$\mathcal{L}(\varepsilon) \in \mathcal{P}_n(0, \sigma^2 \Sigma) \ \text{with} \ \sigma^2 \Sigma \in \mathcal{S}(n), \tag{5.2}$$

where $\mathcal{P}_n(0, \Omega)$ denotes the set of distributions on R^n with mean 0 and covariance matrix Ω, and $\mathcal{S}(n)$ the set of $n \times n$ positive definite matrices. The matrix Σ is assumed to be a continuous function of an unknown one-dimensional parameter θ:

$$\Sigma = \Sigma(\theta),$$

Generalized Least Squares Takeaki Kariya and Hiroshi Kurata
© 2004 John Wiley & Sons, Ltd ISBN: 0-470-86697-7 (PPC)

where

$$\Sigma(\theta)^{-1} = I_n + \lambda_n(\theta)C \quad (\theta \in \Theta). \tag{5.3}$$

Here, C is an $n \times n$ known nonnegative definite matrix with latent roots $0 \le d_1 \le \cdots \le d_n$, Θ is an open interval in R^1 and $\lambda_n = \lambda_n(\theta)$ is a continuous function on Θ. The functional form of λ_n is assumed to be known. Although λ_n is allowed to depend on n, we omit the suffix n for notational simplicity.

A GLSE considered here is given by

$$b(\hat{\Sigma}) = (X'\hat{\Sigma}^{-1}X)^{-1}X'\hat{\Sigma}^{-1}y \text{ with } \hat{\Sigma} = \Sigma(\hat{\theta}), \tag{5.4}$$

where $\hat{\theta} \equiv \hat{\theta}(e)$ is an estimator of θ based on the ordinary least squares (OLS) residual vector

$$e = Ny \text{ with } N = I_n - X(X'X)^{-1}X'. \tag{5.5}$$

A sufficient condition for which $b(\hat{\Sigma})$ has finite second moment is that $\hat{\theta}(e)$ is a continuous scale-invariant function of e:

$$\hat{\theta}(ae) = \hat{\theta}(e) \text{ for any } a > 0. \tag{5.6}$$

The risk matrix of the GLSE $b(\hat{\Sigma})$ is given by

$$R(b(\hat{\Sigma}), \beta) = E\{(b(\hat{\Sigma}) - \beta)(b(\hat{\Sigma}) - \beta)'\}. \tag{5.7}$$

As is shown in Theorem 3.9, the risk matrix is bounded below by the covariance matrix of the Gauss–Markov estimator (GME)

$$b(\Sigma) = (X'\Sigma^{-1}X)^{-1}X'\Sigma^{-1}y,$$

when the distribution $P = \mathcal{L}(\varepsilon)$ of error term ε belongs to the subclass $\mathcal{Q}_n(0, \sigma^2\Sigma)$ of $\mathcal{P}_n(0, \sigma^2\Sigma)$, that is,

$$R(b(\hat{\Sigma}), \beta) \ge \text{Cov}(b(\Sigma)) = \sigma^2(X'\Sigma^{-1}X)^{-1}. \tag{5.8}$$

(For definition of the class $\mathcal{Q}_n(0, \sigma^2\Sigma)$, see (3.86) or Section 5.2). In this chapter, we derive an upper bound for the risk matrix of a GLSE $b(\hat{\Sigma})$ relative to the covariance matrix of the GME $b(\Sigma)$:

$$R(b(\hat{\Sigma}), \beta) \le \gamma_0(b(\hat{\Sigma})) \text{ Cov}(b(\Sigma)), \tag{5.9}$$

where $\gamma_0 = \gamma_0(b(\hat{\Sigma}))$ is a nonrandom real-valued function associated with $b(\hat{\Sigma})$, and is viewed as a measure of the efficiency of $b(\hat{\Sigma})$.

To present these results, this chapter has the following composition:

5.2 Upper Bound for the Risk Matrix of a GLSE

5.3 Upper Bound Problem for a GLSE in the Anderson Model

5.4 Upper Bound Problem for a GLSE in a Two-equation Heteroscedastic Model

5.5 Empirical Example: Automobile Data.

In Section 5.2, an upper bound of the form (5.9) will be obtained under a general setup. The result is applied to typical GLSEs for two specific models in Sections 5.3 and 5.4. The Anderson model and a two-equation heteroscedastic model will be treated. Section 5.5 will be devoted to illustrating an empirical analysis on an automobile data by using the GLSEs in a heteroscedastic model.

Finally, we note that the results in this chapter are applicable to the OLSE

$$b(I_n) = (X'X)^{-1}X'y,$$

since the OLSE is a GLSE with $\hat{\theta}(e) \equiv c$, where c is a constant determined by

$$\lambda(c) = 0.$$

5.2   Upper Bound for the Risk Matrix of a GLSE

This section gives an upper bound of the form (5.9) under a general setup.

Decomposition of a GLSE. We begin by providing a decomposition of the risk matrix of a GLSE. To do so, as before, let Z be any $n \times (n-k)$ matrix such that

$$Z'X = 0, \ Z'Z = I_{n-k} \text{ and } ZZ' = N,$$

where N is defined in (5.5). Transform the matrices X and Z to $\overline{X}$ and $\overline{Z}$ respectively, via

$$\overline{X} = \Sigma^{-1/2}XA^{-1/2} : n \times k,$$
$$\overline{Z} = \Sigma^{1/2}ZB^{-1/2} : n \times (n-k), \tag{5.10}$$

where

$$A = X'\Sigma^{-1}X \in \mathcal{S}(k) \text{ and } B = Z'\Sigma Z \in \mathcal{S}(n-k).$$

Then the matrix Γ defined by

$$\Gamma = \begin{pmatrix} \overline{X}' \\ \overline{Z}' \end{pmatrix}$$

is an $n \times n$ orthogonal matrix. Using this matrix, we define the two uncorrelated vectors $\eta_1 : k \times 1$ and $\eta_2 : (n-k) \times 1$ by

$$\eta = \Gamma\tilde{\varepsilon} = \begin{pmatrix} \overline{X}'\tilde{\varepsilon} \\ \overline{Z}'\tilde{\varepsilon} \end{pmatrix} = \begin{pmatrix} \eta_1 \\ \eta_2 \end{pmatrix}, \tag{5.11}$$

where

$$\tilde{\varepsilon} = \Sigma^{-1/2}\varepsilon.$$

Let

$$\overline{\Sigma} = \Sigma^{-1/2}\hat{\Sigma}\Sigma^{-1/2}. \tag{5.12}$$

Then the GLSE $b(\hat{\Sigma})$ in (5.4) can be decomposed in terms of η_1 and η_2 as

$$\begin{aligned}
b(\hat{\Sigma}) - \beta &= (X'\hat{\Sigma}^{-1}X)^{-1}X'\hat{\Sigma}^{-1}\varepsilon \\
&= A^{-1/2}(\overline{X}'\overline{\Sigma}^{-1}\overline{X})^{-1}\overline{X}'\overline{\Sigma}^{-1}[\overline{X}\eta_1 + \overline{Z}\eta_2] \\
&= A^{-1/2}\eta_1 + A^{-1/2}(\overline{X}'\overline{\Sigma}^{-1}\overline{X})^{-1}\overline{X}'\overline{\Sigma}^{-1}\overline{Z}\eta_2 \tag{5.13} \\
&= [b(\Sigma) - \beta] + [b(\hat{\Sigma}) - b(\Sigma)],
\end{aligned}$$

where the second equality follows by $X = \Sigma^{1/2}\overline{X}A^{1/2}$ and $\varepsilon = \Sigma^{1/2}\Gamma'\eta$. As is shown in Chapter 3, if the distribution $P = \mathcal{L}(\varepsilon)$ is in the class

$$\mathcal{Q}_n(0, \sigma^2\Sigma) = \{P \in \mathcal{P}_n(0, \sigma^2\Sigma) | E_P\{\eta_1|\eta_2\} = 0 \text{ a.s.}\},$$

then the risk matrix $R(b(\hat{\Sigma}), \beta)$ is decomposed as

$$\begin{aligned}
R(b(\hat{\Sigma}), \beta) &= \text{Cov}(b(\Sigma)) + E\{(b(\hat{\Sigma}) - b(\Sigma))(b(\hat{\Sigma}) - b(\Sigma))'\} \\
&= \sigma^2 A^{-1} + A^{-1/2}E(\Delta\Delta')A^{-1/2}, \tag{5.14}
\end{aligned}$$

where

$$\Delta = (\overline{X}'\overline{\Sigma}^{-1}\overline{X})^{-1}\overline{X}'\overline{\Sigma}^{-1}\overline{Z}\eta_2 \quad : k \times 1. \tag{5.15}$$

See Problem 5.2.2. Note that the OLS residual vector e is a function of η_2 as

$$e = ZB^{1/2}\eta_2.$$

The first term of (5.14) is the covariance matrix of the GME and the second term reflects the loss of efficiency due to estimating θ in $\Sigma = \Sigma(\theta)$. The evaluation of the second term is our concern.

Derivation of upper bound. To evaluate the quantity $E(\Delta\Delta')$, decompose the space R^n of y as

$$R^n = B_1 \cup B_2$$

with

$$B_1 = \{y \in R^n | \hat{\lambda} \geq \lambda\} \text{ and } B_2 = \{y \in R^n | \hat{\lambda} < \lambda\},$$

where $\lambda = \lambda(\theta)$ and

$$\hat{\lambda} = \lambda(\hat{\theta}) = \lambda(\hat{\theta}(e)).$$

And let

$$W_1 = 1, \quad W_2 = \frac{(1 + \lambda d_n)^2}{(1 + \hat{\lambda} d_n)^2}. \tag{5.16}$$

Let Ψ be an $n \times n$ orthogonal matrix such that

$$\Psi'C\Psi = D \equiv \begin{pmatrix} d_1 & & 0 \\ & \ddots & \\ 0 & & d_n \end{pmatrix},$$

where $d_1 \leq \cdots \leq d_n$ are the latent roots of C. Then clearly the matrices Σ and $\hat{\Sigma}$ are expressed as

$$\Sigma^{-1} = \Psi(I_n + \lambda D)\Psi'$$

$$= \Psi \begin{pmatrix} 1 + \lambda d_1 & & 0 \\ & \ddots & \\ 0 & & 1 + \lambda d_n \end{pmatrix} \Psi',$$

and

$$\hat{\Sigma}^{-1} = \Psi(I_n + \hat{\lambda} D)\Psi'$$

$$= \Psi \begin{pmatrix} 1 + \hat{\lambda} d_1 & & 0 \\ & \ddots & \\ 0 & & 1 + \hat{\lambda} d_n \end{pmatrix} \Psi',$$

respectively. Hence,

$$\Sigma^{1/2} = \Psi \begin{pmatrix} (1 + \lambda d_1)^{-1/2} & & 0 \\ & \ddots & \\ 0 & & (1 + \lambda d_n)^{-1/2} \end{pmatrix} \Psi',$$

from which the matrix $\overline{\Sigma}^{-1}$ can be rewritten as

$$\overline{\Sigma}^{-1} = \Psi \begin{pmatrix} \frac{1+\hat{\lambda} d_1}{1+\lambda d_1} & & 0 \\ & \ddots & \\ 0 & & \frac{1+\hat{\lambda} d_n}{1+\lambda d_n} \end{pmatrix} \Psi'$$

$$= \Psi \left[I_n + (\hat{\lambda} - \lambda) \begin{pmatrix} \frac{d_1}{1+\lambda d_1} & & 0 \\ & \ddots & \\ 0 & & \frac{d_n}{1+\lambda d_n} \end{pmatrix} \right] \Psi'$$

$$= I_n + (\hat{\lambda} - \lambda)F \quad \text{(say)}. \tag{5.17}$$

Here, of course,

$$F = \Psi \begin{pmatrix} \frac{d_1}{1+\lambda d_1} & & 0 \\ & \ddots & \\ 0 & & \frac{d_n}{1+\lambda d_n} \end{pmatrix} \Psi' : n \times n. \tag{5.18}$$

Define

$$V = \overline{X}' F \overline{Z} : k \times (n - k).$$

Note that the matrix V is nonrandom. The following theorem is due to Toyooka and Kariya (1986).

Theorem 5.1 *For an estimator* $\hat{\theta} = \hat{\theta}(e)$ *of* θ, *let*

$$G_j = \chi_{B_j} \frac{(\hat{\lambda} - \lambda)^2 W_j \eta_2' V' V \eta_2}{\sigma^2} \ (j = 1, 2), \tag{5.19}$$

where χ_B *denotes the indicator function of a set* B. *Then, for the GLSE* $b(\hat{\Sigma})$ *with* $\hat{\Sigma} = \Sigma(\hat{\theta})$, *the following inequality holds:*

$$(b(\hat{\Sigma}) - b(\Sigma))(b(\hat{\Sigma}) - b(\Sigma))' \leq (G_1 + G_2)(\sigma^2 A^{-1}), \tag{5.20}$$

and thus an upper bound for the risk matrix is given by

$$R(b(\hat{\Sigma}), \beta) \leq \gamma_0(b(\hat{\Sigma}))\text{Cov}(b(\Sigma)) \tag{5.21}$$

with

$$\gamma_0(b(\hat{\Sigma})) = 1 + g_1 + g_2, \tag{5.22}$$

where $g_j = E(G_j)$ $(j = 1, 2)$.

   Proof. The vector Δ is rewritten as

$$\Delta = (\hat{\lambda} - \lambda)J^{-1}\overline{X}' F \overline{Z} \eta_2$$
$$= (\hat{\lambda} - \lambda)J^{-1}V\eta_2, \tag{5.23}$$

where

$$J = I_k + (\hat{\lambda} - \lambda)\overline{X}' F \overline{X}. \tag{5.24}$$

By using the Cauchy–Schwarz inequality (2.70), we have, for any $a \in R^k$,

$$a'\Delta\Delta'a = [(\hat{\lambda} - \lambda)a'J^{-1}V\eta_2]^2$$
$$\leq (\hat{\lambda} - \lambda)^2(a'J^{-2}a)(\eta_2'V'V\eta_2). \tag{5.25}$$

On the set $B_1 = \{\hat{\lambda} \geq \lambda\}$, the inequality

$$J \geq I_k$$

holds since $\overline{X}' F \overline{X} \geq 0$. Thus, we have

$$\chi_{B_1} a' J^{-2} a \leq \chi_{B_1} W_1 a' a, \tag{5.26}$$

where $W_1 = 1$. On the other hand, by noting that the function $f(x) = x/(1 + \lambda x)$ $(x \geq 0)$ is increasing in x for any λ, it follows that

$$F \leq \frac{d_n}{1 + \lambda d_n} I_k. \tag{5.27}$$

This implies that on the set $B_2 = \{\hat{\lambda} < \lambda\}$, the matrix J is bounded from below by $W_2^{-1/2} I_k$, because

$$
\begin{aligned}
J &= I_k + (\hat{\lambda} - \lambda) \overline{X}' F \overline{X} \\
&\geq I_k + (\hat{\lambda} - \lambda) \frac{d_n}{(1 + \lambda d_n)} I_k \\
&= W_2^{-1/2} I_k,
\end{aligned}
\tag{5.28}
$$

where $W_2 = (1 + \lambda d_n)^2/(1 + \hat{\lambda} d_n)^2$. This yields

$$\chi_{B_2} a' J^{-2} a \leq \chi_{B_2} W_2 a' a. \tag{5.29}$$

Therefore, from (5.25), (5.26) and (5.29), the following inequality is obtained: for any $a \in R^k$,

$$a' \Delta \Delta' a \leq (\hat{\lambda} - \lambda)^2 \{W_1 \chi_{B_1} + W_2 \chi_{B_2}\}(a' a)(\eta_2' V' V \eta_2), \tag{5.30}$$

or equivalently,

$$\Delta \Delta' \leq (\hat{\lambda} - \lambda)^2 \{W_1 \chi_{B_1} + W_2 \chi_{B_2}\} \eta_2' V' V \eta_2 \, I_k, \tag{5.31}$$

which is further equivalent to

$$\Delta \Delta' \leq (G_1 + G_2) \sigma^2 I_k. \tag{5.32}$$

Hence, we obtain

$$A^{-1/2} \Delta \Delta A^{-1/2} \leq (G_1 + G_2) \sigma^2 A^{-1},$$

which is equivalent to (5.20). Further,

$$\sigma^2 A^{-1} + A^{-1/2} \Delta \Delta A^{-1/2} \leq (1 + G_1 + G_2) \sigma^2 A^{-1}.$$

This completes the proof.

Combining this result with (5.8), we obtain the following inequality:

$$\text{Cov}(b(\Sigma)) \leq R(b(\hat{\Sigma}), \beta) \leq \gamma_0(b(\hat{\Sigma})) \text{Cov}(b(\Sigma)). \tag{5.33}$$

Interpretation of the upper bound as a loss function. The upper bound $\gamma_0 = \gamma_0(b(\hat{\Sigma}))$ thus derived can be viewed as a loss function for choosing an estimator $\hat{\theta}$ in the GLSE $b(\Sigma(\hat{\theta}))$. To see this, let

$$\xi_j = 1 + \lambda d_j \ (j = 1, \cdots, n)$$

and $\hat{\xi}_n = 1 + \hat{\lambda} d_n$. Then $\xi_1 \leq \cdots \leq \xi_n$ are the latent roots of the matrix Σ^{-1}, and $\hat{\xi}_n$ is interpreted as an estimator of ξ_n. By noting that the factor $\eta_2' V' V \eta_2$ in (5.20) is free from $\hat{\theta}$, we rewrite $G_1 + G_2$ as

$$G_1 + G_2 = Q_1 \times Q_2 \tag{5.34}$$

with

$$
\begin{aligned}
Q_1 &= \{\chi_{\{\hat{\lambda} \geq \lambda\}} W_1 + \chi_{\{\hat{\lambda} < \lambda\}} W_2\}(\hat{\lambda} - \lambda)^2 / (1 + \lambda d_n)^2 \\
&= \chi_{\{\hat{\lambda} \geq \lambda\}} \times 1 \times \frac{(\hat{\lambda} - \lambda)^2}{(1 + \lambda d_n)^2} \\
&\quad + \chi_{\{\hat{\lambda} < \lambda\}} \times \frac{(1 + \lambda d_n)^2}{(1 + \hat{\lambda} d_n)^2} \times \frac{(\hat{\lambda} - \lambda)^2}{(1 + \lambda d_n)^2} \\
&= \chi_{\{\hat{\xi}_n \geq \xi_n\}} \frac{(\hat{\xi}_n - \xi_n)^2}{\xi_n^2} + \chi_{\{\hat{\xi}_n < \xi_n\}} \frac{(\hat{\xi}_n - \xi_n)^2}{\hat{\xi}_n^2} \\
&= \chi_{\{\hat{\xi}_n \geq \xi_n\}} (\hat{\xi}_n / \xi_n - 1)^2 + \chi_{\{\hat{\xi}_n < \xi_n\}} (\xi_n / \hat{\xi}_n - 1)^2 \\
&= L(\hat{\xi}_n, \xi_n) \ \text{(say)}
\end{aligned}
\tag{5.35}
$$

and

$$
\begin{aligned}
Q_2 &= \frac{(1 + \lambda d_n)^2 \ \eta_2' V' V \eta_2}{\sigma^2} \\
&= \frac{\xi_n^2 \ \eta_2' V' V \eta_2}{\sigma^2}.
\end{aligned}
\tag{5.36}
$$

Thus, the first factor $Q_1 = L(\hat{\xi}_n, \xi_n)$ in the equality (5.34) is viewed as a loss function for estimating θ via

$$\xi_n = 1 + \lambda(\theta) d_n.$$

As a loss function, L satisfies the following symmetric inverse property:

$$L(\hat{\xi}_n, \xi_n) = L(\hat{\xi}_n^{-1}, \xi_n^{-1}), \tag{5.37}$$

which means that L equally penalizes the underestimate and the overestimate of ξ_n. In the subsequent sections, this property is further investigated in the context of the heteroscedastic model.

Greater but simpler upper bound. Unfortunately, it is, in general, difficult to evaluate the upper bound γ_0 in an explicit way, since in most cases Q_1 and Q_2 in (5.34) are correlated. Hence, we further assume that the error term ε is normally distributed:

$$\mathcal{L}(\varepsilon) = N_n(0, \sigma^2 \Sigma(\theta)), \tag{5.38}$$

and derive a greater but simpler upper bound γ_1. Note that the condition (5.38) is restated in terms of $\eta = \Gamma \Sigma^{-1/2} \varepsilon$ as

$$\mathcal{L}(\eta) = N_n(0, \sigma^2 I_n).$$

By applying the Cauchy–Schwarz inequality (2.79) to g_j's, it follows that

$$
\begin{aligned}
g_j &= E(G_j) \\
&= E\left[\chi_{B_j} W_j (\hat{\lambda} - \lambda)^2 \times \frac{\eta_2' V' V \eta_2}{\sigma^2} \right] \\
&\leq \{ E[\chi_{B_j} W_j^2 (\hat{\lambda} - \lambda)^4] \}^{1/2} \times \left\{ E\left[\left(\frac{\eta_2' V' V \eta_2}{\sigma^2} \right)^2 \right] \right\}^{1/2}. \tag{5.39}
\end{aligned}
$$

The second factor of the above inequality can be calculated by using the following lemma whose proof is fairly straightforward and omitted (Problem 5.2.3).

Lemma 5.2 *Suppose that $\mathcal{L}(v) = N_m(0, I_m)$. Then for any $m \times m$ symmetric matrix C, the following equality holds:*

$$E\{(v'Cv)^2\} = [tr(C)]^2 + 2\, tr(C^2). \tag{5.40}$$

Applying this lemma to η_2/σ with $\mathcal{L}(\eta_2/\sigma) = N_{n-k}(0, I_{n-k})$ yields

$$
\begin{aligned}
E\left\{ \left(\frac{\eta_2' V' V \eta_2}{\sigma^2} \right)^2 \right\} &= [tr(V'V)]^2 + 2\, tr[(V'V)^2] \\
&= \delta^2 \ \text{(say)}, \tag{5.41}
\end{aligned}
$$

which leads to the following theorem due to Toyooka and Kariya (1986).

Theorem 5.3 *When $\mathcal{L}(\varepsilon) = N_n(0, \sigma^2 \Sigma(\theta))$, the inequality*

$$\text{Cov}(b(\Sigma)) \leq R(b(\hat{\Sigma}), \beta) \leq \gamma_1(b(\hat{\Sigma}))\, \text{Cov}(b(\Sigma)) \tag{5.42}$$

holds, where

$$\gamma_1(b(\hat{\Sigma})) = 1 + (\overline{g}_1 + \overline{g}_2)\delta \tag{5.43}$$

with

$$\overline{g}_j = \{ E[\chi_{B_j} W_j^2 (\hat{\lambda} - \lambda)^4] \}^{1/2} \ (j = 1, 2). \tag{5.44}$$

An interpretation of the factor δ. The factor δ in $\gamma_1(b(\hat{\Sigma}))$ defined by (5.41) is important in the sense that $\delta = 0$ implies that

$$E\{(b(\hat{\Sigma}) - b(\Sigma))(b(\hat{\Sigma}) - b(\Sigma))'\} = 0,$$

which is clearly equivalent to the identical equality between the GLSE $b(\hat{\Sigma})$ and the GME $b(\Sigma)$:

$$b(\hat{\Sigma}) = b(\Sigma) \text{ a.s.} \tag{5.45}$$

Here, note that since the quantity $\eta_2' V' V \eta_2$ depends only on X and Σ and is free from the estimator $\hat{\theta}$ of θ, δ does not reflect the efficiency of $\hat{\theta}$. Instead, it is a measure of the deviation of the regression model from the model with *Rao's covariance structure*. As will be investigated in Chapter 8, Rao's covariance structure is a relation between X and Σ under which the OLSE $b(I_n)$ is identically equal to the GME $b(\Sigma)$. More precisely, the identical equality

$$b(I_n) = b(\Sigma) \text{ for any } y \in R^n \tag{5.46}$$

holds if and only if the model satisfies

$$X'\Sigma Z = 0. \tag{5.47}$$

The covariance structure that has the structure (5.47) is called Rao's covariance structure. In the case in which

$$\Sigma^{-1} = I_n + \lambda C,$$

the condition (5.47) holds if and only if

$$\delta = 0,$$

which is in turn equivalent to (5.45).

Proposition 5.4 *Under the covariance structure* $\Sigma(\theta)^{-1} = I_n + \lambda(\theta)C$, $\delta = 0$ *holds if and only if* $X'\Sigma Z = 0$.

Proof. First note that $\Sigma = I_n - \lambda F$ and hence,

$$\Sigma = \Sigma^{-1/2}(I_n - \lambda F)\Sigma^{1/2}$$
$$= I_n - \lambda \Sigma^{-1/2} F \Sigma^{1/2}.$$

From (5.41), $\delta = 0$ holds if and only if $V = 0$, where $V = \overline{X}' F \overline{Z}$. Hence, from the definition of $\overline{X}$ and $\overline{Z}$, the condition $V = 0$ is equivalent to

$$X'\Sigma^{-1/2} F \Sigma^{1/2} Z = 0,$$

and this holds if and only if $\lambda X'\Sigma^{-1/2} F \Sigma^{1/2} Z = 0$, since $\lambda(\theta) = 0$ implies $X'\Sigma^{-1/2} F \Sigma^{1/2} Z = 0$. Now we have

$$\lambda X'\Sigma^{-1/2} F \Sigma^{1/2} Z = X'(I_n - \lambda \Sigma^{-1/2} F \Sigma^{1/2})Z$$
$$= X'\Sigma Z, \tag{5.48}$$

where the first equality is due to $X'Z = 0$. This completes the proof.

Efficiency of the OLSE. Finally, it is remarked that the results in this section are applicable to the OLSE $b(I_n) = (X'X)^{-1}X'y$, since the OLSE is a special case of the GLSE in (5.4). To see this, let $\hat{\theta} = \hat{\theta}(e)$ be a constant function $\hat{\theta}(e) = c$ with c determined by $\lambda(c) = 0$. Then clearly,

$$\hat{\Sigma} = \Sigma(\hat{\theta}) = [I_n + \lambda(c)C]^{-1} = I_n$$

holds, and therefore, the GLSE $b(\hat{\Sigma})$ reduces to the OLSE $b(I_n)$. By using

$$E[\eta_2' V'V \eta_2/\sigma^2] = tr(V'V) \tag{5.49}$$

and by letting $\hat{\lambda} = 0$ in (5.19), the upper bound $\gamma_0(b(I_n))$ is calculated as

$$\gamma_0(b(I_n)) = 1 + \lambda^2[\chi_{\{\lambda\leq 0\}} + \chi_{\{\lambda>0\}}(1 + \lambda d_n)^2]tr(V'V), \tag{5.50}$$

the second term of which attains zero when $\lambda = 0$ (i.e., $\Sigma = I_n$) or $V = 0$ (i.e., the model is of Rao's covariance structure). The greater upper bound $\gamma_1(b(I_n))$ is also obtained from (5.44) as

$$\gamma_1(b(I_n)) = 1 + \lambda^4[\chi_{\{\lambda\leq 0\}} + \chi_{\{\lambda>0\}}(1 + \lambda d_n)^4]\delta, \tag{5.51}$$

the second term of which becomes zero when $\lambda = 0$ or $\delta = 0$. It is noted that the second terms of the bounds in (5.50) and (5.51) do not necessarily go to 0 even when $n \to \infty$.

5.3 Upper Bound Problem for a GLSE in the Anderson Model

In this section, the results developed in the previous sections are applied to the Anderson model. The upper bound γ_1 in Theorem 5.3 is evaluated under normality assumption.

The Anderson model and GLSEs. As is introduced in Example 2.1, the Anderson model is a model (5.1) with covariance structure

$$\text{Cov}(\varepsilon) = \sigma^2 \Sigma(\theta)$$

with

$$\Sigma(\theta)^{-1} = I_n + \lambda(\theta)C \quad (\theta \in \Theta), \tag{5.52}$$

where

$$\sigma^2 > 0, \ \lambda \equiv \lambda(\theta) = \frac{\theta}{(1-\theta)^2}, \ \Theta = (-1, 1)$$

and

$$
C = \begin{pmatrix}
1 & -1 & & & & & 0 \\
-1 & 2 & -1 & & & & \\
& \ddots & \ddots & \ddots & & & \\
& & \ddots & \ddots & \ddots & & \\
& & & \ddots & 2 & -1 \\
0 & & & & -1 & 1
\end{pmatrix}.
$$

The latent roots $d_1 \leq \cdots \leq d_n$ of the matrix C are given by

$$
d_j = 2[1 - \cos(\pi(j-1)/n)] \quad (j = 1, \cdots, n). \tag{5.53}
$$

The matrix $\Sigma(\theta)$ is positive definite on Θ since

$$
-\frac{1}{4} < \lambda(\theta) < \infty \text{ and } d_1 = 0 < d_2 < \cdots < d_n < 4. \tag{5.54}
$$

Here, let us assume that the error term ε is normally distributed:

$$
\mathcal{L}(\varepsilon) = N_n(0, \sigma^2 \Sigma(\theta)).
$$

A GLSE considered here is of the form $b(\hat{\Sigma})$ in (5.4) with

$$
\hat{\Sigma}^{-1} = \Sigma(\hat{\theta})^{-1} = I_n + \hat{\lambda} C, \tag{5.55}
$$

where $\hat{\lambda} \equiv \lambda(\hat{\theta})$ and $\hat{\theta}$ is an estimator of θ based on the OLS residual vector e in (5.5) satisfying the following conditions:

(1) $\hat{\theta}(e) \in \Theta$ a.s.;

(2) $\hat{\theta}(e)$ is an even function in e, that is, $\hat{\theta}(e) = \hat{\theta}(-e)$;

(3) $\hat{\theta}(e)$ is a scale-invariant function in the sense that $\hat{\theta}(ae) = \hat{\theta}(e)$ holds for any $a > 0$;

(4) $\hat{\theta}(e)$ is continuous in e.

These conditions guarantee that the GLSE $b(\hat{\Sigma}) = b(\Sigma(\hat{\theta}))$ with such $\hat{\theta}$ is an unbiased estimator with finite second moment. Under these conditions, the upper bound

$$
\gamma_1(b(\hat{\Sigma})) = 1 + (\bar{g}_1 + \bar{g}_2)\delta
$$

in Theorem 5.3 will be evaluated below. Since it is difficult to obtain an exact expression of γ_1, evaluation will be limited to an asymptotic one. To this end, the following condition is imposed on $\hat{\theta}$ in addition to the conditions (1) to (4):

(5) $\sqrt{n}(\hat{\theta} - \theta) \to_d N(0, 1 - \theta^2)$,

where $\to_d$ denotes convergence in distribution.

Evaluation of the upper bound. To evaluate γ_1 for a GLSE $b(\hat{\Sigma}) = b(\Sigma(\hat{\theta}))$ with $\hat{\theta}$ satisfying the conditions (1) to (5), we use the following two lemmas.

Lemma 5.5 *For any $\hat{\theta}$ such that*

$$\hat{\theta} - \theta = O_p(1/\sqrt{n}), \tag{5.56}$$

the quantity $W_2 = [(1 + \lambda d_n)/(1 + \hat{\lambda} d_n)]^2$ is evaluated as

$$W_2 = 1 + O_p(1/\sqrt{n}). \tag{5.57}$$

Proof. By Taylor's theorem, there exists a random variable θ^* satisfying $|\theta^* - \theta| \le |\hat{\theta} - \theta|$ and

$$\lambda(\hat{\theta}) = \lambda(\theta) + \lambda'(\theta^*)(\hat{\theta} - \theta). \tag{5.58}$$

Here,

$$\lambda'(\theta) = (1 + \theta)/(1 - \theta)^3$$

is continuous in θ, and

$$\theta^* \to_p \theta \text{ as } n \to \infty,$$

where $\to_p$ denotes convergence in probability. Hence, the equation (5.58) is rewritten as

$$\lambda(\hat{\theta}) = \lambda(\theta) + O_p(1/\sqrt{n}). \tag{5.59}$$

Next, by using the following two formulas

$$\cos\left(\pi - \frac{\pi}{n}\right) = \cos \pi \cos(\pi/n) + \sin \pi \sin(\pi/n)$$
$$= -\cos(\pi/n)$$

and

$$\cos x = 1 - \frac{x^2}{2!} + \frac{x^4}{4!} - \frac{x^6}{6!} + \cdots$$
$$= 1 + o(x)(x \to 0),$$

the quantity

$$d_n = 2\left[1 - \cos\left(\pi - \frac{\pi}{n}\right)\right]$$

is expanded as

$$d_n = 4 + o(1/n). \tag{5.60}$$

By using (5.59) and (5.60), we obtain

$$W_2 = \left(\frac{1 + \lambda d_n}{1 + \hat{\lambda} d_n}\right)^2$$

$$= \left[\frac{1 + 4\lambda + o(1/n)}{1 + 4\lambda + O_p(1/\sqrt{n})}\right]^2$$

$$= [1 + O_p(1/\sqrt{n})]^2$$

$$= 1 + O_p(1/\sqrt{n}). \tag{5.61}$$

This completes the proof.

On the other hand, since $\chi_{B_j} \leq 1$,

$$\overline{g}_j \equiv \{E[\chi_{B_j} W_j^2 (\hat{\lambda} - \lambda)^4]\}^{1/2}$$

$$\leq \{E[W_j^2 (\hat{\lambda} - \lambda)^4]\}^{1/2} \equiv \overline{h}_j \quad (j = 1, 2). \tag{5.62}$$

This leads to an upper bound

$$\gamma_2(b(\hat{\Sigma})) = 1 + \delta(\overline{h}_1 + \overline{h}_2), \tag{5.63}$$

which of course satisfies $\gamma_1(b(\hat{\Sigma})) \leq \gamma_2(b(\hat{\Sigma}))$. Using Lemma 5.5 for $\hat{\theta}$ satisfying the condition (5), the quantities $\overline{h}_j$'s are evaluated by the usual delta method with the following condition:

(6) The remainder terms $O_p(1/\sqrt{n})$ in (5.56) and (5.57) satisfy

$$E[O_p(1/\sqrt{n})] = O(1/n).$$

A typical choice for such a $\hat{\theta}$ is

$$\hat{\theta}_0(e) = \frac{e' K e}{e' e}, \tag{5.64}$$

where $K = (k_{ij})$ is an $n \times n$ symmetric matrix such that $k_{ij} = 1/2$ if $|i - j| = 1$ and $k_{ij} = 0$ otherwise. (See Examples 2.5 and 3.1).

Lemma 5.6 *For any $\hat{\theta}$ satisfying conditions (1) to (6),*

$$\overline{h}_j = \frac{1}{n} \frac{\sqrt{3}(1 + \theta)^3}{(1 - \theta)^5} + o(1/n) \quad (j = 1, 2). \tag{5.65}$$

Proof. Since

$$\lambda(\hat{\theta}) - \lambda(\theta) = \lambda'(\theta^*)(\hat{\theta} - \theta)$$

$$= \lambda'(\theta)(\hat{\theta} - \theta) + o_p(1/\sqrt{n})$$

follows from (5.58), the expansion below is obtained from Lemma 5.5:

$$W_j^2[\lambda(\hat{\theta}) - \lambda(\theta)]^4 = [\lambda'(\theta)(\hat{\theta} - \theta)]^4 + o_p(1/n^2) \ (j = 1, 2). \tag{5.66}$$

Since

$$\sqrt{n}\lambda'(\theta)(\hat{\theta} - \theta) \to_d N(0, \ [\lambda'(\theta)]^2(1 - \theta^2)) = N\left(0, \frac{(1 + \theta)^3}{(1 - \theta)^5}\right)$$

follows from the condition (5), it holds that

$$n^2 W_j^2[\lambda(\hat{\theta}) - \lambda(\theta)]^4 \to_d \mathcal{L}(Z^4),$$

where Z is a random variable such that

$$\mathcal{L}(Z) = N\left(0, \frac{(1 + \theta)^3}{(1 - \theta)^5}\right).$$

Hence, we obtain

$$\lim_{n \to \infty} n^2 E\{W_j^2[\lambda(\hat{\theta}) - \lambda(\theta)]^4\} = 3 \times \left[\frac{(1 + \theta)^3}{(1 - \theta)^5}\right]^2,$$

from which

$$\{E\{W_j^2[\lambda(\hat{\theta}) - \lambda(\theta)]^4\}\}^{1/2} = \frac{\sqrt{3}(1 + \theta)^3}{n(1 - \theta)^5} + o(1/n) \tag{5.67}$$

is established. Here, we used the fact that

$$E(Z^3) = 3\sigma^4 \text{ when } \mathcal{L}(Z) = N(0, \sigma^2).$$

This completes the proof.

Thus, the following theorem due to Toyooka and Kariya (1986) is proved.

Theorem 5.7 *The upper bound $\gamma_2(b(\hat{\Sigma}))$ in (5.63) is approximated as*

$$1 + \delta\left[\frac{2\sqrt{3}(1 + \theta)^3}{n(1 - \theta)^5}\right]. \tag{5.68}$$

Here, the quantity δ is defined in (5.41). In most cases, δ is of $O(1)$.

Corollary 5.8 *The result in Theorem 5.7 holds for the GLSE with $\hat{\theta}_0$ in (5.64).*

Proof. It is easy to see that $\hat{\theta}_0$ in (5.64) satisfies (1) to (4). The asymptotic normality (5) of $\hat{\theta}_0$ is shown in the Appendix. Through the result, (6) is also proved there.

5.4   Upper Bound Problem for a GLSE in a Two-equation Heteroscedastic Model

In this section, the results obtained in Section 5.2 are applied to typical GLSEs in a two-equation heteroscedastic model.

The two-equation heteroscedastic model and GLSEs. The model considered here is a two-equation heteroscedastic model, which is defined as a model

$$y = X\beta + \varepsilon \text{ with } \mathcal{L}(\varepsilon) \in \mathcal{P}_n(0, \Omega)$$

with the following structure:

$$y = \begin{pmatrix} y_1 \\ y_2 \end{pmatrix}, \quad X = \begin{pmatrix} X_1 \\ X_2 \end{pmatrix},$$

$$\varepsilon = \begin{pmatrix} \varepsilon_1 \\ \varepsilon_2 \end{pmatrix} \text{ and } \Omega = \begin{pmatrix} \sigma_1^2 I_{n_1} & 0 \\ 0 & \sigma_2^2 I_{n_2} \end{pmatrix}, \tag{5.69}$$

where

$$n = n_1 + n_2, \quad y_j : n_j \times 1,$$

$$X_j : n_j \times k \text{ and } \varepsilon_j : n_j \times 1 (j = 1, 2).$$

For simplicity, the matrices X_j's are assumed to be of full rank:

$$rank \ X_j = k_j \ (j = 1, 2),$$

and the error term ε is normally distributed:

$$\mathcal{L}(\varepsilon) = N_n(0, \sigma^2 \Sigma(\theta)).$$

The covariance matrix Ω can be rewritten as (5.3):

$$\Omega = \sigma^2 \Sigma(\theta)$$

with

$$\Sigma^{-1} = \Sigma(\theta)^{-1} = I_n + \lambda(\theta)D \ (\theta \in \Theta), \tag{5.70}$$

where

$$\sigma^2 = \sigma_1^2, \ \Theta = (0, \infty),$$

$$\theta = \sigma_1^2/\sigma_2^2 \in \Theta, \ \lambda(\theta) = \theta - 1,$$

$$D = \begin{pmatrix} 0 & 0 \\ 0 & I_{n_2} \end{pmatrix}. \tag{5.71}$$

Let us consider a GLSE of the form

$$b(\hat{\Sigma}) = (X'\hat{\Sigma}^{-1}X)^{-1}X'\hat{\Sigma}^{-1}y \text{ with } \hat{\Sigma} = \Sigma(\hat{\theta}), \tag{5.72}$$

where $\hat{\theta} = \hat{\theta}(e)$ is an estimator of θ based on the OLS residual vector e in (5.5). Typical examples are the restricted GLSE and the unrestricted GLSE (see Example 2.6).

Upper bound $\gamma_0(b(\hat{\Sigma}))$. We first derive the upper bound γ_0 obtained in Theorem 5.1. To do so, note that for any $\hat{\theta}$ such that $\hat{\theta} \in \Theta$ a.s., the sets B_1 and B_2 and the quantities W_1, W_2 in (5.16) and F in (5.18) are calculated as

$$B_1 = \{y \in R^n | \hat{\theta} \ge \theta\} \text{ and } B_2 = \{y \in R^n | \hat{\theta} < \theta\},$$

and

$$W_1 = 1, \ W_2 = \theta^2/\hat{\theta}^2, \ F = \theta^{-1}D, \tag{5.73}$$

respectively. Also, let

$$M = (X'\Sigma^{-1}X)^{-1/2}X'\Sigma^{-1/2}F\Sigma^{-1/2}[I_n - X(X'\Sigma^{-1}X)^{-1}X'\Sigma^{-1}]. \tag{5.74}$$

Lemma 5.9 *For any GLSE $b(\hat{\Sigma})$ with $\hat{\Sigma} = \Sigma(\hat{\theta})$, the following inequality holds:*

$$(b(\hat{\Sigma}) - \beta)(b(\hat{\Sigma}) - \beta)' \le (1 + G_1 + G_2)\sigma^2(X'\Sigma(\theta)^{-1}X)^{-1} \tag{5.75}$$

with

$$G_j = \chi_{B_j}(\hat{\theta} - \theta)^2 W_j \varepsilon' M' M \varepsilon/\sigma^2 \ (j = 1, 2). \tag{5.76}$$

And thus,

$$\text{Cov}(b(\hat{\Sigma})) \le \gamma_0(b(\hat{\Sigma}))\text{Cov}(b(\Sigma)) \tag{5.77}$$

holds, where $b(\Sigma) = (X'\Sigma^{-1}X)^{-1}X'\Sigma^{-1}y$ is the GME of β and

$$\gamma_0(b(\hat{\Sigma})) = 1 + g_1 + g_2 \text{ with } g_j = E(G_j).$$

The quadratic form $\varepsilon' M' M \varepsilon$ in (5.76) is the same as $\eta_2' V' V \eta_2$ in (5.19) (Problem 5.4.1).

To give an interpretation of the upper bound γ_0, it is convenient to rewrite the GLSE $b(\hat{\Sigma})$ as a weighted sum of the two OLSEs, say $\hat{\beta}_j$'s, obtained from each homoscedastic regression model $y_j = X_j\beta + \varepsilon_j \ (j = 1, 2)$, that is,

$$b(\hat{\Sigma}) = (X_1'X_1 + \hat{\theta}X_2'X_2)^{-1}(X_1'X_1\hat{\beta}_1 + \hat{\theta}X_2'X_2\hat{\beta}_2), \tag{5.78}$$

where

$$\hat{\beta}_j = (X_j'X_j)^{-1}X_j'y_j \ (j = 1, 2). \tag{5.79}$$

Lemma 5.10 *The upper bound* $1 + G_1 + G_2$ *in (5.75) is decomposed as*

$$1 + G_1 + G_2 = 1 + L(\hat{\theta}, \theta) \, Q(\hat{\beta}_1 - \hat{\beta}_2), \tag{5.80}$$

where the functions L and Q are defined as

$$L(\hat{\theta}, \theta) = \ell(\hat{\theta}/\theta) + \ell(\theta/\hat{\theta}) \tag{5.81}$$

with

$$\ell(t) = (t - 1)^2 \chi_{\{t>1\}},$$

and

$$\begin{aligned} Q(x) &= \frac{\theta^2}{\sigma^2} x' X_1' X_1 (X'\Sigma^{-1}X)^{-1} X_2' X_2 \\ &\quad \times (X'\Sigma^{-1}X)^{-1} X_2' X_2 (X'\Sigma^{-1}X)^{-1} X_1' X_1 x \\ &\equiv \frac{\theta^2}{\sigma^2} x' K x \text{ (say),} \end{aligned}$$

respectively.

 Proof. Proof is available as an exercise (Problem 5.4.2).

 Note that the function $Q(\hat{\beta}_1 - \hat{\beta}_2)$ in (5.80) does not depend on the choice of $\hat{\theta}$. On the other hand, the function $L(\hat{\theta}, \theta)$ can be regarded as a loss function for choosing an estimator of θ, and has the following symmetric inverse property:

$$L(\hat{\theta}, \theta) = L(\hat{\theta}^{-1}, \theta^{-1}). \tag{5.82}$$

The unrestricted GLSE. Now, we treat a class of unbiased GLSEs including the unrestricted GLSE and evaluate the upper bounds for their covariance matrices. The class treated is a class of GLSEs $b(\hat{\Sigma}) = b(\Sigma(\hat{\theta}))$ with $\hat{\theta}$ defined by

$$\hat{\theta} = \hat{\theta}(e; c) = cS_1^2/S_2^2, \ c > 0, \tag{5.83}$$

where the statistics S_j^2's are given by

$$S_j^2 = e_j' N_j e_j,$$

$$N_j = I_{n_j} - X_j (X_j' X_j)^{-1} X_j',$$

$$e = \begin{pmatrix} e_1 \\ e_2 \end{pmatrix} : n \times 1 \text{ with } e_j : n_j \times 1. \tag{5.84}$$

As is stated in Example 2.6, the statistic S_j^2 coincides with the sum of squared residuals calculated from the jth equation:

$$S_j^2 = (y_j - X_j \hat{\beta}_j)'(y_j - X_j \hat{\beta}_j) \ (j = 1, 2), \tag{5.85}$$

and hence, S_j^2/m_j with $m_j = n_j - k_j$ is an unbiased estimator of σ_j^2. The unrestricted GLSE is obtained by letting $c = m_2/m_1$ in (5.83). Since the statistics $\hat{\beta}_1, \hat{\beta}_2, S_1^2$ and S_2^2 are independent, the two factors $L(\hat{\theta}, \theta)$ and $Q(\hat{\beta}_1 - \hat{\beta}_2)$ in (5.80) are also independent. Hence, the quantity $g_1 + g_2$ in Lemma 5.9 is evaluated as

$$g_1 + g_2 = E(G_1 + G_2)$$
$$= E[L(\hat{\theta}, \theta)]E[Q(\hat{\beta}_1 - \hat{\beta}_2)]$$
$$\equiv \rho(c; m_1, m_2) \, E[Q(\hat{\beta}_1 - \hat{\beta}_2)] \text{ (say)}, \qquad (5.86)$$

where the last line defines the function $\rho(c; m_1, m_2) = E[L(\hat{\theta}, \theta)]$. The quantity $E[Q(\hat{\beta}_1 - \hat{\beta}_2)]$ is further calculated as

$$E[Q(\hat{\beta}_1 - \hat{\beta}_2)] = (\theta^2/\sigma^2) \, tr[K \, \text{Cov}(\hat{\beta}_1 - \hat{\beta}_2)]$$
$$= \sum_{i=1}^{k} \frac{\theta r_i}{(1 + \theta r_i)^2}$$
$$\equiv q(r_1, \cdots, r_k; \theta) \text{ (say)}, \qquad (5.87)$$

where $r_1, \cdots, r_k$ are the latent roots of $(X_1'X_1)^{-1/2}X_2'X_2(X_1'X_1)^{-1/2}$ (see Problem 5.4.3).

Next, to describe $\rho(c; m_1, m_2)$, the hypergeometric function defined below is used:

$$_2F_1(a_1, a_2; a_3; z) = \sum_{j=0}^{\infty} \frac{(a_1)_j(a_2)_j}{(a_3)_j} \frac{z^j}{j!} \qquad (5.88)$$

with

$$(a)_j = \prod_{i=0}^{j-1}(a + i) \text{ and } (a)_0 = 1,$$

which converges for $|z| < 1$ (see, for example, Abramowitz and Stegun, 1972). The following result is due to Kurata (2001).

Theorem 5.11 *Let $m_j > 4(j = 1, 2)$. Then, for any $\hat{\theta} = \hat{\theta}(e; c) = cS_1^2/S_2^2 \, (c > 0)$, the following equality holds:*

$$\rho(c; m_1, m_2) = \begin{cases} \rho_1(c; m_1, m_2) & (0 < c < 1) \\ \rho_2(m_1, m_2) & (c = 1) \\ \rho_3(c; m_1, m_2) & (1 < c) \end{cases} \qquad (5.89)$$

holds, where ρ_1, ρ_2 and ρ_3 are given by

$$\rho_1(c; m_1, m_2) = \frac{1}{B\left(\frac{m_1}{2}, \frac{m_2}{2}\right)}[B((m_2 - 4)/2, 3)c^{m_2/2}]$$
$$\times {_2F_1}((m_1 + m_2)/2, (m_2 - 4)/2; (m_2 + 2)/2; -c)$$

$$+B((m_1 - 4)/2, 3)c^{m_2/2}(1 + c)^{-(m_1+m_2)/2}$$

$$\times {}_2F_1((m_1 + m_2)/2, 3; (m_1 + 2)/2; (1 + c)^{-1})], \tag{5.90}$$

$$\rho_2(m_1, m_2) = \frac{1}{2^{(m_1+m_2)/2}B\left(\frac{m_1}{2}, \frac{m_2}{2}\right)}[B((m_2 - 4)/2, 3)$$

$$\times {}_2F_1((m_1 + m_2)/2, 3; (m_2 + 2)/2; 1/2)$$

$$+B((m_1 - 4)/2, 3){}_2F_1((m_1 + m_2)/2, 3; (m_1 + 2)/2; 1/2)], \tag{5.91}$$

and

$$\rho_3(c; m_1, m_2) = \rho_1(1/c; m_2, m_1), \tag{5.92}$$

respectively. Thus, for any GLSE $b(\hat{\Sigma})$ with $\hat{\Sigma} = \Sigma(\hat{\theta})$,

$$\mathrm{Cov}(b(\hat{\Sigma})) \le [1 + \rho(c; m_1, m_2)q(r_1, \cdots, r_k; \theta)]\mathrm{Cov}(b(\Sigma))$$

holds.

Proof. Let $v_j = S_j^2/\sigma_j^2$ so that $\mathcal{L}(v_j) = \chi_{m_j}^2$, where χ_m^2 denotes the χ^2 distribution with degrees of freedom m. Then, by (5.81), it is obtained that

$$\rho(c; m_1, m_2) = E[\ell(cv_1/v_2)] + E[\ell(v_2/cv_1)], \tag{5.93}$$

from which the equality

$$\rho(c; m_1, m_2) = \rho(1/c; m_2, m_1) \tag{5.94}$$

follows. Letting $a = 1/2^{(m_1+m_2)/2}\Gamma(m_1/2)\Gamma(m_2/2)$, the first term of the right-hand side of (5.93) is written as

$$E[\ell(cv_1/v_2)] = a \int\int_{cv_1/v_2 \ge 1} \left(\frac{cv_1}{v_2} - 1\right)^2 v_1^{m_1/2-1} v_2^{m_2/2-1}$$

$$\times \exp\left(-\frac{v_1 + v_2}{2}\right) dv_1\, dv_2.$$

Making transformation $z_1 = v_1$ and $z_2 = v_2/(cv_1)$ with $dv_1 dv_2 = cz_1 dz_1 dz_2$ and integrating with respect to z_1 yields

$$a\, c^{m_2/2} \int_0^1 z_2^{(m_2/2-2)-1}(1 - z_2)^2$$

$$\times \left[\int_0^\infty z_1^{(m_1+m_2)/2-1} \exp\left(-\frac{(1 + cz_2)z_1}{2}\right) dz_1\right] dz_2$$

$$= a' \int_0^1 z_2^{(m_2/2-2)-1}(1 - z_2)^2(1 + cz_2)^{-(m_1+m_2)/2}\, dz_2 \tag{5.95}$$

with

$$a' = a \times c^{m_2/2} 2^{(m_1+m_2)/2} \Gamma((m_1 + m_2)/2) = \frac{c^{m_2/2}}{B(m_1/2, m_2/2)}.$$

To evaluate (5.95) in the case in which $0 < c < 1$, we use the following well-known formula

$$\int_0^1 t^{a_2-1}(1-t)^{a_3-a_2-1}(1-zt)^{-a_1} dt = B(a_2, a_3 - a_2) {}_2F_1(a_1, a_2; a_3; z), \quad (5.96)$$

which is valid for $0 < a_2 < a_3$ and $|z| < 1$. Applying (5.96) to (5.95) proves the first term on the right-hand side of (5.90). When $1 \le c$, applying the following formula

$$\int_0^1 t^{a_2-1}(1-t)^{a_3-a_2-1}(1-zt)^{-a_1} dt$$

$$= (1-z)^{-a_1} B(a_2, a_3 - a_2) {}_2F_1\left(a_1, a_3 - a_2; a_3; \frac{z}{z-1}\right), \quad (5.97)$$

which is valid for $0 < a_2 < a_3$ and $z \le -1$, establishes

$$E[\ell(cv_1/v_2)] = \frac{B((m_2 - 4)/2, 3)}{B(m_1/2, m_2/2)}(1 + c)^{-(m_1+m_2)/2} c^{m_2/2}$$

$$\times {}_2F_1((m_1 + m_2)/2, 3; (m_2 + 2)/2; c/(1 + c)). \quad (5.98)$$

Substituting $c = 1$ into (5.98) yields the first term of (5.91). Next, we consider the second term on the right-hand side of (5.93). By interchanging m_1 and m_2 and replacing c by $1/c$ in (5.95), we obtain

$$E[\ell(v_2/cv_1)] = a'' \int_0^1 z_1^{(m_1/2-2)-1}(1 - z_1)^2$$

$$\times (1 + c^{-1}z_1)^{-(m_1+m_2)/2} dz_1 \quad (5.99)$$

with $a'' = 1/c^{m_1/2} B(m_1/2, m_2/2)$. Since $0 < c \le 1$ is equivalent to $1 \le c^{-1}$, applying (5.97) to the right-hand side of (5.99) establishes the second term of (5.90). The second term of (5.91) is obtained by letting $c = 1$. Thus, (5.90) and (5.91) are proved. Finally, (5.92) is obtained from (5.94). This completes the proof.

In Table 5.1, we treat the unrestricted GLSE ($c = m_2/m_1$) and summarize the values of $\rho(m_2/m_1; m_1, m_2)$ for $m_1, m_2 = 5, 10, 15, 20, 25, 50$. The table is symmetric in m_1 and m_2, which is a consequence of (5.94). It can be observed that the upper bound monotonically decreases in m_1 and m_2.

Comparison with the upper bound in Chapter 4. Next, let us consider a relation between the upper bound

$$\gamma_0(b(\hat{\Sigma})) = 1 + \rho(c; m_1, m_2) \, q(r_1, \cdots, r_k; \theta)$$

$$= 1 + E[L(\hat{\theta}, \theta)] \, E[Q(\hat{\beta}_1 - \hat{\beta}_2)]$$

Table 5.1   Values of $\rho(c; m_1, m_2)$ with $c = m_2/m_1$.

| | | | m_1 | | | |
m_2	5	10	15	20	25	50
5	18.3875	8.8474	7.7853	7.3423	7.0956	6.6356
10	8.8474	1.8203	1.2826	1.0879	0.9873	0.8138
15	7.7853	1.2826	0.8156	0.6518	0.5687	0.4288
20	7.3423	1.0879	0.6518	0.5013	0.4258	0.3001
25	7.0956	0.9873	0.5687	0.4258	0.3544	0.2369
50	6.6356	0.8138	0.4288	0.3001	0.2369	0.1349

Source: Kurata (2001) with permission.

and the upper bound

$$\alpha_0(b(\hat{\Sigma})) = 1 + E[a(\hat{\theta}, \theta)]$$

with

$$a(\hat{\theta}, \theta) = \frac{[(\hat{\theta}/\theta) - 1]^2}{4(\hat{\theta}/\theta)} \tag{5.100}$$

treated in Chapter 4. The relation between the two upper bounds is indefinite. More precisely,

Proposition 5.12 *The relation between $L(\hat{\theta}, \theta)$ and $a(\hat{\theta}, \theta)$ is given by*

$$\frac{1}{4} \times L(\hat{\theta}, \theta) \geq a(\hat{\theta}, \theta). \tag{5.101}$$

The range of the function q is given by

$$0 < q(r_1, \cdots, r_k; \theta) \leq \frac{k}{4},$$

and its maximum is attained when $r_1 = \cdots = r_k = 1/\theta$. As $r_i\theta$ goes to either 0 or ∞ $(i = 1, \cdots, k)$, the function q converges to 0.

Proof. The inequality (5.101) is proved as

$$L(\hat{\theta}, \theta) = (\hat{\theta}/\theta - 1)^2 \chi_{\{\hat{\theta}/\theta > 1\}} + (\theta/\hat{\theta} - 1)^2 \chi_{\{\theta/\hat{\theta} > 1\}}$$

$$\geq \frac{(\hat{\theta}/\theta - 1)^2}{\hat{\theta}/\theta} \chi_{\{\hat{\theta}/\theta > 1\}} + \frac{(\theta/\hat{\theta} - 1)^2}{\theta/\hat{\theta}} \chi_{\{\theta/\hat{\theta} > 1\}}$$

$$= \frac{(\hat{\theta}/\theta - 1)^2}{\hat{\theta}/\theta} \chi_{\{\hat{\theta}/\theta > 1\}} + \frac{(\hat{\theta}/\theta - 1)^2}{\hat{\theta}/\theta} \chi_{\{\hat{\theta}/\theta < 1\}}$$

$$= \frac{(\hat{\theta}/\theta - 1)^2}{\hat{\theta}/\theta}$$

$$= 4 \times a(\hat{\theta}, \theta).$$

The rest is evident from (5.87). This completes the proof.

The inequality (5.101) clearly implies that

$$1 + \rho(c; m_1, m_2) \times \frac{1}{4} \geq 1 + E[a(\hat{\theta}, \theta)].$$

However, the factor $q(r_1, \cdots, r_k; \theta)$ can be so small (or large) that

$$1 + \rho(c; m_1, m_2)q(r_1, \cdots, r_k; \theta) \leq (\text{or} \geq) 1 + E[a(\hat{\theta}, \theta)]$$

holds. While the upper bound $1 + \rho(c; m_1, m_2)q(r_1, \cdots, r_k; \theta)$ depends on the regressor matrix X through r_i's, the alternative upper bound $1 + E[a(\hat{\theta}, \theta)]$ ignores the information contained in X.

5.5   Empirical Example: Automobile Data

Table 5.2 shows the famous data on automobile speed (in miles per hour) and distance (in feet) covered to come to a standstill after braking (Sen and Srivastava, 1990; Ezekiel and Fox, 1959). This section gives an example of analysis on this data by using the unrestricted GLSE in a two-equation heteroscedastic model.

Two-equation heteroscedastic model. Let $speed_j$ and $distance_j$ ($j = 1, \cdots, 62$) be the automobile speed and the distance covered to come to a standstill after braking. It may be reasonable to assume that the relation between the two variables, speed and distance, are determined by

$$distance_j = \beta_1 speed_j + \beta_2(speed_j)^2 + \varepsilon_j,$$

where ε_j is an error term. It can be said that the scatter plot in Figure 5.1 may support this assumption. However, it can also be observed that the figure indicates a possible heteroscedasticity among $\varepsilon_1, \ldots, \varepsilon_{62}$. A possible model for this data is a model with the assumption that the variance of ε_j is a function of $speed_j$:

$$Var(\varepsilon_j) = \sigma^2 \times f(speed_j) \text{ with } \sigma^2 > 0.$$

Here, a typical choice for f is $f(x) = x$, or x^2. Sen and Srivastava (1990) analyze this data from this viewpoint. In this book, in order to apply the results in previous sections and chapters, we divide the data into two sets and consider a two-equation heteroscedastic model. Let $\{(speed_1, distance_1), \ldots, (speed_{35}, distance_{35})\}$ and $\{(speed_{36}, distance_{36}), \ldots, (speed_{62}, distance_{62})\}$ be the first and second groups respectively. Suppose the data of each group is generated by a homoscedastic regression model. This assumption leads to the following two-equation heteroscedastic model:

$$y = X\beta + \varepsilon \tag{5.102}$$

Table 5.2   Automobile data.

Data Number	Speed	Distance	Data Number	Speed	Distance
1	4	4	32	18	29
2	5	2	33	18	34
3	5	4	34	18	47
4	5	8	35	19	30
5	5	8	36	20	48
6	7	7	37	21	39
7	7	7	38	21	42
8	8	8	39	21	55
9	8	9	40	24	56
10	8	11	41	25	33
11	8	13	42	25	48
12	9	5	43	25	56
13	9	5	44	25	59
14	9	13	45	26	39
15	10	8	46	26	41
16	10	14	47	27	57
17	10	17	48	27	78
18	12	11	49	28	64
19	12	19	50	28	84
20	12	21	51	29	54
21	13	15	52	29	68
22	13	18	53	30	60
23	13	27	54	30	67
24	14	14	55	30	101
25	14	16	56	31	77
26	15	16	57	35	85
27	16	14	58	35	107
28	16	19	59	36	79
29	16	34	60	39	138
30	17	22	61	40	110
31	17	29	62	40	134

Source: Methods of Correlation and Regression Analysis, Ezekiel and Fox.
1959. © John Wiley & Sons Limited. Reproduced with permission.

with $n_1 = 35$, $n_2 = 27$, $n = n_1 + n_2 = 62$, $k = 2$ and

$$y = \begin{pmatrix} y_1 \\ y_2 \end{pmatrix} : n \times 1, \quad X = \begin{pmatrix} X_1 \\ X_2 \end{pmatrix} : n \times k,$$

$$\varepsilon = \begin{pmatrix} \varepsilon_1 \\ \varepsilon_2 \end{pmatrix} : n \times 1,$$

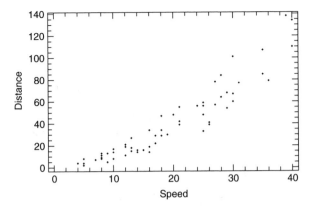

Figure 5.1   Scatter plot of speed and distance.

where

$$y_1 = \begin{pmatrix} \text{distance}_1 \\ \vdots \\ \text{distance}_{35} \end{pmatrix} : 35 \times 1, \quad y_2 = \begin{pmatrix} \text{distance}_{36} \\ \vdots \\ \text{distance}_{62} \end{pmatrix} : 27 \times 1,$$

$$X_1 = \begin{pmatrix} \text{speed}_1 & (\text{speed}_1)^2 \\ \vdots & \vdots \\ \text{speed}_{35} & (\text{speed}_{35})^2 \end{pmatrix} : 35 \times 2,$$

$$X_2 = \begin{pmatrix} \text{speed}_{36} & (\text{speed}_{36})^2 \\ \vdots & \vdots \\ \text{speed}_{62} & (\text{speed}_{62})^2 \end{pmatrix} : 27 \times 2.$$

We assume that

$$\mathcal{L}(\varepsilon) = N_n(0, \Omega) \text{ with } \Omega = \begin{pmatrix} \sigma_1^2 I_{n_1} & 0 \\ 0 & \sigma_2^2 I_{n_2} \end{pmatrix}. \tag{5.103}$$

The matrix Ω can be rewritten as

$$\Omega = \sigma^2 \Sigma(\theta)$$

with

$$\Sigma(\theta)^{-1} = I_n + \lambda(\theta)D \ (\theta \in \Theta), \tag{5.104}$$

where $\sigma^2 = \sigma_1^2$, $\lambda(\theta) = \theta - 1$, $\Theta = (0, \infty)$ and

$$\theta = \sigma_1^2 / \sigma_2^2. \tag{5.105}$$

The definition of the matrix D is given in (5.71).

The estimate of $\beta = (\beta_1, \beta_2)'$ by the OLSE

$$b(I_n) = (X'X)^{-1}X'y \tag{5.106}$$

is given by $b(I_n) = (0.577, 0.0621)'$. Hence, we get

$$\text{distance} = 0.577 \text{ speed} + 0.0621 \text{ (speed)}^2. \tag{5.107}$$

However, since the estimate of the variance ratio $\theta = \sigma_1^2/\sigma_2^2$ is close to zero, the efficiency of the OLSE is expected to be low. In fact, the estimates of σ_1^2, σ_2^2 and θ are obtained by

$$\hat{\sigma}_1^2 = S_1^2/m_1 = 29.959, \tag{5.108}$$

$$\hat{\sigma}_2^2 = S_2^2/m_2 = 193.373, \tag{5.109}$$

$$\hat{\theta} = \hat{\sigma}_1^2/\hat{\sigma}_2^2 = cS_1^2/S_2^2 = 0.1549, \tag{5.110}$$

respectively, where $c = m_2/m_1$, $m_1 = n_1 - k = 33$ and $m_2 = n_2 - k = 25$. As is shown in Section 4.5 of Chapter 4, the upper bound $\alpha_0(b(I_n))$ is given by

$$\alpha_0(b(I_n)) = \frac{(\sigma_1^2 + \sigma_2^2)^2}{4\sigma_1^2\sigma_2^2} = \frac{(1+\theta)^2}{4\theta}, \tag{5.111}$$

which is estimated as

$$\alpha_0(b(I_n)) = \frac{(1+\hat{\theta})^2}{4\hat{\theta}} = \frac{(1+0.1549)^2}{4 \times 0.1549} = 2.1524. \tag{5.112}$$

On the other hand, the unrestricted GLSE defined by

$$b(\Sigma(\hat{\theta})) = (X'\Sigma(\hat{\theta})^{-1}X)^{-1}X'\Sigma(\hat{\theta})^{-1}y \text{ with } \hat{\theta} \text{ in } (5.110) \tag{5.113}$$

estimates the model as

$$\text{distance} = 0.569 \text{ speed} + 0.0623 \text{ (speed)}^2. \tag{5.114}$$

The two regression lines obtained from the OLSE and the unrestricted GLSE are given in Figure 5.2.

The upper bound $\alpha_0(b(\Sigma(\hat{\theta})))$ is calculated as

$$\alpha_0(b(\Sigma(\hat{\theta}))) = 1 + \frac{1}{2(m_1 - 2)} + \frac{1}{2(m_2 - 2)}$$

$$= 1 + \frac{1}{2(33 - 2)} + \frac{1}{2(25 - 2)}$$

$$= 1.0379. \tag{5.115}$$

Clearly,

$$\alpha_0(b(\Sigma(\hat{\theta}))) < \alpha_0(b(I_n))$$

holds and hence the unrestricted GLSE seems to be reasonable in this case.

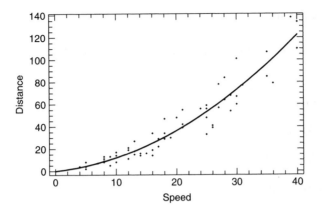

Figure 5.2   Regression lines obtained from the OLSE and the unrestricted GLSE (in dotted line). The two lines almost lie one upon the other.

Next, let us examine the upper bound $\gamma_0(b(\Sigma(\hat{\theta})))$ derived in Theorem 5.11. The upper bound $\gamma_0(b(\Sigma(\hat{\theta})))$ is of the form

$$\gamma_0(b(\Sigma(\hat{\theta}))) = 1 + g_1 + g_2 \text{ with } g_j = E(G_j)$$

$$= 1 + \rho(c; m_1, m_2)\, q(r_1, r_2; \theta). \tag{5.116}$$

For the definition of G_j and $\rho(c; m_1, m_2)$, see Section 5.4. In this case, the function q becomes

$$q(r_1, r_2; \theta) = \frac{\theta r_1}{(1 + \theta r_1)^2} + \frac{\theta r_2}{(1 + \theta r_2)^2}, \tag{5.117}$$

where r_1 and r_2 are latent roots of the matrix $(X_1'X_1)^{-1/2}X_2'X_2(X_1'X_1)^{-1/2}$. The latent roots r_1 and r_2 are calculated as

$$r_1 = 105.9558, r_2 = 0.4834. \tag{5.118}$$

From this, the value of the function $q(r_1, r_2; \theta)$ is estimated by

$$q(r_1, r_2; \hat{\theta}) = \frac{\hat{\theta} r_1}{(1 + \hat{\theta} r_1)^2} + \frac{\hat{\theta} r_2}{(1 + \hat{\theta} r_2)^2} \tag{5.119}$$

$$= 0.1189.$$

On the other hand, the function $\rho(c; m_1, m_2)$ with $m_1 = 33$, $m_2 = 25$ and $c = m_2/m_1 = 0.7576(< 1)$ is evaluated as

$$\rho(c; m_1, m_2) = \rho(0.7576; 33, 25) = 0.2937. \tag{5.120}$$

Thus, from (5.116), we obtain the following estimate of $\gamma_0(b(\Sigma(\hat{\theta})))$:

$$\gamma_0(b(\Sigma(\hat{\theta}))) = 1 + 0.1189 \times 0.2937 = 1.0349, \tag{5.121}$$

which is almost the same as $\alpha_0(b(\Sigma(\hat{\theta})))$.

5.6 Problems

5.2.1 Show that the following three conditions are equivalent:

(1) $E(u_1|u_2) = 0$;

(2) $E(\eta_1|\eta_2) = 0$;

(3) $E[b(\Sigma)|e] = \beta$.

5.2.2 Establish (5.14).

5.2.3 Establish Lemma 5.2.

5.2.4 Under the assumption of Proposition 5.4, show that $\delta = 0$ is also equivalent to each of the following two conditions:

1. $X'\Sigma^{-1}Z = 0$;

2. $X'\hat{\Sigma}^{-1}\Sigma Z = 0$.

5.3.1 In the Anderson model treated in Section 5.3:

1. Find the inverse matrix Σ of Σ^{-1} in (5.52).

2. By using this, find an expression of $\{\varepsilon_j\}$ like that of AR(1) process:

$$\varepsilon_j = \theta_j \varepsilon_{j-1} + \xi_j \ (j = 1, \cdots, n).$$

5.3.2 Analyze the CO_2 emission data given in Section 2.5 by evaluating the upper bounds for the covariance matrix of the GLSE treated in Section 5.3.

5.4.1 The quadratic form $\varepsilon'M'M\varepsilon$ in (5.76) is the same as $\eta_2'V'V\eta_2$ in (5.19).

5.4.2 Establish Lemma 5.10. See Kurata (2001).

5.4.3 Establish the equality (5.87). See Kurata (2001).

6

Bounds for Normal Approximation to the Distributions of GLSP and GLSE

6.1 Overview

In application, the distribution of a generalized least squares predictor (GLSP) is often approximated by the distribution of the Gauss–Markov predictor (GMP), which is normal if the error term is normally distributed. Such an approximation is based on the fact that the GLSPs are asymptotically equivalent to the GMP under appropriate regularity conditions. In this chapter, we evaluate with a uniform bound, goodness of the normal approximations to the probability density function (pdf) and the cumulative distribution function (cdf) of a GLSP. The results are applied to the case in which the distribution of a generalized least squares estimator (GLSE) is approximated by that of the Gauss–Markov estimator (GME). To describe the problem, let a general linear regression model be

$$\begin{pmatrix} y \\ y_0 \end{pmatrix} = \begin{pmatrix} X \\ X_0 \end{pmatrix} \beta + \begin{pmatrix} \zeta \\ \zeta_0 \end{pmatrix} \tag{6.1}$$

where

$$y : n \times 1, \quad y_0 : m \times 1, \quad X : n \times k, \quad X_0 : m \times k, \quad rank X = k,$$

Generalized Least Squares   Takeaki Kariya and Hiroshi Kurata
© 2004 John Wiley & Sons, Ltd   ISBN: 0-470-86697-7 (PPC)

and the error term is assumed to be normally distributed:

$$\mathcal{L}\left(\begin{pmatrix} \zeta \\ \zeta_0 \end{pmatrix}\right) = N_{n+m}\left(0, \sigma^2 \Phi\right) \tag{6.2}$$

with

$$\Phi = \begin{pmatrix} \Sigma & \Delta' \\ \Delta & \Sigma_0 \end{pmatrix}.$$

Here, $\Sigma \in \mathcal{S}(n)$, $\Sigma_0 : m \times m$ and $\Delta : m \times n$ are unknown, and $\mathcal{S}(n)$ denotes the set of $n \times n$ positive definite matrices. The submodel $y = X\beta + \zeta$ in (6.1) corresponds to the observable part of the model, while $y_0 = X_0\beta + \zeta_0$ is to be predicted.

As stated in Chapter 3, the problem of estimating β is a particular case of the prediction problem. In fact, by letting

$$m = k, \quad X_0 = I_k, \quad \Delta = 0 \quad \text{and} \quad \Sigma_0 = 0, \tag{6.3}$$

the predictand y_0 in (6.1) reduces to $y_0 = \beta$. Hence, the problem is clearly equivalent to estimation of β in

$$y = X\beta + \zeta \quad \text{with} \quad \mathcal{L}(\zeta) = N_n(0, \sigma^2\Sigma). \tag{6.4}$$

By transforming $\zeta = \varepsilon$ and $\zeta_0 = \Delta\Sigma^{-1}\varepsilon + \varepsilon_0$ in (6.1), the model is rewritten as

$$\begin{pmatrix} y \\ y_0 \end{pmatrix} = \begin{pmatrix} X \\ X_0 \end{pmatrix}\beta + \begin{pmatrix} I_n & 0 \\ \Delta\Sigma^{-1} & I_m \end{pmatrix}\begin{pmatrix} \varepsilon \\ \varepsilon_0 \end{pmatrix}, \tag{6.5}$$

where

$$\mathcal{L}\left(\begin{pmatrix} \varepsilon \\ \varepsilon_0 \end{pmatrix}\right) = N_{n+m}(0, \sigma^2\Omega) \tag{6.6}$$

with

$$\Omega = \begin{pmatrix} \Sigma & 0 \\ 0 & \Sigma_0 - \Delta\Sigma^{-1}\Delta' \end{pmatrix}.$$

When condition (6.3) holds, the reduced model is equivalent to

$$y = X\beta + \varepsilon \quad \text{with} \quad \mathcal{L}(\varepsilon) = N_n(0, \sigma^2\Sigma). \tag{6.7}$$

The GMP y_0^* of y_0 is given by

$$y_0^* \equiv \tilde{y}_0(\Sigma, \Delta) = X_0 b(\Sigma) + \Delta\Sigma^{-1}[y - Xb(\Sigma)], \tag{6.8}$$

where $b(\Sigma)$ is the GME of β:

$$b(\Sigma) = (X'\Sigma^{-1}X)^{-1}X'\Sigma^{-1}y. \tag{6.9}$$

The GMP is the best linear unbiased predictor when Σ and Δ are known (see Section 3.2 of Chapter 3). The GLSP $\hat{y}_0$ of y_0 is defined by replacing unknown Σ and Δ by their estimators $\hat{\Sigma}$ and $\hat{\Delta}$ respectively:

$$\hat{y}_0 \equiv \tilde{y}_0(\hat{\Sigma}, \hat{\Delta}) = X_0 b(\hat{\Sigma}) + \hat{\Delta}\hat{\Sigma}^{-1}[y - Xb(\hat{\Sigma})], \tag{6.10}$$

where $b(\hat{\Sigma})$ is a GLSE of β:

$$b(\hat{\Sigma}) = (X'\hat{\Sigma}^{-1}X)^{-1}X'\hat{\Sigma}^{-1}y. \tag{6.11}$$

Here, the quantities $\hat{\Sigma} \equiv \hat{\Sigma}(e)$ and $\hat{\Delta} \equiv \hat{\Delta}(e)$ are functions of the ordinary least squares (OLS) residual vector e:

$$e = Ny \quad \text{with} \quad N = I_n - X(X'X)^{-1}X'. \tag{6.12}$$

As stated in Proposition 3.2, the GLSP $\hat{y}_0$ in (6.10) can be rewritten as

$$\hat{y}_0 = C(e)y \quad \text{with} \quad C(e)X = X_0 \tag{6.13}$$

by letting

$$C(e) = X_0(X'\hat{\Sigma}^{-1}X)^{-1}X'\hat{\Sigma}^{-1}$$
$$+ \hat{\Delta}\hat{\Sigma}^{-1}[I_n - X(X'\hat{\Sigma}^{-1}X)^{-1}X'\hat{\Sigma}^{-1}]. \tag{6.14}$$

Throughout this chapter, we assume that the function $C(e)$ in (6.13) satisfies the following conditions:

(1) $C(e)$ is continuous in e;

(2) $C(e)$ is scale-invariant in the sense that $C(ae) = C(e)$ for any $a > 0$;

(3) $C(e)$ is an even function of e, that is, $C(-e) = C(e)$.

These conditions guarantee that the GLSP $\hat{y}_0 = C(e)y$ is an unbiased predictor with a finite second moment (Proposition 3.3).

In this chapter, the distribution of a GLSP $\hat{y}_0$ is approximated by that of the GMP y_0^*. To describe this approximation, we reproduce the decomposition of a GLSP described in Section 3.2 of Chapter 3: for a GLSP $\hat{y}_0$ satisfying conditions (1) to (3), let Y be the prediction error of $\hat{y}_0$, namely,

$$Y = \hat{y}_0 - y_0,$$

and decompose Y as

$$Y = Y_1 + Y_2, \tag{6.15}$$

where

$$Y_1 \equiv y_0^* - y_0 = \tilde{y}_0(\Sigma, \Delta) - y_0,$$
$$Y_2 \equiv \hat{y}_0 - y_0^* = \tilde{y}_0(\hat{\Sigma}, \hat{\Delta}) - \tilde{y}_0(\Sigma, \Delta).$$

Note that the first term Y_1 of Y in (6.15) is the prediction error of the GMP. Moreover, the two terms Y_1 and Y_2 are independent. To see this, let Z be an $n \times (n - k)$ matrix such that

$$N = ZZ' \quad \text{and} \quad Z'Z = I_{n-k}, \tag{6.16}$$

and make a transformation

$$\begin{pmatrix} u_1 \\ u_2 \\ \varepsilon_0 \end{pmatrix} = \begin{pmatrix} A^{-1/2}X'\Sigma^{-1} & 0 \\ B^{-1/2}Z' & 0 \\ 0 & I_m \end{pmatrix} \begin{pmatrix} \varepsilon \\ \varepsilon_0 \end{pmatrix} \tag{6.17}$$

with

$$A \equiv X'\Sigma^{-1}X \in S(k) \quad \text{and} \quad B \equiv Z'\Sigma Z \in S(n - k).$$

Then the distribution of $(u_1', u_2', \varepsilon_0')'$ is obtained as

$$\mathcal{L}\left(\begin{pmatrix} u_1 \\ u_2 \\ \varepsilon_0 \end{pmatrix}\right) = N_{n+m}\left(\begin{pmatrix} 0 \\ 0 \\ 0 \end{pmatrix}, \sigma^2 \begin{pmatrix} I_k & 0 & 0 \\ 0 & I_{n-k} & 0 \\ 0 & 0 & \Sigma_0 - \Delta\Sigma^{-1}\Delta' \end{pmatrix}\right). \tag{6.18}$$

On the other hand, Y_1 and Y_2 in (6.15) are rewritten as

$$Y_1 \equiv Y_{11} + Y_{12},$$
$$Y_2 \equiv [C(ZB^{1/2}u_2) - \Delta\Sigma^{-1}]\Sigma ZB^{-1/2}u_2, \tag{6.19}$$

where

$$Y_{11} = (X_0 - \Delta\Sigma^{-1}X)A^{-1/2}u_1 \quad \text{and} \quad Y_{12} = -\varepsilon_0.$$

It can be observed that the identity

$$e = ZB^{1/2}u_2$$

is used in (6.19). Thus, Y_{11}, Y_{12} and Y_2 are independent and the distribution of $Y_1 = Y_{11} + Y_{12}$ is given by

$$\mathcal{L}(Y_1) = N_m(0, V_1) \quad \text{with} \quad V_1 = E(Y_1Y_1') = V_{11} + V_{12}, \tag{6.20}$$

where

$$V_{11} \equiv E(Y_{11}Y_{11}') = \sigma^2(X_0 - \Delta\Sigma^{-1}X)A^{-1}(X_0 - \Delta\Sigma^{-1}X)',$$
$$V_{12} \equiv E(Y_{12}Y_{12}') = \sigma^2(\Sigma_0 - \Delta\Sigma^{-1}\Delta'), \tag{6.21}$$

and the conditional distribution of $Y = Y_1 + Y_2$ given Y_2 is

$$\mathcal{L}(Y|Y_2) = N_m(Y_2, V_1). \tag{6.22}$$

The distribution of Y_2 in (6.19) may be complicated. But it follows:

Lemma 6.1 *Under conditions (1) to (3), Y_2 has a finite second moment and*

$$E(Y_2) = 0. \tag{6.23}$$

Proof. Denote

$$Y_2 = Y_2(u_2)$$

as a function of u_2. Then $Y_2(u_2)$ is continuous in u_2, since $C(\cdot)$ is continuous by condition (1). For a moment, let

$$\mathbf{c} = \|u_2\| \quad \text{and} \quad \mathbf{v} = u_2/\|u_2\|.$$

By using condition (2), the function $Y_2(u_2)$ is expressed as

$$
\begin{aligned}
Y_2(u_2) &= Y_2(\mathbf{cv}) \\
&= [C(ZB^{1/2}(\mathbf{cv})) - \Delta\Sigma^{-1}]\Sigma Z B^{-1/2}(\mathbf{cv}) \\
&= \mathbf{c} \times [C(ZB^{1/2}\mathbf{v}) - \Delta\Sigma^{-1}]\Sigma Z B^{-1/2}\mathbf{v} \\
&= \|u_2\| \times Y_2(u_2/\|u_2\|).
\end{aligned}
$$

Here, the quantities $\mathbf{c} = \|u_2\|$ and $\mathbf{v} = u_2/\|u_2\|$ are independent, $\mathcal{L}(\mathbf{c}^2/\sigma^2) = \chi^2_{n-k}$ and $\mathbf{v}$ is distributed as the uniform distribution on the unit sphere $\mathcal{U}(n-k) = \{u \in R^{n-k} \mid \|u\| = 1\}$ (see Proposition 1.4). Hence,

$$
\begin{aligned}
V_2 &= E(Y_2 Y_2') \\
&= E[\mathbf{c}^2 Y_2(\mathbf{v}) Y_2(\mathbf{v})'] \\
&= E(\mathbf{c}^2) \times E[Y_2(\mathbf{v}) Y_2(\mathbf{v})'] \\
&= (n-k)\sigma^2 \times E[Y_2(\mathbf{v}) Y_2(\mathbf{v})']. \tag{6.24}
\end{aligned}
$$

Since $Y_2(\mathbf{v})$ is a continuous function on a compact set $\mathcal{U}(n-k)$, it is bounded, and hence V_2 is finite.

Next, to show $E(Y_2) = 0$, note that the function $Y_2(u_2)$ is an odd function of u_2:

$$Y_2(-u_2) = -Y_2(u_2),$$

since $C(\cdot)$ is an even function. Hence, by using $\mathcal{L}(u_2) = \mathcal{L}(-u_2)$, we obtain

$$E[Y_2(u_2)] = E[Y_2(-u_2)] = -E[Y_2(u_2)],$$

from which (6.23) follows. This completes the proof.

Let a pdf of $Y = \hat{y}_0 - y_0$ be $f(x)$. The function $f(x)$ is approximated by

$$\phi(x; V_1) = \frac{1}{(2\pi)^{m/2}|V_1|^{1/2}} \exp\left(-\frac{1}{2}x' V_1^{-1} x\right),$$

which is a pdf of $Y_1 = y_0^* - y_0$ in (6.20). We are interested in evaluating the goodness of this approximation by deriving a bound for

$$\sup_{x \in R^m} |f(x) - \phi(x; V_1)|. \tag{6.25}$$

We develop the theory in the following order:

6.2 Uniform Bounds for Normal Approximations to the Probability Density Functions

6.3 Uniform Bounds for Normal Approximations to the Cumulative Distribution Functions.

In Section 6.2, we obtain a bound for (6.25) under the conditions described above. The results are applied to typical GLSPs and GLSEs in the AR(1) error model and a seemingly unrelated regression (SUR) model. Similar analyses are pursued for the cdf of a GLSP (GLSE) in Section 6.3.

6.2   Uniform Bounds for Normal Approximations to the Probability Density Functions

In this section, we first discuss a GLSP $\tilde{y}_0 = C(e)y$ that satisfies conditions (1) to (3), and evaluate a bound for the quantity in (6.25) under a general setup. As a particular case of interest, a bound for the pdf of a GLSE $b(\hat{\Sigma})$ is also obtained.

The main theorem. Let $\xi(t)$ be the characteristic function of Y:

$$\xi(t) = E[\exp(it'Y)] \quad (t \in R^m). \tag{6.26}$$

The function $\xi(t)$ is the product of the characteristic functions of Y_j $(j = 1, 2)$:

$$\xi(t) = \xi_1(t)\xi_2(t) \tag{6.27}$$

where $\xi_j(t) = E\{\exp(it'Y_j)\}$, and therefore, $\xi_1(t)$ is given by

$$\xi_1(t) = \exp\left(-\frac{t'V_1 t}{2}\right),$$

since $\mathcal{L}(Y_1) = N_m(0, V_1)$ (see Problem 1.2.1). By using the inversion formula (see e.g., Theorem 2.6.3 of Anderson, 1984), the pdf $f(x)$ of Y is expressed as

$$f(x) = \frac{1}{(2\pi)^m} \int_{R^m} \exp(-it'x)\xi(t)dt. \tag{6.28}$$

Similarly,

$$\phi(x; V_1) = \frac{1}{(2\pi)^m} \int_{R^m} \exp(-it'x)\xi_1(t)dt. \tag{6.29}$$

Here is the main theorem.

Theorem 6.2 *Suppose that a GLSP $\hat{y}_0 = C(e)y$ in (6.13) satisfies conditions (1) to (3) in Section 6.1. Then the following inequality holds:*

$$\sup_{x \in R^m} |f(x) - \phi(x; V_1)| \le \frac{tr(V_2 V_1^{-1})}{\sqrt{2}(2\pi)^{m/2}|V_1|^{1/2}}. \tag{6.30}$$

Proof. By Taylor's theorem,

$$\cos x = 1 - \cos(\theta_1 x)\frac{x^2}{2} \quad \text{for some } \theta_1 \in [0, 1],$$

$$\sin x = x - \sin(\theta_2 x)\frac{x^2}{2} \quad \text{for some } \theta_2 \in [0, 1].$$

Hence, it follows that

$$\exp(ix) = \cos x + i \sin x$$

$$= 1 + ix - [\cos(\theta_1 x) + i \sin(\theta_2 x)]\frac{x^2}{2}. \tag{6.31}$$

Therefore, the function $\xi_2(t)$ is evaluated as

$$\xi_2(t) = E[\exp(it'Y_2)]$$

$$= 1 + it'E(Y_2) - E\left\{[\cos(\theta_1 t'Y_2) + i \sin(\theta_2 t'Y_2)]\frac{(t'Y_2)^2}{2}\right\},$$

where θ_j's are random variables such that $\theta_j \in [0, 1]$ a.s. Since $E(Y_2) = 0$ (see Lemma 6.1),

$$\xi_2(t) = 1 - E\left[\{\cos(\theta_1 t'Y_2) + i \sin(\theta_2 t'Y_2)\}\frac{(t'Y_2)^2}{2}\right]$$

$$= 1 + v(t) \text{ (say)}. \tag{6.32}$$

This yields $\xi(t) = \xi_1(t)[1 + v(t)]$, and hence

$$\xi(t) - \xi_1(t) = \xi_1(t)v(t) \tag{6.33}$$

Now using the inversion formula (6.28) and (6.29) entails

$$|f(x) - \phi(x; V_1)| = \left|\frac{1}{(2\pi)^m}\int_{R^m} \exp(-it'x)[\xi(t) - \xi_1(t)]dt\right|$$

$$= \left|\frac{1}{(2\pi)^m}\int_{R^m} \exp(-it'x)\xi_1(t)v(t)dt\right| \quad \text{(from (6.33))}$$

$$\le \frac{1}{(2\pi)^m}\int_{R^m} \left|\exp(-it'x)\xi_1(t)v(t)\right|dt$$

$$= \frac{1}{(2\pi)^m}\int_{R^m} \xi_1(t)|v(t)| \, dt, \tag{6.34}$$

where $|\exp(-it'x)| = 1$ is used in the last line. Here, the factor $|v(t)|$ is bounded above by $t'V_2t/\sqrt{2}$, since

$$
\begin{aligned}
|v(t)| &\le \frac{1}{2}\, E[|\cos(\theta_1 t'Y_2) + i\sin(\theta_2 t'Y_2)|\,(t'Y_2)^2] \\
&= \frac{1}{2}\, E\{[\cos^2(\theta_1 t'Y_2) + \sin^2(\theta_2 t'Y_2)]^{1/2}(t'Y_2)^2\} \\
&\le \frac{1}{2}\, E[\sqrt{2}(t'Y_2)^2] \\
&= \frac{1}{\sqrt{2}} t' E(Y_2 Y_2')t,
\end{aligned}
$$

where $\cos^2(x) + \sin^2(y) \le 2$ is used in the third line. Hence, the extreme right-hand side of (6.34) is bounded above by

$$
\frac{1}{\sqrt{2}(2\pi)^m} \int_{R^m} \xi_1(t)(t'V_2t)dt = \frac{1}{\sqrt{2}(2\pi)^m} \int_{R^m} \exp\left(-\frac{1}{2}t'V_1t\right)(t'V_2t)dt.
$$

To evaluate this integral, it is convenient to regard t as a normally distributed random vector such that $\mathcal{L}(t) = N_m(0, U_1)$ with $U_1 = V_1^{-1}$. In fact,

$$
\begin{aligned}
&\frac{1}{\sqrt{2}(2\pi)^m} \int_{R^m} \exp\left(-\frac{1}{2}t'V_1t\right)(t'V_2t)dt \\
&= \frac{|U_1|^{1/2}}{\sqrt{2}(2\pi)^{m/2}} \times \frac{1}{(2\pi)^{m/2}|U_1|^{1/2}} \int_{R^m} (t'V_2t)\exp\left(-\frac{1}{2}t'U_1^{-1}t\right)dt \\
&= \frac{|U_1|^{1/2}}{\sqrt{2}(2\pi)^{m/2}} \times E\{t'V_2t\} \\
&= \frac{|U_1|^{1/2}}{\sqrt{2}(2\pi)^{m/2}} \times tr(V_2U_1) \\
&= \frac{tr(V_2V_1^{-1})}{\sqrt{2}(2\pi)^{m/2}|V_1|^{1/2}}.
\end{aligned}
$$

Since the extreme right-hand side of the above equality does not depend on x, this completes the proof.

This theorem is due to Kariya and Toyooka (1992) and Usami and Toyooka (1997b). In applications, it is often the case that for any $a \in R^m$, $a'V_2a \to 0$ as $n \to \infty$. Hence, the bound in this theorem will be effective.

Estimation version of Theorem 6.2. As stated in the previous section, the problem of estimating β is a particular case of prediction problem. Put $m = k$, $X_0 = I_k$,

$\Delta = 0$ and $\Sigma_0 = 0$ in (6.5) and consider the model (6.7). Then a GLSP $\hat{y}_0 = C(e)y$ with $C(e)X = X_0$ in (6.13) reduces to a GLSE

$$\hat{\beta} = C(e)y \text{ with } C(e)X = I_k,$$

which is decomposed into two independent parts as

$$\overline{Y} \equiv \hat{\beta} - \beta$$
$$= [b(\Sigma) - \beta] + [\hat{\beta} - b(\Sigma)]$$
$$= A^{-1/2}u_1 + C(ZB^{1/2}u_2)\Sigma ZB^{-1/2}u_2$$
$$\equiv \overline{Y}_1 + \overline{Y}_2 \text{ (say).} \tag{6.35}$$

Corollary 6.3 *Suppose that a GLSE $\hat{\beta} = C(e)y$ satisfies conditions (1) to (3), and let the pdf of $\hat{\beta} - \beta$ be $f(x)$. Then the following inequality holds:*

$$\sup_{x \in R^k} |f(x) - \phi(x; \overline{V}_1)| \le \frac{tr(\overline{V}_2 \overline{V}_1^{-1})}{\sqrt{2}(2\pi)^{k/2}|\overline{V}_1|^{1/2}}, \tag{6.36}$$

where

$$\overline{V}_1 = E(\overline{Y}_1 \overline{Y}_1') = E[(b(\Sigma) - \beta)(b(\Sigma) - \beta)']$$
$$= \text{Cov}(b(\Sigma)) = \sigma^2 A^{-1},$$
$$\overline{V}_2 = E(\overline{Y}_2 \overline{Y}_2') = E[(\hat{\beta} - b(\Sigma))(\hat{\beta} - b(\Sigma))'], \tag{6.37}$$

and $\phi(x; \overline{V}_1)$ denotes the pdf of $\mathcal{L}(\overline{Y}_1) = N_k(0, \overline{V}_1)$.

Proof. The proof is available as an exercise (Problem 6.2.1).

Evaluation by using the upper bounds in Chapters 4 and 5. To evaluate the bound in the right-hand side of (6.36), we can use the results of Chapters 4 and 5, in which an upper bound for the covariance matrix of a GLSE $\hat{\beta} \equiv b(\hat{\Sigma})$ is obtained as

$$\text{Cov}(b(\hat{\Sigma})) \le (1 + \gamma) \text{ Cov}(b(\Sigma)) \text{ with } \gamma = \gamma(b(\hat{\Sigma})). \tag{6.38}$$

This inequality is rewritten in terms of $\overline{V}_1$ and $\overline{V}_2$ as

$$\overline{V}_1 + \overline{V}_2 \le (1 + \gamma)\overline{V}_1,$$

which is in turn equivalent to

$$\overline{V}_1^{-1/2}\overline{V}_2\overline{V}_1^{-1/2} \le \gamma I_k. \tag{6.39}$$

Hence, we obtain

Corollary 6.4 *For any GLSE $\hat{\beta} = b(\hat{\Sigma})$ satisfying conditions (1) to (3), the pdf of $\hat{\beta} - \beta$ is approximated by the normal pdf to the extent*

$$\sup_{x \in R^k} |f(x) - \phi(x; \overline{V}_1)| \leq \frac{tr(\overline{V}_2 \overline{V}_1^{-1})}{\sqrt{2}(2\pi)^{k/2}|\overline{V}_1|^{1/2}} \leq \frac{k\gamma(b(\hat{\Sigma}))}{\sqrt{2}(2\pi)^{k/2}|\overline{V}_1|^{1/2}}. \qquad (6.40)$$

Practical use of the uniform bounds. Note that the uniform bounds derived in Theorem 6.2 and its corollaries depend on the units of y and X. This is not curious, since the values of the functions $f(x)$ and $\phi(x; V_1)$ in the left-hand side of (6.30) depend on V_1 and V_2. Hence, for practical use of the uniform bounds, we need to modify the bounds from this viewpoint. For this purpose, we propose to use the following inequality for practical use:

- For the inequality in (6.30), we divide both sides of (6.30) by

$$\phi(0; V_1) = \phi(x; V_1)\Big|_{x=0} = \frac{1}{(2\pi)^{m/2}|V_1|^{1/2}},$$

and use the following inequality:

$$\sup_{x \in R^m} |f(x) - \phi(x; V_1)| \Big/ \phi(0; V_1) \leq \frac{tr(V_2 V_1^{-1})}{\sqrt{2}}. \qquad (6.41)$$

Similarly,

- For the inequality (6.36) in Corollary 6.3, we use

$$\sup_{x \in R^k} |f(x) - \phi(x; \overline{V}_1)| \Big/ \phi(0; \overline{V}_1) \leq \frac{tr(\overline{V}_2 \overline{V}_1^{-1})}{\sqrt{2}}. \qquad (6.42)$$

- For the inequality (6.40) in Corollary 6.4,

$$\sup_{x \in R^k} |f(x) - \phi(x; \overline{V}_1)| \Big/ \phi(0; \overline{V}_1) \leq \frac{tr(\overline{V}_2 \overline{V}_1^{-1})}{\sqrt{2}} \leq \frac{k\gamma(b(\hat{\Sigma}))}{\sqrt{2}}. \qquad (6.43)$$

Examples. Here are three simple examples.

Example 6.1 (SUR model) For simplicity, let us consider the two-equation SUR model. As is proved in (3) of Theorem 4.10, an upper bound for the covariance matrix of the unrestricted Zellner estimator (UZE) $b(S \otimes I_m)$ is given by

$$\text{Cov}(b(S \otimes I_m)) \leq \left[1 + \frac{2}{q-3}\right] \text{Cov}(b(\Sigma \otimes I_m)), \qquad (6.44)$$

where $\mathcal{L}(S) = W_2(\Sigma, q)$, the Wishart distribution with mean $q\Sigma$ and degrees of freedom q.

The UZE can be rewritten as

$$b(S \otimes I_m) = C(e)y \quad \text{with } C(e) = (X'(S^{-1} \otimes I_m)X)^{-1}X'(S^{-1} \otimes I_m),$$

and the function $C(e)$ satisfies conditions (1) to (3). Recall that S is a function of e. Hence, by letting

$$\gamma(b(S \otimes I_m)) = \frac{2}{q-3},$$

and letting f be the pdf of $b(S \otimes I_m) - \beta$, it holds that

$$\sup_{x \in R^k} |f(x) - \phi(x; \overline{V}_1)| \leq \frac{\sqrt{2}k}{(2\pi)^{k/2}|\overline{V}_1|^{1/2}(q-3)}, \tag{6.45}$$

where $\phi(x; \overline{V}_1)$ is the pdf of $b(\Sigma \otimes I_m) - \beta$, namely, the pdf of $N_k(0, \overline{V}_1)$. Here

$$\overline{V}_1 = (X'(\Sigma^{-1} \otimes I_m)X)^{-1}.$$

Example 6.2 (CO$_2$ emission data) In Section 4.6 of Chapter 4, by using the UZE $b(S \otimes I_m)$, we fitted the following two-equation SUR model to the CO$_2$ emission data of Japan and the USA:

$$\log(CO2_1) = 4.673 + 0.378 \, \log(GNP_1) \quad : \text{Japan}$$

$$\log(CO2_2) = 6.057 + 0.280 \, \log(GNP_2) \quad : \text{USA}. \tag{6.46}$$

The data is given in Section 2.5 of Chapter 2. When the matrix $\Sigma \in \mathcal{S}(2)$ is estimated by

$$S/q = \frac{1}{24} \begin{pmatrix} 0.06013 & 0.02925 \\ 0.02925 & 0.02563 \end{pmatrix}$$

$$= \begin{pmatrix} 0.002505 & 0.001219 \\ 0.001219 & 0.001068 \end{pmatrix}, \tag{6.47}$$

an estimate of the matrix

$$\overline{V}_1 = (X'(\Sigma^{-1} \otimes I_m)X)^{-1}$$

is obtained as

$$\hat{\overline{V}}_1 = (X'[(S/q)^{-1} \otimes I_m]X)^{-1}$$

$$= \begin{pmatrix} 0.036447 & * & * & * \\ -0.0063078 & 0.0010945 & * & * \\ 0.039136 & -0.0067826 & 0.076814 & * \\ -0.0045964 & 0.00079751 & -0.0090273 & 0.0010614 \end{pmatrix},$$

which is a symmetric matrix. From this, it follows

$$|\hat{\overline{V}}_1| = 8.587 \times 10^{-16}.$$

Hence, by letting $k = 4$ and $q = 24$, the uniform bound in (6.45) is estimated by

$$\frac{\sqrt{2k}}{(2\pi)^{k/2}|\hat{\overline{V}}_1|^{1/2}(q-3)} = 232855.9.$$

However, since the value depends on the unit of y and X, we use (6.43) and obtain

$$\sup_{x \in R^k} |f(x) - \phi(x; \overline{V}_1)|/\phi(0; \overline{V}_1) \le \frac{k\gamma(b(\hat{\Sigma}))}{\sqrt{2}} = \frac{4 \times [2/(24-3)]}{\sqrt{2}} = 0.2694.$$

Example 6.3 (Anderson model)   In the Anderson model described in Section 5.3 of Chapter 5, Theorem 5.7 shows that if an estimator $\hat{\theta} = \hat{\theta}(e)$ satisfies the conditions of Theorem 5.7, then the corresponding GLSE $b(\hat{\Sigma})$ with $\hat{\Sigma} = \Sigma(\hat{\theta})$ approximately satisfies

$$\text{Cov}(b(\hat{\Sigma})) \le \left\{1 + \delta \left[\frac{2}{n} \frac{\sqrt{3}(1+\theta)^3}{(1-\theta)^5}\right]\right\} \text{Cov}(b(\Sigma)), \qquad (6.48)$$

where the definition of δ is given in (5.41). When we rewrite $b(\hat{\Sigma})$ as

$$b(\hat{\Sigma}) = C(e)y \quad \text{with } C(e) = (X'\hat{\Sigma}^{-1}X)^{-1}X'\hat{\Sigma}^{-1},$$

the function $C(e)$ satisfies conditions (1) to (3). Hence, by Corollaries 6.3 and 6.4, the following approximation is obtained for the pdf $f(x)$ of $b(\hat{\Sigma}) - \beta$:

$$\sup_{x \in R^k} |f(x) - \phi(x; \overline{V}_1)| \le \frac{k\delta}{\sqrt{2}(2\pi)^{k/2}|\overline{V}_1|^{1/2}} \left[\frac{2}{n} \frac{\sqrt{3}(1+\theta)^3}{(1-\theta)^5}\right], \qquad (6.49)$$

where $\phi(x; \overline{V}_1)$ is the pdf of $b(\Sigma) - \beta$, namely, the pdf of $N_k(0, \overline{V}_1)$ with

$$\overline{V}_1 = \sigma^2 (X'\Sigma(\theta)^{-1}X)^{-1}.$$

Some complementary results are provided as exercise.

6.3   Uniform Bounds for Normal Approximations to the Cumulative Distribution Functions

In this section, analyses similar to those of Section 6.2 are pursued for the cdf of a GLSP $\hat{y}_0$. Bounds for the normal approximation to the cdf of a linear combination of the form

$$a'(\hat{y}_0 - y_0) \quad \text{with } a \in R^m \qquad (6.50)$$

are obtained.

Bound for the pdf of $a'(\hat{y}_0 - y_0)$. Fix a nonnull vector $a \in R^m$. For any GLSP $\hat{y}_0 = C(e)y$ in (6.13) satisfying conditions (1) to (3), consider a linear combination of the form

$$w = a'Y \quad \text{with } Y = \hat{y}_0 - y_0. \tag{6.51}$$

By using the decomposition $Y = Y_1 + Y_2$ in (6.15), we write w in (6.51) as

$$w = w_1 + w_2 \quad \text{with } w_1 = a'Y_1 \text{ and } w_2 = a'Y_2, \tag{6.52}$$

where the two terms w_1 and w_2 are clearly independent. Then it readily follows from (6.22) that

$$\mathcal{L}(w|w_2) = N(w_2, v_1) \quad \text{with } v_1 = a'V_1 a, \tag{6.53}$$

where $V_1 = E(Y_1 Y_1')$. The pdf of w is approximated by $\phi(x; v_1)$, which is the pdf of $\mathcal{L}(w_1) = N(0, v_1)$.

Theorem 6.5 *Let h be the pdf of w. Then the following inequality holds:*

$$\sup_{x \in R^1} |h(x) - \phi(x; v_1)| \leq \frac{v_2}{2\sqrt{\pi} v_1^{3/2}}, \tag{6.54}$$

where $v_2 = a'V_2 a$ with $V_2 = E(Y_2 Y_2')$.

Proof. The proof is essentially the same as that of Theorem 6.2. Let the characteristic functions of w, w_1 and w_2 be $\zeta(s)$, $\zeta_1(s)$ and $\zeta_2(s)$ respectively, where $s \in R^1$:

$$\zeta(s) = E[\exp(isw)] = E[\exp(is(w_1 + w_2))] = \zeta_1(s)\zeta_2(s),$$

$$\zeta_1(s) = E[\exp(isw_1)] \quad \text{and} \quad \zeta_1(s) = E[\exp(isw_2)]$$

Then these functions are written in terms of ξ, ξ_1 and ξ_2 as

$$\zeta(s) = \xi(sa), \quad \zeta_1(s) = \xi_1(sa) = \exp(-s^2 v_1/2) \quad \text{and} \quad \zeta_2(s) = \xi_2(sa),$$

where ξ, ξ_1 and ξ_2 are respectively the characteristic functions of Y, Y_1 and Y_2. Hence, we obtain from (6.27)

$$\zeta(s) = \zeta_1(s)\zeta_2(s) = \exp(-s^2 v_1/2)E[\exp(isw_2)]. \tag{6.55}$$

Thus, from the proof of Theorem 6.2, the function ζ is expressed as

$$\zeta(s) = \zeta_1(s)[1 + v^*(s)],$$

where

$$v^*(s) = v(sa) = -E\{[\cos(\theta_1 sw_2) + i\sin(\theta_2 sw_2)](sw_2)^2\}/2$$

(see (6.32)).

By the inversion formula, the two pdf's h and ϕ are expressed as

$$h(x) = \frac{1}{2\pi} \int_{-\infty}^{\infty} \exp(-isx)\zeta(s)ds$$

and

$$\phi(x; v_1) = \frac{1}{2\pi} \int_{-\infty}^{\infty} \exp(-isx)\zeta_1(s)ds$$

respectively. Hence, for any $x \in R^1$,

$$|h(x) - \phi(x; v_1)| = \left| \frac{1}{2\pi} \int_{R^1} \exp(-isx)[\zeta(s) - \zeta_1(s)]ds \right|$$

$$\leq \frac{1}{2\pi} \int_{R^1} \zeta_1(s)|v^*(s)|ds$$

$$\leq \frac{v_2}{2\sqrt{2\pi}} \int_{R^1} \zeta_1(s)s^2 ds$$

$$= \frac{v_2}{2\sqrt{2\pi}} \frac{(2\pi)^{1/2}}{v_1^{3/2}},$$

where

$$|v^*(s)| \leq s^2 v_2 / \sqrt{2}$$

and

$$\int_{-\infty}^{\infty} s^2 \exp(-s^2 v_1/2)ds = 2 \int_0^{\infty} s^2 \exp(-s^2 v_1/2)ds = (2\pi/v_1^3)^{1/2}$$

are used. This completes the proof.

Arguing in the same way as in Corollary 6.3 yields the following estimation version of Theorem 6.5 whose proof is available as an exercise (Problem 6.3.1).

Corollary 6.6 *For any fixed vector $a \in R^k$ with $a \neq 0$, let g be a pdf of the linear combination $w = a'(\hat{\beta} - \beta)$, where $\hat{\beta}$ is a GLSE of the form $\hat{\beta} = C(e)y$ with $C(e)X = I_k$. Suppose that the function $C(e)$ satisfies conditions (1) to (3). Then the following inequality holds:*

$$\sup_{x \in R^1} |g(x) - \phi(x; \bar{v}_1)| \leq \frac{\bar{v}_2}{2\sqrt{\pi}\bar{v}_1^{3/2}}, \tag{6.56}$$

where $\phi(x; \bar{v}_1)$ denotes the pdf of $\mathcal{L}(\bar{w}_1) = N(0, \bar{v}_1)$,

$$\bar{w}_1 = a'[b(\Sigma) - \beta],$$
$$\bar{w}_2 = a'[\hat{\beta} - b(\Sigma)],$$

and

$$\bar{v}_1 = a'\bar{V}_1 a = E(\bar{w}_1^2) = \sigma^2 a'(X'\Sigma^{-1}X)^{-1}a,$$
$$\bar{v}_2 = a'\bar{V}_2 a = E(\bar{w}_2^2).$$

Bounds for the cdf of $w = a'(\hat{y}_0 - y_0)$. Next, we shall derive a uniform bound for the approximation to the cdf of $w = a'Y = a'(\hat{y}_0 - y_0)$. This result is more important because it directly evaluates the error of the normal approximation.

To this end, we need the following smoothing lemma in the theory of Fourier transform (see, for example, Feller, 1966, p 512)).

Lemma 6.7 *Let G be a cdf satisfying*

$$\int_{R^1} x \, dG(x) = 0, \tag{6.57}$$

and let $\psi(s) = \int_{R^1} \exp(isx)dG(x)$ be the characteristic function of G. Let G_0 be a function such that

$$\lim_{x \to \pm\infty} [G(x) - G_0(x)] = 0, \tag{6.58}$$

and that G_0 has a bounded derivative $g_0 = G_0'$, that is, there exists a positive constant $c > 0$ such that

$$|g_0(x)| \le c \text{ for any } x \in R^1. \tag{6.59}$$

Let ψ_0 be the Fourier transform of g_0: $\psi_0(s) = \int_{R^1} \exp(isx)g_0(x)dx$. Suppose further that ψ_0 is continuously differentiable and satisfies

$$\psi_0(0) = 1 \text{ and } \psi_0'(0) = 0. \tag{6.60}$$

Then for all x and $T > 0$, the inequality

$$|G(x) - G_0(x)| \le \frac{1}{\pi} \int_{-T}^{T} \left| \frac{\psi(s) - \psi_0(s)}{s} \right| ds + \frac{24c}{\pi T} \tag{6.61}$$

holds.

Since the left-hand side of the inequality (6.61) is free from $T > 0$, this lemma implies that

$$|G(x) - G_0(x)| \le \frac{1}{\pi} \int_{-\infty}^{\infty} \left| \frac{\psi(s) - \psi_0(s)}{s} \right| ds. \tag{6.62}$$

Let H be the cdf of w, namely,

$$H(x) = P(w \le x), \tag{6.63}$$

and let

$$\Phi(x; v_1) = \int_{-\infty}^{x} \phi(w_1; v_1)dw_1,$$

which is the cdf of $\mathcal{L}(w_1) = N(0, v_1)$. Note that (6.58) is trivial when G_0 is a cdf.

To apply Lemma 6.7 to the cdf's H and Φ, we check the conditions (6.57) to (6.60). Since the GLSP $\hat{y}_0$ in (6.50) is unbiased, it holds that

$$E(w) = \int_{-\infty}^{\infty} w \, dH(w) = 0,$$

proving (6.57). Next, the derivative $\phi(x; v_1) = \Phi'(x; v_1)$ is clearly bounded, and its Fourier transform, which is given by $\zeta_1(s) = \exp(-s^2 v_1/2)$, is continuously differentiable and satisfies

$$\zeta_1(0) = 1 \quad \text{and} \quad \zeta_1'(0) = -v_1 s \exp(-s^2 v_1/2)|_{s=0} = 0,$$

proving (6.59) and (6.60). Hence, by Lemma 6.7 and (6.62), we have

$$|H(x) - \Phi(x; v_1)| \leq \frac{1}{\pi} \int_{-\infty}^{\infty} \left| \frac{\zeta(s) - \zeta_1(s)}{s} \right| ds. \tag{6.64}$$

As shown in the proof of Theorem 6.5, it holds that $|\zeta(s) - \zeta_1(s)| = |v^*(s)|\zeta_1(s)$ and $|v^*(s)| \leq s^2 v_2/\sqrt{2}$. Thus, the right-hand side of (6.64) is further bounded above by

$$\frac{v_2}{\sqrt{2\pi}} \int_{-\infty}^{\infty} |s|\zeta_1(s)ds = \frac{v_2}{\sqrt{2\pi}} \int_{-\infty}^{\infty} |s| \exp(-s^2 v_1/2)ds$$

$$= \frac{v_2}{\sqrt{2\pi}} \times 2 \int_0^{\infty} s \exp(-s^2 v_1/2)ds$$

$$= \frac{\sqrt{2}v_2}{\pi v_1}, \tag{6.65}$$

where the last equality is obtained from $\int_0^{\infty} s \exp(-s^2 v_1/2)ds = 1/v_1$. Summarizing the results above yields

Theorem 6.8 *For any fixed $a \in R^m$ with $a \neq 0$ and any GLSP $\hat{y}_0 = C(e)y$ with $C(e)X = X_0$ satisfying conditions (1) to (3), the normal approximation of the cdf $H(x)$ of $w = a'(\hat{y}_0 - y_0)$ is evaluated as*

$$\sup_{x \in R^1} |H(x) - \Phi(x; v_1)| \leq \frac{\sqrt{2}v_2}{\pi v_1}. \tag{6.66}$$

Corollary 6.9 *For any $a \in R^k$ with $a \neq 0$, let G be the cdf of the linear combination $\bar{w} = a'(\hat{\beta} - \beta)$ of a GLSE $\hat{\beta} = C(e)y$ with $C(e)X = I_k$, and suppose that the function $C(e)$ satisfies conditions (1) to (3). Then the following inequality holds:*

$$\sup_{x \in R^1} |G(x) - \Phi(x; \bar{v}_1)| \leq \frac{\sqrt{2}\bar{v}_2}{\pi \bar{v}_1}, \tag{6.67}$$

where $\bar{v}_1$ and $\bar{v}_2$ are defined in Corollary 6.6.

As done in the previous section, we evaluate the bound obtained above by using the results of Chapters 4 and 5. If an upper bound for the covariance matrix of a GLSE $b(\hat{\Sigma})$ is given by

$$\text{Cov}(b(\hat{\Sigma})) \leq (1 + \gamma)\,\text{Cov}(b(\Sigma)) \quad \text{with } \gamma = \gamma(b(\hat{\Sigma})), \tag{6.68}$$

then clearly

$$\bar{v}_2 \leq \gamma \bar{v}_1 \tag{6.69}$$

follows, since

$$\text{Cov}(b(\hat{\Sigma})) = \overline{V}_1 + \overline{V}_2, \quad \text{Cov}(b(\Sigma)) = \overline{V}_1 \quad \text{and} \quad \bar{v}_i = a'\overline{V}_i a \quad (i = 1, 2).$$

Thus, we obtain

$$\sup_{x \in R^1} |G(x) - \Phi(x; \bar{v}_1)| \leq \frac{\sqrt{2}\bar{v}_2}{\pi \bar{v}_1} \leq \frac{\sqrt{2}\gamma}{\pi}. \tag{6.70}$$

Example 6.4 (SUR model) Let us consider the UZE $b(S \otimes I_m)$ in the SUR model discussed in Example 6.1. Let G be the cdf of $a'[b(S \otimes I_m) - \beta]$. Then from (6.44), we obtain

$$\bar{v}_2 \leq \frac{2}{q-3}\bar{v}_1,$$

from which

$$\sup_{x \in R^1} |G(x) - \Phi(x; \bar{v}_1)| \leq \frac{2\sqrt{2}}{\pi(q-3)} \tag{6.71}$$

follows.

Example 6.5 (Anderson model) Consider the Anderson model described in Example 6.3. For a GLSE $b(\hat{\Sigma})$ with $\hat{\Sigma} = \Sigma(\hat{\theta})$ treated there, let G be the cdf of $a'[b(\hat{\Sigma}) - \beta]$. Then, by (6.48) and (6.69), an approximation

$$\bar{v}_2 \leq \delta \left[\frac{2}{n} \frac{\sqrt{3}(1+\theta)^3}{(1-\theta)^5} \right] \bar{v}_1, \tag{6.72}$$

is obtained, where

$$\bar{v}_1 = \sigma^2 a'(X'\Sigma(\theta)^{-1}X)^{-1}a.$$

Hence, by Corollary 6.9, we obtain the following approximation:

$$\sup_{x \in R^1} |G(x) - \Phi(x; \bar{v}_1)| \leq \frac{\sqrt{2}\delta}{\pi} \left(\frac{2}{n} \frac{\sqrt{3}(1+\theta)^3}{(1-\theta)^5} \right). \tag{6.73}$$

Numerical studies. Usami and Toyooka (1997b) numerically evaluated the uniform bounds given above by using a Monte Carlo simulation. The model considered by them is the following:

$$y_j = \beta_1 + \beta_2 j + \varepsilon_j \quad (j = 1, \cdots, n)$$

$$\varepsilon_j = \theta \varepsilon_{j-1} + \xi_j, \tag{6.74}$$

where $|\theta| < 1$, and ξ_j's are identically and independently distributed as the normal distribution $N(0, \sigma^2)$. In this model, one-step-ahead prediction of y_{n+1} is made. Hence, this model is written by the model (6.1) with $m = 1$, $k = 2$,

$$X = \begin{pmatrix} 1 & 1 \\ 1 & 2 \\ \vdots & \vdots \\ 1 & n \end{pmatrix} : n \times 2,$$

$$X_0 = (1, n + 1) : 1 \times 2,$$

$$y_0 = y_{n+1} \quad \text{(The predictand is } y_{n+1}\text{)}$$

and

$$\Phi = \begin{pmatrix} \Sigma & \Delta' \\ \Delta & \Sigma_0 \end{pmatrix} \in S(n + 1),$$

where

$$\Sigma = \Sigma(\theta) = \frac{1}{1 - \theta^2} (\theta^{|i-j|}) \in S(n),$$

$$\Delta = \Delta(\theta) = \frac{1}{1 - \theta^2} (\theta^n, \theta^{n-1}, \cdots, \theta) : 1 \times n,$$

$$\Sigma_0 = \Sigma_0(\theta) = \frac{1}{1 - \theta^2}.$$

The GMP and GLSP are given by

$$y_{n+1}^* = \tilde{y}_{n+1}(\Sigma, \Delta) = X_0 b(\Sigma) + \Delta \Sigma^{-1}[y - X b(\Sigma)]$$

and

$$\hat{y}_{n+1} = \tilde{y}_{n+1}(\hat{\Sigma}, \hat{\Delta}) = X_0 b(\hat{\Sigma}) + \hat{\Delta} \hat{\Sigma}^{-1}[y - X b(\hat{\Sigma})]$$

respectively, where $\hat{\Sigma} = \Sigma(\hat{\theta})$, $\hat{\Delta} = \Delta(\hat{\theta})$ and

$$\hat{\theta} = \hat{\theta}(e) = \frac{\sum_{j=2}^{n} e_j e_{j-1}}{\sum_{j=1}^{n} e_j^2}$$

and

$$b(\hat{\Sigma}) = (X'\hat{\Sigma}^{-1}X)^{-1}X'\hat{\Sigma}^{-1}y.$$

In this model, the uniform bound obtained in Theorem 6.2 is given by

$$\text{UB}_{\text{pdf}}(\hat{y}_{n+1}) = \frac{tr(V_2V_1^{-1})}{\sqrt{2}(2\pi)^{1/2}|V_1|^{1/2}}$$

$$= \frac{V_2}{2\sqrt{\pi}V_1^{3/2}}, \qquad (6.75)$$

where $V_1 = E[(y^*_{n+1} - y_{n+1})^2]$ and $V_2 = E[(\hat{y}_{n+1} - y^*_{n+1})^2]$. Let

$$b(\hat{\Sigma}) = \begin{pmatrix} b_1(\hat{\Sigma}) \\ b_2(\hat{\Sigma}) \end{pmatrix}.$$

Corresponding to Corollaries 6.3 and 6.6, we define

$$\text{UB}_{\text{pdf}}(b(\hat{\Sigma})) = \frac{tr(\overline{V}_2\overline{V}_1^{-1})}{2\sqrt{2}\pi|V_1|^{1/2}},$$

$$\text{UB}_{\text{pdf}}(b_i(\hat{\Sigma})) = \frac{\overline{v}_2}{2\sqrt{\pi}\overline{v}_1^{3/2}} \quad (i = 1, 2), \qquad (6.76)$$

respectively. The latter is obtained by letting

$$a = (1, 0)', \quad (0, 1)'$$

in Corollary 6.6. Similarly, let $\text{UB}_{\text{cdf}}(\hat{y}_{n+1})$ and $\text{UB}_{\text{cdf}}(b_i(\hat{\Sigma}))$ $(i = 1, 2)$ be the uniform bounds for the cdf's of $\hat{y}_{n+1}$ and $b_i(\hat{\Sigma})$ $(i = 1, 2)$ respectively. By using Theorem 6.8 and Corollary 6.9, these quantities are written by

$$\text{UB}_{\text{cdf}}(\hat{y}_{n+1}) = \frac{\sqrt{2}v_2}{\pi v_1},$$

$$\text{UB}_{\text{cdf}}(b_i(\hat{\Sigma})) = \frac{\sqrt{2}\overline{v}_2}{\pi\overline{v}_1} \quad (i = 1, 2),$$

respectively.

In Usami and Toyooka (1997b), the cases in which

$$\beta = (10, 2)', \quad \sigma^2 = 10^2,$$

$$\theta = 0.1, \ 0.3, \ 0.5, \ 0.7, \ 0.9,$$

$$n = 30, \ 50$$

are treated. Under this setup, the quantities V_1 and $\overline{V}_1$ are exactly determined. On the other hand, the values of V_2 and $\overline{V}_2$ are evaluated by a Monte Carlo simulation with replication 3000. The results are summarized in Tables 6.1 and 6.2.

Further, in Tables 6.3 and 6.4, the values of V_1, V_2, $\overline{V}_1$ and $\overline{V}_2$ are shown. Figures 6.1, 6.2 and 6.3 show $\phi(x; V_1)$, $\phi(x; V_1) - \mathrm{UB}_{\mathrm{pdf}}(\hat{y}_{n+1})$ (in dashed line)

Table 6.1   Uniform bounds for pdf ($n = 30$).

θ	$\mathrm{UB}_{\mathrm{pdf}}(\hat{y}_{n+1})$	$\mathrm{UB}_{\mathrm{pdf}}(b(\hat{\Sigma}))$	$\mathrm{UB}_{\mathrm{pdf}}(b_1(\hat{\Sigma}))$	$\mathrm{UB}_{\mathrm{pdf}}(b_2(\hat{\Sigma}))$
0.1	0.00101	0.00190	0.00032	0.00691
0.3	0.00124	0.00179	0.00038	0.00821
0.5	0.00168	0.00172	0.00050	0.01080
0.7	0.00272	0.00163	0.00073	0.01612
0.9	0.00596	0.00103	0.00095	0.02369

Source: Usami and Toyooka (1997b) with permission.

Table 6.2   Uniform bounds for pdf ($n = 50$).

θ	$\mathrm{UB}_{\mathrm{pdf}}(\hat{y}_{n+1})$	$\mathrm{UB}_{\mathrm{pdf}}(b(\hat{\Sigma}))$	$\mathrm{UB}_{\mathrm{pdf}}(b_1(\hat{\Sigma}))$	$\mathrm{UB}_{\mathrm{pdf}}(b_2(\hat{\Sigma}))$
0.1	0.00057	0.00219	0.00018	0.00630
0.3	0.00066	0.00209	0.00022	0.00759
0.5	0.00085	0.00197	0.00028	0.00980
0.7	0.00130	0.00183	0.00041	0.01459
0.9	0.00294	0.00138	0.00072	0.02746

Source: Usami and Toyooka (1997b) with permission.

Table 6.3   The values of V_2, V_1, $\overline{V}_2$ and $\overline{V}_1$ ($n = 30$).

θ	V_2	V_1	$\overline{V}_2$	$\overline{V}_1$
0.1	4.346	113.91	$\begin{pmatrix} 0.0792 & -0.0046 \\ -0.0046 & 0.0003 \end{pmatrix}$	$\begin{pmatrix} 16.993 & -0.833 \\ -0.833 & 0.054 \end{pmatrix}$
0.3	5.317	113.62	$\begin{pmatrix} 0.1854 & -0.0107 \\ -0.0107 & 0.0007 \end{pmatrix}$	$\begin{pmatrix} 26.663 & -1.294 \\ -1.294 & 0.083 \end{pmatrix}$
0.5	7.153	113.11	$\begin{pmatrix} 0.5824 & -0.0330 \\ -0.0330 & 0.0022 \end{pmatrix}$	$\begin{pmatrix} 47.753 & -2.274 \\ -2.274 & 0.147 \end{pmatrix}$
0.7	11.435	112.09	$\begin{pmatrix} 2.9405 & -0.1610 \\ -0.1610 & 0.0104 \end{pmatrix}$	$\begin{pmatrix} 109.332 & -4.986 \\ -4.986 & 0.322 \end{pmatrix}$
0.9	24.046	109.03	$\begin{pmatrix} 35.9805 & -1.6216 \\ -1.6216 & 0.1045 \end{pmatrix}$	$\begin{pmatrix} 486.336 & -17.936 \\ -17.936 & 1.157 \end{pmatrix}$

Source: Usami and Toyooka (1997b) with permission.

Table 6.4   The values of V_2, V_1, $\overline{V}_2$ and $\overline{V}_1$ ($n = 50$).

θ	V_2	V_1	$\overline{V}_2$	$\overline{V}_1$
0.1	2.270	108.21	$\begin{pmatrix} 0.020400 & -0.000718 \\ -0.000718 & 0.000028 \end{pmatrix}$	$\begin{pmatrix} 10.066 & -0.298 \\ -0.298 & 0.012 \end{pmatrix}$
0.3	2.641	108.10	$\begin{pmatrix} 0.049833 & -0.001738 \\ -0.001738 & 0.000068 \end{pmatrix}$	$\begin{pmatrix} 16.124 & -0.475 \\ -0.475 & 0.019 \end{pmatrix}$
0.5	3.374	107.92	$\begin{pmatrix} 0.160991 & -0.005566 \\ -0.005566 & 0.000219 \end{pmatrix}$	$\begin{pmatrix} 29.902 & -0.871 \\ -0.871 & 0.034 \end{pmatrix}$
0.7	5.125	107.53	$\begin{pmatrix} 0.914683 & -0.030884 \\ -0.030884 & 0.001209 \end{pmatrix}$	$\begin{pmatrix} 73.471 & -2.084 \\ -2.084 & 0.082 \end{pmatrix}$
0.9	11.395	106.12	$\begin{pmatrix} 20.477519 & -0.611951 \\ -0.611951 & 0.023725 \end{pmatrix}$	$\begin{pmatrix} 400.795 & -9.950 \\ -9.950 & 0.390 \end{pmatrix}$

Source: Usami and Toyooka (1997b) with permission.

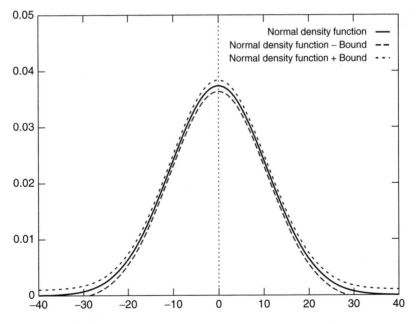

Figure 6.1   :$\phi(x; V_1)$, $\phi(x; V_1) - \text{UB}_{\text{pdf}}(\hat{y}_{n+1})$ (in dashed line) and $\phi(x; V_1) + \text{UB}_{\text{pdf}}(\hat{y}_{n+1})$ (in dotted line). $\theta = 0.1$ and $n = 30$. Source: Usami and Toyooka (1997b) with permission.

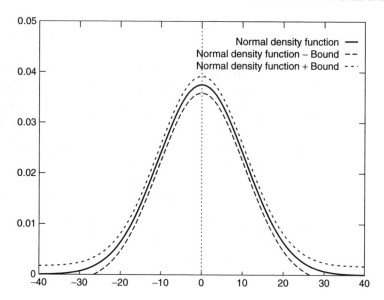

Figure 6.2 : $\phi(x; V_1)$, $\phi(x; V_1) - \mathrm{UB}_{\mathrm{pdf}}(\hat{y}_{n+1})$ (in dashed line) and $\phi(x; V_1) + \mathrm{UB}_{\mathrm{pdf}}(\hat{y}_{n+1})$ (in dotted line). $\theta = 0.5$ and $n = 30$. Source: Usami and Toyooka (1997b) with permission.

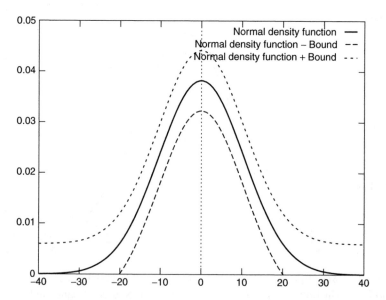

Figure 6.3 : $\phi(x; V_1)$, $\phi(x; V_1) - \mathrm{UB}_{\mathrm{pdf}}(\hat{y}_{n+1})$ (in dashed line) and $\phi(x; V_1) + \mathrm{UB}_{\mathrm{pdf}}(\hat{y}_{n+1})$ (in dotted line). $\theta = 0.9$ and $n = 30$. Source: Usami and Toyooka (1997b) with permission.

Table 6.5   Uniform bounds for the cdf ($n = 30$).

θ	$\mathrm{UB_{cdf}}(\hat{y}_{n+1})$	$\mathrm{UB_{cdf}}(b_1(\hat{\Sigma}))$	$\mathrm{UB_{cdf}}(b_2(\hat{\Sigma}))$
0.1	0.01717	0.00210	0.00255
0.3	0.02107	0.00313	0.00378
0.5	0.02847	0.00549	0.00660
0.7	0.04592	0.01211	0.01459
0.9	0.09928	0.03330	0.04066

Source: Usami and Toyooka (1997b) with permission.

Table 6.6   Uniform bounds for the cdf ($n = 50$).

θ	$\mathrm{UB_{cdf}}(\hat{y}_{n+1})$	$\mathrm{UB_{cdf}}(b_1(\hat{\Sigma}))$	$\mathrm{UB_{cdf}}(b_2(\hat{\Sigma}))$
0.1	0.00944	0.00091	0.00109
0.3	0.01100	0.00139	0.00165
0.5	0.01407	0.00242	0.00289
0.7	0.02145	0.00560	0.00666
0.9	0.04834	0.02300	0.02737

Source: Usami and Toyooka (1997b) with permission.

and $\phi(x; V_1) + \mathrm{UB_{pdf}}(\hat{y}_{n+1})$ (in dotted line) for the case in which $\theta = 0.1$, 0.5 and 0.9 respectively ($n = 30$). From these tables and figures, Usami and Toyooka (1997b) observed that

(1) As θ increases, the uniform bounds for the pdf's of $\hat{y}_{n+1}$ and $b_i(\hat{\Sigma})$s become larger;

(2) For each θ, the bounds in the case of $n = 30$ are greater than those in the case of $n = 50$;

(3) However, points (1) and (2) do not hold for $\mathrm{UB_{pdf}}(b(\hat{\Sigma}))$.

They also evaluated the uniform bounds $\mathrm{UB_{cdf}}(\hat{y}_{n+1})$ and $\mathrm{UB_{cdf}}(b_i(\hat{\Sigma}))$ ($i = 1, 2$) for the cdf's of $\hat{y}_{n+1}$ and $b_i(\hat{\Sigma})$ ($i = 1, 2$). The results are summarized in Tables 6.5 and 6.6.

6.4   Problems

6.1.1 Let P_j ($j = 1, 2$) be two probability distributions, and suppose that P_js have pdf p_js with respect to the Lebesgue measure on R^n: $\mathrm{d}P_j/\mathrm{d}x = p_j(x)$ ($j = 1, 2$). Let $\mathcal{B}$ be the Borel σ algebra on R^n. The variational distance between P_1 and P_2

is defined by

$$d(P_1, P_2) = 2 \sup_{B \in \mathcal{B}} |P_1(B) - P_2(B)|.$$

Establish the following identity:

$$d(P_1, P_2) = \int_{R^n} |p_1(x) - p_2(x)| dx = 2 \int_{R^n} \left(\frac{p_1(x)}{p_2(x)} - 1 \right)^+ p_2(x) dx,$$

from which

$$d(P_1, P_2) \leq \sup_{x \in R^n} \left(\frac{p_1(x)}{p_2(x)} - 1 \right)^+$$

follows, where for a function p,

$$p^+(x) = \max(p(x), 0).$$

See, for example, Chapters 7, 8 and 9 of Eaton (1989).

6.2.1 Prove Corollary 6.3.

6.2.2 The OLSE $b(I_n) = (X'X)^{-1}X'y$ also satisfies conditions (1) to (3) required in Corollaries 6.3 and 6.4.

   (1) Evaluate the bound in the right-hand side of (6.36) in Corollary 6.3 for the OLSE.

   (2) In the Anderson model described in Example 6.3, derive the bound for the pdf of the OLSE.

   (3) In the two-equation SUR model described in Example 6.1, derive the bound for the pdf of the OLSE.

6.2.3 Using the results in Chapters 4 and 5, derive a bound for the pdf of the unrestricted GLSE in the two-equation heteroscedastic model.

6.3.1 Prove Corollary 6.6.

6.3.2 Let us consider the distribution of the OLSE $b(I_n)$.

   (1) Evaluate the bound in the right-hand side of (6.67) in Corollary 6.9 for the OLSE.

   (2) In the Anderson model described in Example 6.5, derive the bound for the cdf of the OLSE.

   (3) In the two-equation SUR model described in Example 6.4, derive the bound for the cdf of the OLSE.

6.3.3 Using the results in Chapters 4 and 5, derive a bound for the cdf of the unrestricted GLSE in the two-equation heteroscedastic model.

7

A Maximal Extension of the Gauss–Markov Theorem and Its Nonlinear Versions

7.1 Overview

In this chapter, we first provide a maximal extension of the Gauss–Markov theorem (GMT, Theorem 2.1) in its linear framework, in particular, the maximal class of distributions (of the error term) under which the GMT holds is derived. The result suggests some further extension of the nonlinear GMT (Theorem 3.9) established in Chapter 3 for nonlinear estimators including generalized least square estimators (GLSEs) and location-equivariant estimators. An application to elliptically symmetric distributions is also given. The results in this chapter are due to Kariya and Kurata (2002).

Criterion of optimality. To state the problem more precisely, let the general linear regression model be

$$y = X\beta + \varepsilon, \tag{7.1}$$

where

$$y : n \times 1, \ X : n \times k, \ \text{and} \ rank \, X = k.$$

The GMT states that if the model (7.1) satisfies the condition

$$\mathcal{L}(\varepsilon) \in \tilde{\mathcal{P}}_n(0, \Sigma) \ \text{with} \ \Sigma \in \mathcal{S}(n), \tag{7.2}$$

Generalized Least Squares Takeaki Kariya and Hiroshi Kurata
© 2004 John Wiley & Sons, Ltd ISBN: 0-470-86697-7 (PPC)

the Gauss–Markov estimator (GME) defined by

$$b(\Sigma) = (X'\Sigma^{-1}X)^{-1}X'\Sigma^{-1}y \tag{7.3}$$

minimizes the risk matrix

$$R_P(\hat{\beta}, \beta) = E_P[(\hat{\beta} - \beta)(\hat{\beta} - \beta)'] \tag{7.4}$$

in the class C_0 of linear unbiased estimators:

$$C_0 = \{\hat{\beta} = Cy \mid C \text{ is a } k \times n \text{ matrix satisfying } CX = I_k\}. \tag{7.5}$$

Here, $P \equiv \mathcal{L}(\varepsilon)$ denotes the distribution of the error term ε, $\mathcal{S}(n)$ the set of $n \times n$ positive definite matrices, and for $\mu \in R^n$ and $\Phi \in \mathcal{S}(n)$,

$$\tilde{\mathcal{P}}_n(\mu, \Phi) = \bigcup_{\gamma > 0} \mathcal{P}_n(\mu, \gamma\Phi),$$

where $\mathcal{P}_n(\mu, \gamma\Phi)$ denotes the class of distributions with mean μ and covariance matrix $\gamma\Phi$. The risk matrix in (7.4) is the covariance matrix of $\hat{\beta} \in C_0$ under (7.2).

It should be noted here that the optimality of the GME depends on the class $\tilde{\mathcal{P}}_n(0, \Sigma)$ of distributions in (7.2) as well as the class C_0 of estimators. In fact, the class C_0 is implied by unbiasedness when $\tilde{\mathcal{P}}_n(0, \Sigma)$ is first fixed. Hence, it is appropriate to begin with the notion of optimality adopted in this chapter.

Definition 7.1 *For a class C of estimators and for a class $\mathcal{P}$ of distributions of ε, an estimator $\hat{\beta}^o$ is said to be $(C, \mathcal{P})$-optimal, if $\hat{\beta}^o \in C$ and*

$$R_P(\hat{\beta}^o, \beta) \le R_P(\hat{\beta}, \beta) \quad for \text{ all } \hat{\beta} \in C \text{ and } P \in \mathcal{P}. \tag{7.6}$$

In the case of the GMT, the statement along this definition becomes

Theorem 7.2 (Gauss–Markov theorem)

$$b(\Sigma) \text{ is } (C_0, \tilde{\mathcal{P}}_n(0, \Sigma))\text{-optimal}. \tag{7.7}$$

A maximal extension of the GMT and its nonlinear version. In this chapter, on the basis of Definition 7.1, the GMT above will be strengthened in several ways. The organization of this chapter is as follows:

7.2 An Equivalence Relation on $\mathcal{S}(n)$

7.3 A Maximal Extension of the Gauss–Markov Theorem

7.4 Nonlinear Versions of the Gauss–Markov Theorem.

In Section 7.3, for the given $\Sigma \in \mathcal{S}(n)$ and for the class C_0, we will derive a maximal class, say $\tilde{\mathcal{P}}_{max}(\Sigma)$, of the distributions of ε in the sense that the following two conditions hold:

(1) $b(\Sigma)$ is $(\mathcal{C}_0, \tilde{\mathcal{P}}_{max}(\Sigma))$-optimal;

(2) There is no $(v, \Psi) \in R^n \times \mathcal{S}(n)$ such that $b(\Sigma)$ is $(\mathcal{C}_0, \mathcal{P}^*)$-optimal, where
$\mathcal{P}^* = \tilde{\mathcal{P}}_{max}(\Sigma) \cup \tilde{\mathcal{P}}_n(v, \Psi)$.

The class $\tilde{\mathcal{P}}_{max}(\Sigma)$ derived in this chapter contains a class $\tilde{\mathcal{P}}_n(\mu, \Phi)$ with some $\mu(\neq 0)$ and $\Phi(\neq \Sigma)$ as well as the class $\tilde{\mathcal{P}}_n(0, \Sigma)$ for which the original GMT (Theorem 7.2) holds. Note that (1) and (2) imply that $b(\Sigma)$ is in particular $(\mathcal{C}_0, \tilde{\mathcal{P}}_n$ $(\mu, \Phi))$-optimal as long as $\tilde{\mathcal{P}}_n(\mu, \Phi) \subset \tilde{\mathcal{P}}_{max}(\Sigma)$. Then $\mathcal{C}_0$ is not necessarily the class of linear unbiased estimators under $P \in \tilde{\mathcal{P}}_n(\mu, \Phi)$, though it is a class of linear estimators.

In Section 7.4, by applying the results in Section 7.3, we will extend the class $\mathcal{C}_0$ of linear estimators to the class $\mathcal{C}_2$ of the location-equivariant estimators, which is introduced in Chapter 2 as

$$\mathcal{C}_2 = \{\hat{\beta} = \hat{\beta}(y) \mid \hat{\beta}(y) = b(I_n) + d(e), \ d \text{ is a } k \times 1 \text{ vector-valued}$$

$$\text{measurable function on } R^n\}, \tag{7.8}$$

where

$$b(I_n) = (X'X)^{-1}X'y$$

is the ordinary least squares estimator (OLSE) and e is the OLS residual vector defined by

$$e = Ny \quad \text{with} \quad N = I_n - X(X'X)^{-1}X'. \tag{7.9}$$

As shown in Proposition 2.5, the class $\mathcal{C}_2$ contains the class $\mathcal{C}_1$ of GLSEs of the form

$$b(\hat{\Sigma}) = (X'\hat{\Sigma}^{-1}X)^{-1}X'\hat{\Sigma}^{-1}y \quad \text{with} \quad \hat{\Sigma} = \hat{\Sigma}(e), \tag{7.10}$$

where $\hat{\Sigma}(e)$ is an estimator of Σ. Section 7.4 is also devoted to finding a subclass $\tilde{\mathcal{Q}}_n(\Sigma)$ of $\tilde{\mathcal{P}}_{max}(\Sigma)$ for which

$$b(\Sigma) \text{ is } (\mathcal{C}_2, \tilde{\mathcal{Q}}_n(\Sigma))\text{-optimal.} \tag{7.11}$$

Further, it is shown that $\tilde{\mathcal{Q}}_n(\Sigma)$ contains a class of elliptically symmetric distributions. To develop the theory in the sequel, we equip an equivalence relation $\sim$ on the set $\mathcal{S}(n)$ of $n \times n$ positive definite matrices by

$$\Sigma \sim \Phi \quad \text{if and only if} \quad b(\Sigma) = b(\Phi) \text{ for all } y \in R^n, \tag{7.12}$$

which is of course equivalent to

$$B(\Sigma) = B(\Psi)$$

with

$$B(\Omega) = (X'\Omega^{-1}X)^{-1}X'\Omega^{-1}.$$

The relation plays an important role in describing and interpreting the classes of distributions treated in the following sections.

7.2   An Equivalence Relation on $\mathcal{S}(n)$

In this section, we characterize the equivalence relation $\sim$ introduced in (7.12). Consider the linear regression model

$$y = X\beta + \varepsilon \tag{7.13}$$

with $X : n \times k$ and $rank\,X = k$.

An equivalence relation. Now fix $\Sigma \in \mathcal{S}(n)$ and let us consider the problem of characterizing the matrices Ψ's that satisfy

$$\Psi \sim \Sigma.$$

To do so, let

$$\overline{X}_\Sigma = \Sigma^{-1/2} X (X'\Sigma^{-1}X)^{-1/2} : n \times k,$$

$$\overline{Z}_\Sigma = \Sigma^{1/2} Z (Z'\Sigma Z)^{-1/2} : n \times (n-k), \tag{7.14}$$

and form the $n \times n$ orthogonal matrix

$$\Gamma_\Sigma = \begin{pmatrix} \overline{X}'_\Sigma \\ \overline{Z}'_\Sigma \end{pmatrix}, \tag{7.15}$$

where Z is an $n \times (n-k)$ matrix such that

$$X'Z = 0, \quad ZZ' = N \quad \text{and} \quad Z'Z = I_{n-k}$$

and it is fixed throughout. Further, let

$$\eta = \Sigma^{-1/2}\varepsilon$$

and define the two vectors $\tilde{\eta}_1 : k \times 1$ and $\tilde{\eta}_2 : (n-k) \times 1$ by

$$\tilde{\eta} = \Gamma_\Sigma\,\eta = \begin{pmatrix} \overline{X}'_\Sigma \eta \\ \overline{Z}'_\Sigma \eta \end{pmatrix} = \begin{pmatrix} \tilde{\eta}_1 \\ \tilde{\eta}_2 \end{pmatrix}. \tag{7.16}$$

Then it is easy to see that for $\Sigma \in \mathcal{S}(n)$ fixed,

$$b(\Sigma) - \beta = (X'\Sigma^{-1}X)^{-1}X'\Sigma^{-1}\varepsilon$$
$$= (X'\Sigma^{-1}X)^{-1/2}\tilde{\eta}_1, \tag{7.17}$$

and for any $\Psi \in \mathcal{S}(n)$,

$$b(\Psi) - \beta = (X'\Psi^{-1}X)^{-1}X'\Psi^{-1}\varepsilon$$
$$= (X'\Psi^{-1}X)^{-1}X'\Psi^{-1}[X(X'\Sigma^{-1}X)^{-1}X'\Sigma^{-1} + \Sigma Z(Z'\Sigma Z)^{-1}Z']\varepsilon$$
$$= (X'\Sigma^{-1}X)^{-1/2}\tilde{\eta}_1 + (X'\Psi^{-1}X)^{-1}X'\Psi^{-1}\Sigma Z(Z'\Sigma Z)^{-1/2}\tilde{\eta}_2$$
$$= [b(\Sigma) - \beta] + [b(\Psi) - b(\Sigma)], \tag{7.18}$$

where the matrix identity

$$X(X'\Sigma^{-1}X)^{-1}X'\Sigma^{-1} + \Sigma Z(Z'\Sigma Z)^{-1}Z' = I_n \tag{7.19}$$

is used in the second line of (7.18). The following theorem characterizes the equivalence relation in (7.12).

Theorem 7.3 *For Σ and Ψ in $\mathcal{S}(n)$, the following three statements are equivalent:*

(1) $\Psi \sim \Sigma$;

(2) $X'\Psi^{-1}\Sigma Z = 0$;

(3) $\Psi \in \mathcal{R}(\Sigma)$, *where*

$$\mathcal{R}(\Sigma) = \{\Phi \in \mathcal{S}(n) \mid \Phi = X\Upsilon X' + \Sigma Z\Delta Z'\Sigma, \ \Upsilon \in \mathcal{S}(k), \ \Delta \in \mathcal{S}(n-k)\}$$
$$= \{\Phi \in \mathcal{S}(n) \mid \Phi^{-1} = \Sigma^{-1}X\overline{\Upsilon}X'\Sigma^{-1} + Z\overline{\Delta}Z',$$
$$\overline{\Upsilon} \in \mathcal{S}(k), \ \overline{\Delta} \in \mathcal{S}(n-k)\}. \tag{7.20}$$

Proof. From (7.17) and (7.18), the equality $b(\Psi) = b(\Sigma)$ for all $y \in R^n$ holds if and only if

$$(X'\Psi^{-1}X)^{-1}X'\Psi^{-1}\Sigma Z(Z'\Sigma Z)^{-1/2}\tilde{\eta}_2 = 0 \quad \text{for any } \tilde{\eta}_2 \in R^{n-k}. \tag{7.21}$$

Hence, it is shown that a necessary and sufficient condition for (7.21) is that $X'\Psi^{-1}\Sigma Z = 0$. Thus, the equivalence between (1) and (2) follows. The rest is available as an exercise (see Problem 7.2.1). This completes the proof.

Rao's covariance structure and some related results. Special attention has been accorded in the literature to the condition under which the GME is identical to the OLSE. Below, we briefly review the several relevant results on this problem.

To do so, consider the following general linear regression model

$$y = X\beta + \varepsilon \quad \text{with } P \equiv \mathcal{L}(\varepsilon) \in \tilde{\mathcal{P}}_n(0, \Sigma), \tag{7.22}$$

where

$$y : n \times 1, \ X : n \times k, \ rankX = k.$$

Then it follows

Theorem 7.4 *The identical equality*

$$b(I_n) = b(\Sigma) \quad \text{for any } y \in R^n \tag{7.23}$$

holds if and only if Σ is of the form

$$\Sigma = X\Upsilon X' + Z\Delta Z' \quad \text{for some } \Upsilon \in \mathcal{S}(k) \text{ and } \Delta \in \mathcal{S}(n-k). \tag{7.24}$$

Proof. By using the following expression

$$b(I_n) - b(\Sigma) = (X'X)^{-1}X'\Sigma Z(Z'\Sigma Z)^{-1}Z'y,$$

we see that a necessary and sufficient condition for (7.23) is

$$X'\Sigma Z = 0. \tag{7.25}$$

On the other hand, let

$$G = (X, Z) \in \mathcal{G}\ell(n)$$

whose inverse is given by

$$G^{-1} = \begin{pmatrix} (X'X)^{-1}X' \\ Z' \end{pmatrix}.$$

In general, any $\Sigma \in \mathcal{S}(n)$ is expressed as

$$\Sigma = GG^{-1}\Sigma G'^{-1}G'$$
$$= (X, Z) \begin{pmatrix} \Upsilon & \Xi \\ \Xi' & \Delta \end{pmatrix} \begin{pmatrix} X' \\ Z' \end{pmatrix}, \tag{7.26}$$

where

$$\Upsilon = (X'X)^{-1}X'\Sigma X(X'X)^{-1} \in \mathcal{S}(k),$$
$$\Delta = Z'\Sigma Z \in \mathcal{S}(n-k),$$
$$\Xi = (X'X)^{-1}X'\Sigma Z : k \times (n-k).$$

Here, condition (7.24) is clearly equivalent to

$$\Xi = 0, \tag{7.27}$$

which is further equivalent to (7.25). Thus, it is proved that (7.23) and (7.24) are equivalent. This completes the proof.

The structure (7.24) is called *Rao's covariance structure*. It was originally established by Rao (1967) and Geisser (1970), and has been fully investigated by many papers. Among others, Kruskal (1968) and Zyskind (1967) obtained other characterizations of the equality (7.23).

Proposition 7.5 *The equality (7.23) holds if and only if one of the following two conditions is satisfied:*

(1) *(Kruskal (1968))* $L(\Sigma X) = L(X)$;

(2) *(Zyskind (1967))* $L(X)$ *is spanned by* k *latent vectors of* Σ.

Here, $L(A)$ denotes the linear subspace spanned by the column vectors of matrix A. In Puntanen and Styan (1989), an excellent survey on this topic is given on the basis of approximately 100 references.

Proof. First let us treat Kruskal's condition (1). For Σ in (7.26),

$$\Sigma X = (X, Z) \begin{pmatrix} \Upsilon & \Xi \\ \Xi' & \Delta \end{pmatrix} \begin{pmatrix} X' \\ Z' \end{pmatrix} X$$

$$= X \Upsilon X' X + Z \Xi' X' X.$$

Hence, the condition $L(\Sigma X) = L(X)$ holds if and only if

$$\Xi' X' X = 0,$$

which is equivalent to $X' \Sigma Z = 0$. Hence, (1) is equivalent to the equality (7.23).

Next, consider Zyskind's condition (2). Suppose that the equality (7.23) holds, or equivalently, Σ has Rao's covariance structure. Then Σ is expressed as

$$\Sigma = (X, Z) \begin{pmatrix} \Upsilon & 0 \\ 0 & \Delta \end{pmatrix} \begin{pmatrix} X' \\ Z' \end{pmatrix}$$

$$= (X(X'X)^{-1/2}, Z) \begin{pmatrix} (X'X)^{1/2}\Upsilon(X'X)^{1/2} & 0 \\ 0 & \Delta \end{pmatrix} \begin{pmatrix} (X'X)^{-1/2}X' \\ Z' \end{pmatrix}$$

$$= (\tilde{X}, \tilde{Z}) \begin{pmatrix} \tilde{\Upsilon} & 0 \\ 0 & \tilde{\Delta} \end{pmatrix} \begin{pmatrix} \tilde{X}' \\ \tilde{Z}' \end{pmatrix} \quad \text{(say)}.$$

Here, $(\tilde{X}, \tilde{Z})$ is an $n \times n$ orthogonal matrix. To give spectral decompositions (Lemma 1.10) of $\tilde{\Upsilon}$ and $\tilde{\Delta}$, let Γ_1 and Γ_2 be $k \times k$ and $(n-k) \times (n-k)$ orthogonal matrices such that

$$\tilde{\Upsilon} = \Gamma_1 \Lambda_1 \Gamma_1' \quad \text{and} \quad \tilde{\Delta} = \Gamma_2 \Lambda_2 \Gamma_2',$$

respectively, where $\Lambda_1 : k \times k$ and $\Lambda_2 : (n-k) \times (n-k)$ are diagonal. Then Σ is further rewritten as

$$\Sigma = (\tilde{X}, \tilde{Z}) \begin{pmatrix} \Gamma_1 & 0 \\ 0 & \Gamma_2 \end{pmatrix} \begin{pmatrix} \Lambda_1 & 0 \\ 0 & \Lambda_2 \end{pmatrix} \begin{pmatrix} \Gamma_1' & 0 \\ 0 & \Gamma_2' \end{pmatrix} \begin{pmatrix} \tilde{X}' \\ \tilde{Z}' \end{pmatrix}$$

$$= (\tilde{X}\Gamma_1, \tilde{Z}\Gamma_2) \begin{pmatrix} \Lambda_1 & 0 \\ 0 & \Lambda_2 \end{pmatrix} \begin{pmatrix} \Gamma_1'\tilde{X}' \\ \Gamma_2'\tilde{Z}' \end{pmatrix},$$

where $(\tilde{X}\Gamma_1, \tilde{Z}\Gamma_2)$ is also an $n \times n$ orthogonal matrix. Thus, the column vectors of

$$(\tilde{X}\Gamma_1, \tilde{Z}\Gamma_2) : n \times n$$

are nothing but the latent vectors of Σ. Here,

$$L(X) = L(\tilde{X}\Gamma_1)$$

holds, which is equivalent to Zyskind's condition (2). Hence, it is shown that the equality (7.23) implies the condition (2). Conversely, suppose that Zyskind's condition (2) holds. Let

$$\Sigma = \Psi \Lambda \Psi'$$

be a spectral decomposition of Σ, where Ψ and Λ are $n \times n$ orthogonal and diagonal matrices respectively. By assumption, we can assume without loss of generality that

$$\Psi_1 = XG \text{ for some } G \in \mathcal{G}\ell(k),$$

where

$$\Psi = (\Psi_1, \Psi_2) \text{ with } \Psi_1 : n \times k \text{ and } \Psi_2 : n \times (n-k)$$

and

$$\Lambda = \begin{pmatrix} \Lambda_1 & 0 \\ 0 & \Lambda_2 \end{pmatrix} \text{ with } \Lambda_1 : k \times k \text{ and } \Lambda_2 : (n-k) \times (n-k).$$

Then, by letting $\Upsilon = G\Lambda_1 G'$, $\Delta = \Lambda_2$ and $Z = \Psi_2$, the matrix Σ is expressed as

$$\Sigma = XG\Lambda_1 G'X' + \Psi_2\Lambda_2\Psi_2'$$
$$= X\Upsilon X' + Z\Delta Z'.$$

This shows that Zyskind's condition implies Rao's covariance structure. This completes the proof.

In page 53 of Eaton (1989), another aspect of Rao's covariance structure is described by using group invariance.

Some results obtained from different points of view are also found in the literature. Among others, McElroy (1967) obtained a necessary and sufficient condition under which the equality (7.23) holds for *all* X such that the first column vector of X is $1_n = (1, \cdots, 1)'$. As an extension of McElroy's theorem, Zyskind (1969) derived the covariance structure (7.28) under which the equality (7.23) holds for all X such that $L(X) \supset L(U)$, where U is a fixed $n \times p$ matrix such that $p \leq k$:

$$\Sigma = \lambda I_n + U\Delta U' \text{ for some } \lambda > 0 \text{ and } \Delta : p \times p, \tag{7.28}$$

Mathew (1983) treated the case in which Σ may be incorrectly specified.

Next, let

$$\mathcal{L}(\varepsilon) \in \mathcal{P}_n(0, \sigma^2\Sigma),$$

where $\mathcal{P}_n(0, \sigma^2\Sigma)$ is the set of the distributions on R^n with mean 0 and covariance matrix $\sigma^2\Sigma$. Kariya (1980) showed that the following two equalities

$$b(\Sigma) = b(I_n) \text{ and } s^2(\Sigma) = s^2(I_n) \text{ for all } y \in R^n \tag{7.29}$$

simultaneously hold for given X if and only if Σ has the following structure:

$$\Sigma = X\Upsilon X' + N \text{ for some } \Upsilon \in \mathcal{S}(k), \tag{7.30}$$

where $N = ZZ'$. Here, the statistics $s^2(\Sigma)$ and $s^2(I_n)$ are defined by

$$s^2(\Sigma) = (y - Xb(\Sigma))'\Sigma^{-1}(y - Xb(\Sigma))/m, \qquad (7.31)$$

and

$$s^2(I_n) = (y - Xb(I_n))'(y - Xb(I_n))/m$$

respectively, where $m = n$ or $n - k$. When the problem of estimating σ^2 is considered, it is assumed without loss of generality that $|\Sigma| = 1$ for the sake of identifiability. The result (7.30) was further extended by Kurata (1998).

7.3   A Maximal Extension of the Gauss–Markov Theorem

In the model (7.1), we fix a matrix $\Sigma \in \mathcal{S}(n)$ and derive the class $\tilde{\mathcal{P}}_{max}(\Sigma)$ of distributions of ε for which $b(\Sigma)$ is $(\mathcal{C}_0, \tilde{\mathcal{P}}_{max}(\Sigma))$-optimal. The class $\tilde{\mathcal{P}}_{max}(\Sigma)$ is maximal in the sense of (1) and (2) in Section 7.1

A maximal extension of the GMT. Let $P = \mathcal{L}(\varepsilon)$ be the distribution of ε. As before, $\hat{\beta} = Cy$ in $\mathcal{C}_0$ is decomposed as

$$
\begin{aligned}
\hat{\beta} - \beta &= C\varepsilon \\
&= C[X(X'\Sigma^{-1}X)^{-1}X'\Sigma^{-1} + \Sigma Z(Z'\Sigma Z)^{-1}Z']\varepsilon \\
&= (X'\Sigma^{-1}X)^{-1}X'\Sigma^{-1}\varepsilon + C\Sigma Z(Z'\Sigma Z)^{-1}Z'\varepsilon \\
&= A^{-1/2}\tilde{\eta}_1 + H\tilde{\eta}_2, \qquad (7.32)
\end{aligned}
$$

where $\tilde{\eta} = (\tilde{\eta}_1', \tilde{\eta}_2')'$ is defined in (7.16),

$$A = X'\Sigma^{-1}X, \quad H = C\Sigma Z(Z'\Sigma Z)^{-1/2} \qquad (7.33)$$

and the matrix identity (7.19) is used in the first line of (7.32). Then the risk matrix of $\hat{\beta}$ is expressed as

$$
\begin{aligned}
R_P(\hat{\beta}, \beta) &= E_P[(\hat{\beta} - \beta)(\hat{\beta} - \beta)'] \\
&= A^{-1/2}E_P(\tilde{\eta}_1\tilde{\eta}_1')A^{-1/2} + HE_P(\tilde{\eta}_2\tilde{\eta}_2')H' \\
&\quad + A^{-1/2}E_P(\tilde{\eta}_1\tilde{\eta}_2')H' + HE_P(\tilde{\eta}_2\tilde{\eta}_1')A^{-1/2} \\
&= V_{11} + V_{22} + V_{12} + V_{21} \text{ (say)}. \qquad (7.34)
\end{aligned}
$$

Here we note that $V_{12} = V_{21}'$,

$$R_P(b(\Sigma), \beta) = V_{11}$$

and V_{ij}'s depend on P. Clearly, if $P \in \tilde{\mathcal{P}}_n(\mu, \Phi)$, then $E_P(\varepsilon \varepsilon') = \mu \mu' + \gamma \Phi$ for some $\gamma > 0$, and hence, we see by direct calculation that

$$V_{11} = (X' \Sigma^{-1} X)^{-1} X' \Sigma^{-1} [\mu \mu' + \gamma \Phi] \Sigma^{-1} X (X' \Sigma^{-1} X)^{-1},$$

$$V_{12} = (X' \Sigma^{-1} X)^{-1} X' \Sigma^{-1} [\mu \mu' + \gamma \Phi] Z (Z' \Sigma Z)^{-1} Z' \Sigma C' = V_{21}',$$

$$V_{22} = C \Sigma Z (Z' \Sigma Z)^{-1} Z' [\mu \mu' + \gamma \Phi] Z (Z' \Sigma Z)^{-1} Z' \Sigma C'. \tag{7.35}$$

Theorem 7.6 *Fix* $\Sigma \in \mathcal{S}(n)$. *A necessary and sufficient condition for* $b(\Sigma)$ *to be* $(\mathcal{C}_0, \tilde{\mathcal{P}}_n(\mu, \Phi))$-*optimal is that*

(1) $E_P(\tilde{\eta}_1 \tilde{\eta}_2') = 0$ *holds for any* $P \in \tilde{\mathcal{P}}_n(\mu, \Phi)$,

which is equivalent to

(2) $(\mu, \Phi) \in \mathcal{M}(\Sigma) \times \mathcal{R}(\Sigma)$, *where*

$$\mathcal{M}(\Sigma) = L(X) \cup L(\Sigma Z). \tag{7.36}$$

Proof. Suppose first that (1) holds. Then for any $\hat{\beta} \in \mathcal{C}_0$ and any $P \in \tilde{\mathcal{P}}_n(\mu, \Phi)$, the risk matrix in (7.34) is expressed as

$$R_P(\hat{\beta}, \beta) = V_{11} + V_{22},$$

which is greater than $V_{11} = R_P(b(\Sigma), \beta)$, proving the sufficiency of (1).

Conversely, suppose that $b(\Sigma)$ is $(\mathcal{C}_0, \tilde{\mathcal{P}}_n(\mu, \Phi))$-optimal. Since for any $a \in R$ and any $F : (n - k) \times k$ the estimator of the form

$$\hat{\beta} = Cy \text{ with } C = (X' \Sigma^{-1} X)^{-1} X' \Sigma^{-1} + a F' Z'$$

belongs to $\mathcal{C}_0$, the risk matrix of this estimator is bounded below by that of $b(\Sigma)$, that is,

$$R_P(\hat{\beta}, \beta) = V_{11} + V_{12} + V_{21} + V_{22} \geq V_{11} = R_P(b(\Sigma), \beta)$$

holds for any $a \in R$, $F : (n - k) \times k$, $P \in \tilde{\mathcal{P}}_n(\mu, \Phi)$.

For $c \in R^k$, we set

$$c' R_P(\hat{\beta}, \beta) c = a^2 f_2 + 2a f_1 + f_0,$$

where f_i's are defined as

$$f_0 = c' U_{11} c = c' R_P(b(\Sigma), \beta) c,$$

$$f_1 = c' \{U_{12} + U_{21}\} c,$$

$$f_2 = c' U_{22} c$$

with $U_{11} = V_{11}$, $U_{12} = V_{12}/a = U'_{21}$ and $U_{22} = V_{22}/a^2$. More specifically, if the covariance matrix of $P = \mathcal{L}(\varepsilon) \in \tilde{\mathcal{P}}_n(\mu, \Phi)$ is given by $\gamma \Phi$, the matrices V_{ij}'s and U_{ij}'s are of the form

$$V_{11} = (X'\Sigma^{-1}X)^{-1}X'\Sigma^{-1}[\mu\mu' + \gamma\Phi]\Sigma^{-1}X(X'\Sigma^{-1}X)^{-1}$$

$$\equiv U_{11},$$

$$V_{12} = (X'\Sigma^{-1}X)^{-1}X'\Sigma^{-1}[\mu\mu' + \gamma\Phi]Z(Z'\Sigma Z)^{-1}Z'\Sigma C'$$

$$= a(X'\Sigma^{-1}X)^{-1}X'\Sigma^{-1}[\mu\mu' + \gamma\Phi]ZF$$

$$\equiv aU_{12},$$

$$V_{22} = C\Sigma Z(Z'\Sigma Z)^{-1}Z'[\mu\mu' + \gamma\Phi]Z(Z'\Sigma Z)^{-1}Z'\Sigma C'$$

$$= a^2 F'Z'[\mu\mu' + \gamma\Phi]ZF$$

$$\equiv a^2 U_{22}.$$

Then $(\mathcal{C}_0, \tilde{\mathcal{P}}_n(\mu, \Phi))$-optimality of $b(\Sigma)$ is equivalent to

$$a^2 f_2 + 2af_1 + f_0 \geq f_0$$

holds for any $a \in R^1$, $c \in R^k$, $\gamma > 0$ and $F : (n - k) \times k$. Note that f_i's are free from a, though they depend on c, γ and F. Here, it is easy to see that if $f_1 \neq 0$ for some c, γ and F, then we can choose $a \in R$ such that $a^2 f_2 + 2af_1 < 0$. This clearly contradicts the optimality of $b(\Sigma)$. Hence, it follows that

$$f_1 = 0 \text{ for any } c, \ \gamma, \text{ and } F. \tag{7.37}$$

The condition (7.37) is equivalent to

$$U_{12} + U_{21} = U_{12} + U'_{12} = 0 \text{ for any } \gamma \text{ and } F,$$

which implies $U_{12} = 0$ for any $\gamma > 0$ (see Problem 7.3.1), completing the proof of (1).

Next, to show that (1) and (2) are equivalent, let

$$E_P(\varepsilon) = \mu \text{ and } \text{Cov}_P(\varepsilon) = \gamma\Phi \text{ with } \gamma > 0.$$

Then $E_P(\tilde{\eta}_1 \tilde{\eta}'_2)$ is directly calculated as

$$E_P(\tilde{\eta}_1 \tilde{\eta}'_2) = A^{-1/2}X'\Sigma^{-1}[\mu\mu' + \gamma\Phi]Z(Z'\Sigma Z)^{-1/2}.$$

The condition (1) is equivalent to

$$X'\Sigma^{-1}[\mu\mu' + \gamma\Phi]Z = 0 \text{ for any } \gamma > 0,$$

which is in turn equivalent to

$$-\gamma X'\Sigma^{-1}\Phi Z = X'\Sigma^{-1}\mu\mu'Z \text{ for any } \gamma > 0. \tag{7.38}$$

Since the right-hand side of (7.38) does not depend on $\gamma > 0$, it is equivalent to

$$X'\Sigma^{-1}\Phi Z = 0 \text{ and } X'\Sigma^{-1}\mu\mu'Z = 0.$$

By Theorem 7.3, $X'\Sigma^{-1}\Phi Z = 0$ is equivalent to $\Phi \in \mathcal{R}(\Sigma)$. On the other hand, $X'\Sigma^{-1}\mu\mu'Z = 0$ holds if and only if

$$X'\Sigma^{-1}\mu = 0 \text{ or } Z'\mu = 0,$$

proving equivalence between (1) and (2). This completes the proof.

Now the following result is obvious.

Corollary 7.7 *For fixed* $\Sigma \in \mathcal{S}(n)$, *let*

$$\tilde{\mathcal{P}}_{max}(\Sigma) = \bigcup_{(\mu,\Phi)\in\mathcal{M}(\Sigma)\times\mathcal{R}(\Sigma)} \tilde{\mathcal{P}}_n(\mu, \Phi). \tag{7.39}$$

Then

$$b(\Sigma) \text{ is } (\mathcal{C}_0, \tilde{\mathcal{P}}_{max}(\Sigma))\text{-optimal}, \tag{7.40}$$

and the class $\tilde{\mathcal{P}}_{max}(\Sigma)$ *is maximal.*

Examples. The following two examples illustrate the effect of the condition $\mu \in \mathcal{M}(\Sigma)$. In Example 7.2, it is shown that $b(\Sigma)$ itself is a biased estimator.

Example 7.1   Let

$$\mu = \Sigma Z d \text{ and } \Phi = X\Upsilon X' + \Sigma Z\Delta Z'\Sigma, \tag{7.41}$$

where $d \in R^{n-k}$, $\Upsilon \in \mathcal{S}(k)$ and $\Delta \in \mathcal{S}(n-k)$. Suppose

$$P = \mathcal{L}(\varepsilon) \in \tilde{\mathcal{P}}_n(\mu, \Phi) \text{ with } \text{Cov}_P(\varepsilon) = \gamma\Phi.$$

Then $b(\Sigma) = b(\Phi)$. And $b(\Sigma)$ is unbiased, since

$$\begin{aligned} E_P[b(\Sigma)] &= \beta + (X'\Sigma^{-1}X)^{-1}X'\Sigma^{-1}E_P(\varepsilon) \\ &= \beta + (X'\Sigma^{-1}X)^{-1}X'\Sigma^{-1}\Sigma Z d \\ &= \beta. \end{aligned}$$

The risk matrix of $b(\Sigma)$ is

$$\begin{aligned} R_P(b(\Sigma), \beta) &= (X'\Sigma^{-1}X)^{-1}X'\Sigma^{-1}E_P(\varepsilon\varepsilon')\Sigma^{-1}X(X'\Sigma^{-1}X)^{-1} \\ &= (X'\Sigma^{-1}X)^{-1}X'\Sigma^{-1}[\mu\mu' + \gamma\Phi]\Sigma^{-1}X(X'\Sigma^{-1}X)^{-1}. \end{aligned}$$

Here, substituting (7.41) for μ and Φ in the above equality yields

$$R_P(b(\Sigma), \beta) = \gamma \Upsilon.$$

Further, for $\hat{\beta} = Cy = \beta + C\varepsilon$ in $\mathcal{C}_0$,

$$E_P(\hat{\beta}) = \beta + C\Sigma Zd.$$

Hence, $\mathcal{C}_0$ contains some biased estimators in general. However, for any $\hat{\beta} = Cy \in \mathcal{C}_0$, its risk matrix is greater than that of $b(\Sigma)$:

$$
\begin{aligned}
R_P(\hat{\beta}, \beta) &= C E_P(\varepsilon\varepsilon')C' \\
&= C[\mu\mu' + \gamma\Phi]C' \\
&= C\Sigma Zdd'Z'\Sigma C' + \gamma\Upsilon + \gamma C\Sigma Z\Delta Z'\Sigma C' \\
&\geq \gamma\Upsilon.
\end{aligned}
$$

Therefore, $b(\Sigma)$ minimizes $R_P(\hat{\beta}, \beta)$ among $\mathcal{C}_0$, and the minimum is $\gamma\Upsilon$.

Example 7.2   Let

$$\mu = Xc \text{ and } \Phi = X\Upsilon X' + \Sigma Z\Delta Z'\Sigma,$$

and suppose

$$\mathcal{L}(\varepsilon) \in \tilde{P}_n(\mu, \Phi) \text{ with } \mathrm{Cov}_P(\varepsilon) = \gamma\Phi,$$

where $c \in R^k$, $\Upsilon \in \mathcal{S}(k)$ and $\Delta \in \mathcal{S}(n-k)$. Then $b(\Sigma) = b(\Phi)$, but $b(\Sigma)$ is biased:

$$
\begin{aligned}
E_P[b(\Sigma)] &= \beta + (X'\Sigma^{-1}X)^{-1}X'\Sigma^{-1}Xc \\
&= \beta + c.
\end{aligned}
$$

The risk matrix of $b(\Sigma)$ is evaluated as

$$R_P(b(\Sigma), \beta) = cc' + \gamma\Upsilon.$$

Further, for $\hat{\beta} = Cy$ in $\mathcal{C}_0$,

$$
\begin{aligned}
R_P(\hat{\beta}, \beta) &= C[\mu\mu' + \gamma\Phi]C' \\
&= cc' + \gamma\Upsilon + \gamma C\Sigma Z\Delta Z'\Sigma C' \\
&\geq cc' + \gamma\Upsilon.
\end{aligned}
$$

7.4   Nonlinear Versions of the Gauss–Markov Theorem

In this section, we strengthen the results of the previous section by enlarging the class $\mathcal{C}_0$ to $\mathcal{C}_2$.

Risk matrix of a location-equivariant estimator. A location-equivariant estimator $\hat{\beta} = b(I_n) + d(e) \in \mathcal{C}_2$ is expressed as

$$
\begin{aligned}
\hat{\beta} - \beta &= (X'X)^{-1}X'\varepsilon + d(e) \\
&= (X'X)^{-1}X'[X(X'\Sigma^{-1}X)^{-1}X'\Sigma^{-1} + \Sigma Z(Z'\Sigma Z)^{-1}Z']\varepsilon + d(e) \\
&= (X'\Sigma^{-1}X)^{-1}X'\Sigma^{-1}\varepsilon + [(X'X)^{-1}X'\Sigma Z(Z'\Sigma Z)^{-1}Z'\varepsilon + d(e)] \\
&= A^{-1/2}\tilde{\eta}_1 + h_{\hat{\beta}}(\tilde{\eta}_2) \text{ (say)}, &(7.42)
\end{aligned}
$$

where

$$
\begin{aligned}
h_{\hat{\beta}}(\tilde{\eta}_2) &= (X'X)^{-1}X'\Sigma Z(Z'\Sigma Z)^{-1}Z'\varepsilon + d(e) \\
&= (X'X)^{-1}X'\Sigma Z(Z'\Sigma Z)^{-1/2}\tilde{\eta}_2 + d(Z(Z'\Sigma Z)^{1/2}\tilde{\eta}_2). &(7.43)
\end{aligned}
$$

Here, it is noted that the OLS residual vector e in (7.9) is a function of $\tilde{\eta}_2$:

$$
e = ZZ'\varepsilon = Z(Z'\Sigma Z)^{1/2}\tilde{\eta}_2.
$$

The risk matrix of $\hat{\beta} \in \mathcal{C}_2$ is decomposed as

$$
\begin{aligned}
R_P(\hat{\beta}, \beta) &= A^{-1/2}E_P(\tilde{\eta}_1\tilde{\eta}_1')A^{-1/2} + E_P[h_{\hat{\beta}}(\tilde{\eta}_2)\, h_{\hat{\beta}}(\tilde{\eta}_2)'] \\
&\quad + A^{-1/2}E_P[\tilde{\eta}_1\, h_{\hat{\beta}}(\tilde{\eta}_2)'] + E_P[h_{\hat{\beta}}(\tilde{\eta}_2)\, \tilde{\eta}_1']A^{-1/2} \\
&= V_{11} + V_{22} + V_{12} + V_{21} \text{ (say)}, &(7.44)
\end{aligned}
$$

as long as the four terms are finite. Here, since $\mathcal{C}_2$ includes $\mathcal{C}_0$, it is necessary to assume $(\mu, \Phi) \in \mathcal{M}(\Sigma) \times \mathcal{R}(\Sigma)$ and to let $P = \mathcal{L}(\varepsilon)$ move over $\tilde{\mathcal{P}}_{max}(\Sigma)$, so that the linear result in Section 7.3 should hold in this nonlinear extension. Hence, we impose the moment condition on $\mathcal{C}_2$: $E_P(\hat{\beta}'\hat{\beta}) < \infty$ for any $P \in \tilde{\mathcal{P}}_{max}(\Sigma)$, which holds if and only if $\hat{\beta}$ belongs to

$$
\mathcal{D} = \{\hat{\beta} \in \mathcal{C}_2 \mid E_P\{\hat{\beta}'\hat{\beta}\} < \infty \text{ for any } P \in \tilde{\mathcal{P}}_{max}(\Sigma)\}. \quad (7.45)
$$

Clearly, $\mathcal{D}$ is the maximal class of estimators in $\mathcal{C}_2$ that have the finite second moments for any $P \in \tilde{\mathcal{P}}_{max}(\Sigma)$, and it contains all the GLSEs with finite second moments.

Theorem 7.8 *For fixed* $\Sigma \in \mathcal{S}(n)$, *let*

$$\tilde{Q}_n^1(\Sigma) = \{P \in \tilde{\mathcal{P}}_{max}(\Sigma) \mid E_P[\tilde{\eta}_1 \, h_{\hat{\beta}}(\tilde{\eta}_2)'] = 0 \ \ for \ any \ \hat{\beta} \in \mathcal{D}\}. \quad (7.46)$$

Then

$$b(\Sigma) \ is \ (\mathcal{D}, \tilde{Q}_n^1(\Sigma))\text{-}optimal. \quad (7.47)$$

The proof is clear from (7.44) and the definitions of $\mathcal{D}$ and $\tilde{Q}_1(\Sigma)$. Clearly,

$$\tilde{Q}_n^1(\Sigma) \subset \tilde{\mathcal{P}}_{max}(\Sigma), \quad \text{though} \ \ \mathcal{D} \supset \mathcal{C}_0.$$

Therefore, the nonlinear version of the GMT in the above theorem is not completely stronger than the GMT in Theorem 7.2.

Subclasses of $\tilde{Q}_n^1(\Sigma)$. We describe two important subclasses of $\tilde{Q}_1(\Sigma)$ in (7.46).

Corollary 7.9 *For fixed* $\Sigma \in \mathcal{S}(n)$, *let*

$$\tilde{Q}_n^2(\Sigma) = \{P \in \tilde{\mathcal{P}}_{max}(\Sigma) \mid E_P(\tilde{\eta}_1 | \tilde{\eta}_2) = 0 \ a.s. \ \tilde{\eta}_2\}, \quad (7.48)$$

$$\tilde{Q}_n^3(\Sigma) = \{P \in \tilde{\mathcal{P}}_{max}(\Sigma) \mid \mathcal{L}_P(-\tilde{\eta}_1, \tilde{\eta}_2) = \mathcal{L}_P(\tilde{\eta}_1, \tilde{\eta}_2)\}, \quad (7.49)$$

where $\mathcal{L}_P(\cdot)$ denotes the distribution of $\cdot$ under P. Then

$$\tilde{Q}_n^3(\Sigma) \subset \tilde{Q}_n^2(\Sigma) \subset \tilde{Q}_n^1(\Sigma) \quad (7.50)$$

holds, and hence

$$b(\Sigma) \ is \ (\mathcal{D}, \tilde{Q}_n^i(\Sigma)) - optimal \ (i = 2, 3). \quad (7.51)$$

Proof. For any $P \in \tilde{Q}_n^2(\Sigma)$ and any $\hat{\beta} \in \mathcal{D}$, it holds that

$$E_P\left(\tilde{\eta}_1 \, h(\tilde{\eta}_2)'\right) = E\left[E\left(\tilde{\eta}_1 | \tilde{\eta}_2\right) h(\tilde{\eta}_2)'\right] = 0.$$

Thus, $\tilde{Q}_n^2(\Sigma) \subset \tilde{Q}_n^1(\Sigma)$. It is easy to see that $\tilde{Q}_n^3(\Sigma) \subset \tilde{Q}_n^2(\Sigma)$. This completes the proof.

In the following sections, we will show that $\tilde{Q}_n^3(\Sigma)$ includes a class of elliptically symmetric distributions. Other examples will be found in Kariya and Kurata (2002).

Elliptically symmetric distributions. In the model (7.1), suppose that

$$P \equiv \mathcal{L}(\varepsilon) \in \tilde{\mathcal{E}}_n(\mu, \Phi), \quad (7.52)$$

where

$$\tilde{\mathcal{E}}_n(\mu, \Phi) \equiv \bigcup_{\gamma > 0} \mathcal{E}_n(\mu, \gamma \Phi),$$

and $\mathcal{E}_n(\mu, \gamma\Phi)$ is the class of elliptically symmetric distributions with mean $\mu \in R^n$ and covariance matrix $\gamma\Phi$ (see Section 1.3 of Chapter 1).

Fix $\Sigma \in \mathcal{S}(n)$. Then a sufficient condition for $b(\Sigma)$ to be $(\mathcal{D}, \tilde{\mathcal{E}}_n(\mu, \Phi))$-optimal is that

$$\tilde{\mathcal{E}}_n(\mu, \Phi) \subset \tilde{\mathcal{Q}}_n^3(\Sigma).$$

Theorem 7.10 *Assume* $\mu \in L(\Sigma Z)$. *Then* $\tilde{\mathcal{E}}_n(\mu, \Phi) \subset \tilde{\mathcal{Q}}_n^3(\Sigma)$ *holds if and only if* $\Phi \in \mathcal{R}(\Sigma)$.

Proof. Suppose first that $\tilde{\mathcal{E}}_n(\mu, \Phi) \subset \tilde{\mathcal{Q}}_n^3(\Sigma)$. Since $\tilde{\mathcal{Q}}_n^3(\Sigma)$ is a subclass of $\tilde{\mathcal{P}}_{max}(\Sigma)$, the matrix Φ is clearly in $\mathcal{R}(\Sigma)$.

Conversely, suppose $\Phi \in \mathcal{R}(\Sigma)$ and let

$$\mu = \Sigma Z d.$$

Then $\tilde{\mathcal{E}}_n(\mu, \Phi) \subset \tilde{\mathcal{P}}_{max}(\Sigma)$. For any $P \in \tilde{\mathcal{E}}_n(\mu, \Phi)$, the distribution of

$$\tilde{\eta} = \Gamma_\Sigma \Sigma^{-1/2} \varepsilon = \begin{pmatrix} \tilde{\eta}_1 \\ \tilde{\eta}_2 \end{pmatrix}$$

under P satisfies

$$\mathcal{L}_P(\tilde{\eta}) \in \tilde{\mathcal{E}}_n(\tilde{\mu}, Q).$$

Here, $\tilde{\mu}$ and Q are given by

$$\begin{aligned}
\tilde{\mu} &= \Gamma_\Sigma \Sigma^{-1/2} \mu \\
&= \begin{pmatrix} A^{-1/2} X' \Sigma^{-1/2} \Sigma^{-1/2} \Sigma Z d \\ (Z'\Sigma Z)^{-1/2} Z' \Sigma^{1/2} \Sigma^{-1/2} \Sigma Z d \end{pmatrix} \\
&= \begin{pmatrix} 0 \\ (Z'\Sigma Z)^{1/2} d \end{pmatrix} \\
&= \begin{pmatrix} 0 \\ \tilde{\mu}_2 \end{pmatrix} \quad \text{(say)},
\end{aligned} \tag{7.53}$$

and

$$\begin{aligned}
Q &= \Gamma_\Sigma \Sigma^{-1/2} \Phi \Sigma^{-1/2} \Gamma'_\Sigma \\
&= \begin{pmatrix} Q_{11} & Q_{12} \\ Q_{21} & Q_{22} \end{pmatrix},
\end{aligned}$$

respectively, where

$$\begin{aligned}
Q_{11} &= A^{-1/2} X' \Sigma^{-1} \Phi \Sigma^{-1} X A^{-1/2} \\
Q_{12} &= A^{-1/2} X' \Sigma^{-1} \Phi Z (Z'\Sigma Z)^{-1/2} \\
Q_{21} &= Q'_{12} \\
Q_{22} &= (Z'\Sigma Z)^{-1/2} Z' \Phi Z (Z'\Sigma Z)^{-1/2}.
\end{aligned}$$

Since $\Phi \in \mathcal{R}(\Sigma)$ implies $X'\Sigma^{-1}\Phi Z = 0$, the off-diagonal blocks of Q are zero:

$$Q_{12} = Q'_{21} = 0.$$

From this,

$$\mathcal{L}_P(-\tilde{\eta}_1, \tilde{\eta}_2) = \mathcal{L}_P(\tilde{\eta}_1, \tilde{\eta}_2)$$

follows. To see this, since $\mathcal{L}_P(\tilde{\eta}) \in \tilde{\mathcal{E}}_n(\tilde{\mu}, Q)$, and since

$$\Gamma_0 = \begin{pmatrix} -I_k & 0 \\ 0 & I_{n-k} \end{pmatrix} \in \mathcal{O}(n),$$

it holds that

$$\mathcal{L}_P \begin{pmatrix} Q_{11}^{-1/2}\tilde{\eta}_1 \\ Q_{22}^{-1/2}(\tilde{\eta}_2 - \tilde{\mu}_2) \end{pmatrix} = \mathcal{L}_P(Q^{-1/2}(\tilde{\eta} - \tilde{\mu}))$$

$$= \mathcal{L}_P(\Gamma_0 Q^{-1/2}(\tilde{\eta} - \tilde{\mu}))$$

$$= \mathcal{L}_P \begin{pmatrix} -Q_{11}^{-1/2}\tilde{\eta}_1 \\ Q_{22}^{-1/2}(\tilde{\eta}_2 - \tilde{\mu}_2) \end{pmatrix}.$$

Thus, we see that

$$\mathcal{L}_P(u_1, u_2) = \mathcal{L}_P(-u_1, u_2),$$

where $u_1 = Q_{11}^{-1/2}\tilde{\eta}_1$ and $u_2 = Q_{22}^{-1/2}(\tilde{\eta}_2 - \tilde{\mu}_2)$. Let

$$f(u_1, u_2) = \begin{pmatrix} Q_{11}^{1/2}u_1 \\ Q_{22}^{1/2}u_2 + \tilde{\mu}_2 \end{pmatrix}.$$

Then we have

$$\mathcal{L}_P(\tilde{\eta}_1, \tilde{\eta}_2) = \mathcal{L}_P(f(u_1, u_2)) = \mathcal{L}_P(f(-u_1, u_2)) = \mathcal{L}_P(-\tilde{\eta}_1, \tilde{\eta}_2),$$

proving $P \in \tilde{\mathcal{Q}}_3(\Sigma)$. This completes the proof.

Corollary 7.11 *For a fixed* $\Sigma \in \mathcal{S}(n)$,

$$b(\Sigma) \text{ is } (\mathcal{D}, \tilde{\mathbf{E}}_n(\Sigma))\text{-optimal},$$

where

$$\tilde{\mathbf{E}}_n(\Sigma) = \bigcup_{(\mu, \Phi) \in L(\Sigma Z) \times \mathcal{R}(\Sigma)} \tilde{\mathcal{E}}_n(\mu, \Phi).$$

7.5   Problems

7.2.1 Show that (2) and (3) of Theorem 7.3 are equivalent.

7.3.1 In the proof of Theorem 7.6, show that the statement

$$U_{12} + U'_{12} = 0 \quad \text{for any } \gamma \text{ and } F$$

implies $U_{12} = 0$ for any $\gamma > 0$.
Hint: Fix $\gamma > 0$ and let

$$R = (X'\Sigma^{-1}X)^{-1}X'\Sigma^{-1}\left[\mu\mu' + \gamma\Phi\right]Z.$$

Then the condition is equivalent to

$$RF + F'R' = 0 \quad \text{for any } F.$$

By replacing F by some appropriate matrices, it is shown that $R = 0$.

7.4.1 In Corollary 7.9, show that $\tilde{Q}_n^3(\Sigma) \subset \tilde{Q}_n^2(\Sigma)$.

8

Some Further Extensions

8.1 Overview

In this chapter, we complement and extend the arguments made in the previous chapters, and treat the three topics: the concentration inequality for the Gauss–Markov estimator (GME), the relaxation of the normality assumption in evaluating an upper bound in a seemingly unrelated regression (SUR) model and the degeneracy of the distributions of some generalized least squares estimators (GLSEs).

In Section 8.2, it is shown that under some appropriate conditions, the GME $b(\Sigma)$ maximizes the probability that $b(\Sigma) - \beta$ lies in any symmetric convex set among a class of GLSEs. The probability in question is often called *concentration probability*. The result in Section 8.2 can be viewed as a partial extension of the nonlinear version of the Gauss–Markov theorem established in Chapter 3. In fact, maximizing the concentration probability is a stronger criterion than that of minimizing the risk matrix. However, to establish such a maximization result, some additional assumption such as unimodality is required on the distribution of the error term. The results in Section 8.2 are essentially due to Berk and Hwang (1989) and Eaton (1987, 1988). Some related facts will be found in Hwang (1985), Kuritsyn (1986), Andrews and Phillips (1987), Ali and Ponnapalli (1990), Jensen (1996) and Lu and Shi (2000).

In Section 8.3, an extension of the results provided in Chapter 4 is given. In Chapter 4, we have observed that in a general linear regression model with normally distributed error, several typical GLSEs such as the unrestricted Zellner estimator (UZE) in an SUR model have a simple covariance structure. On the basis of this structure, we obtained upper bounds for the covariance matrices of these GLSEs. The approach adopted in Chapter 4 depends on the normality of the error term. This section is devoted to relaxing the normality assumption by treating the case in which the distribution of the error term is elliptically symmetric. For

Generalized Least Squares   Takeaki Kariya and Hiroshi Kurata
© 2004 John Wiley & Sons, Ltd   ISBN: 0-470-86697-7 (PPC)

simplicity, we limit our consideration to the SUR model. The results derived here are also valid in a heteroscedastic model. This section is due to Kurata (1999). Some related results on the inference of SUR models under nonnormal distributions will be found, for example, in Srivastava and Maekawa (1995), Hasegawa (1995), Ng (2000, 2002) and so on. See also Wu and Perlman (2000).

Section 8.4 is concerned with degeneracy of the distribution of a GLSE. We first introduce the results of Usami and Toyooka (1997a), in which it is shown that in general linear regression models with a certain covariance structure, the distribution of the quantity $b(\hat{\Sigma}) - b(\Sigma)$, the difference between a GLSE and the GME, degenerates into a linear subspace of R^k. Next, some extensions of their results are provided.

8.2 Concentration Inequalities for the Gauss–Markov Estimator

In this section, it is proved that the GME is most concentrated in a class of estimators including GLSEs.

Being most concentrated. In this section, a general linear regression model of the form

$$y = X\beta + \varepsilon \tag{8.1}$$

is considered, where

$$y : n \times 1, \ X : n \times k \text{ and } rank X = k.$$

Let $P \equiv \mathcal{L}(\varepsilon)$ be the distribution of ε, and suppose that P belongs to the class $\mathbf{E}_n(0, \Sigma)$ of elliptically symmetric distributions with location vector 0 and scale matrix $\Sigma \in \mathcal{S}(n)$. Here, recall that $\mathcal{L}(\varepsilon) \in \mathbf{E}_n(0, \Sigma)$ if and only if

$$\mathcal{L}(\Gamma \Sigma^{-1/2}\varepsilon) = \mathcal{L}(\Sigma^{-1/2}\varepsilon) \text{ for any } \Gamma \in \mathcal{O}(n), \tag{8.2}$$

where $\mathcal{S}(n)$ is the set of $n \times n$ positive definite matrices and $\mathcal{O}(n)$, the group of $n \times n$ orthogonal matrices. The class $\mathbf{E}_n(0, \Sigma)$ contains some heavy-tailed distributions without moments such as the multivariate Cauchy distribution (see Section 1.3 of Chapter 1). Note that

$$\mathbf{E}_n(0, \Sigma) \supset \mathcal{E}_n(0, \sigma^2\Sigma)$$

for any $\sigma^2 > 0$, where $\mathcal{E}_n(0, \sigma^2\Sigma)$ is the class of elliptically symmetric distributions with mean 0 and covariance matrix $\sigma^2\Sigma$.

Let $b(\Sigma)$ be the GME of β:

$$b(\Sigma) = (X'\Sigma^{-1}X)^{-1}X'\Sigma^{-1}y. \tag{8.3}$$

To make the notion of "being most concentrated" clear, let C be an arbitrarily given class of estimators of β, and let $\mathcal{K}$ be a set of symmetric convex sets in R^k. Here, it is formally assumed that ϕ and R^k are always in $\mathcal{K}$. Further, a set K ($\subset R^k$) is called *symmetric* if K satisfies

$$-K = K \quad \text{with} \quad -K = \{-x | x \in K\},$$

K is said to be convex if

$$x, y \in K \text{ implies } ax + (1 - a)y \in K \text{ for any } a \in [0, 1],$$

and K is *symmetric convex* if it is symmetric and convex.

An estimator $\hat{\beta}^* \in C$ is called *most concentrated* with respect to $\mathcal{K}$ in the class C under P, if

$$P(\hat{\beta}^* - \beta \in K) \geq P(\hat{\beta} - \beta \in K) \tag{8.4}$$

for any $\beta \in R^k$, $K \in \mathcal{K}$ and $\hat{\beta} \in C$. We often refer to (8.4) as concentration inequality. It should be noted that in the inequality (8.4), the probability P is fixed. If there exists a class, say tentatively $\mathcal{P}$, of distributions under which (8.4) holds for any $P \in \mathcal{P}$, we say that $\hat{\beta}^*$ is most concentrated under $\mathcal{P}$.

Equivalence theorem. In showing that the GME is most concentrated to β, the following two theorems play an essential role:

(1) Anderson's theorem, which provides a technical basis for (8.4);

(2) The equivalence theorem, which shows that being most concentrated is equivalent to being optimal with respect to a class of loss functions.

First, the equivalence theorem due to Berk and Hwang (1989) is introduced, in which a particular class of loss functions is specified.

Theorem 8.1 (Equivalence theorem) *Let C and $\mathcal{K}$ be given classes of estimators and symmetric convex sets in R^n respectively. An estimator $\hat{\beta}^* \in C$ is most concentrated with respect to $\mathcal{K}$ in C, if and only if $\hat{\beta}^*$ satisfies*

$$E\{g(\hat{\beta}^* - \beta)\} \leq E\{g(\hat{\beta} - \beta)\} \tag{8.5}$$

for any $\hat{\beta} \in C$ and for any nonnegative function g such that

$$\{x \in R^k \mid g(x) \leq c\} \in \mathcal{K} \quad \text{for any } c \in [0, \infty). \tag{8.6}$$

Proof. Suppose first that $\hat{\beta}^* \in C$ satisfies the condition (8.5). For each $K \in \mathcal{K}$, let

$$g(x) = 1 - \chi_K(x),$$

where χ denotes the indicator function of the set K. Then the function g is non-negative, and

$$\{x \in R^k \mid g(x) \le c\} = \begin{cases} R^k \ (\in \mathcal{K}) & (1 \le c) \\ K \ (\in \mathcal{K}) & (0 \le c < 1) \end{cases}$$

Hence, by (8.5), we obtain

$$P(\hat{\beta}^* - \beta \in K) \ge P(\hat{\beta} - \beta \in K)$$

for any $\hat{\beta} \in \mathbf{C}$ and $\beta \in R^k$.

Conversely, suppose that $\hat{\beta}^* \in \mathbf{C}$ is most concentrated. We use the notion of stochastic order $\le_{st}$ whose definition and requisite facts are provided in Problem 8.2.1. For any nonnegative function g that satisfies the condition (8.6) and for any $c \in [0, \infty)$, let

$$K = \{x \in R^k \mid g(x) \le c\}.$$

Then by assumption, for any $\hat{\beta} \in \mathbf{C}$ and $\beta \in R^k$

$$\begin{aligned} P(g(\hat{\beta}^* - \beta) \le c) &= P(\hat{\beta}^* - \beta \in K) \\ &\ge P(\hat{\beta} - \beta \in K) \\ &= P(g(\hat{\beta} - \beta) \le c) \end{aligned}$$

holds, which shows that

$$g(\hat{\beta}^* - \beta) \le_{st} g(\hat{\beta} - \beta),$$

where $A \le_{st} B$ means that A is stochastically no greater than B (for details, see Problem 8.2.1). Then by Problem 8.2.2, it follows that

$$E\{g(\hat{\beta}^* - \beta)\} \le E\{g(\hat{\beta} - \beta)\}.$$

This completes the proof.

In Berk and Hwang (1989), the conditions imposed on the function g are more general than those of the above theorem. Several versions of such equivalence theorems can be found in the literature, some of which are closely related to the notion of stochastic order in the context described in Lehmann (1986, page 84): see Problems 8.2.1, 8.2.2 and 8.2.3, Hwang (1985), Andrews and Phillips (1987) and Berk and Hwang (1989).

The following corollary states that being most concentrated implies being optimal in terms of the risk matrix. More specifically,

Corollary 8.2 *Suppose that* $P \equiv \mathcal{L}(\varepsilon)$ *has finite second moments and let* $\mathbf{C}$ *be a class of estimators with the finite second moment under* P. *Let* $\mathcal{K}$ *be the class of*

all symmetric convex sets in R^k. If an estimator $\hat{\beta}^* \in \mathbf{C}$ is most concentrated with respect to $\mathcal{K}$ in the class $\mathbf{C}$, then $\hat{\beta}^*$ is optimal in terms of risk matrix, that is,

$$R(\hat{\beta}^*, \beta) \leq R(\hat{\beta}, \beta) \quad \text{for any } \hat{\beta} \in \mathbf{C} \text{ and } \beta \in R^k, \tag{8.7}$$

where $R(\hat{\beta}, \beta)$ is the risk matrix of $\hat{\beta}$:

$$R(\hat{\beta}, \beta) = E[(\hat{\beta} - \beta)(\hat{\beta} - \beta)'].$$

Proof. The inequality (8.7) (to be shown) is equivalent to

$$a' R(\hat{\beta}^*, \beta)a \leq a' R(\hat{\beta}, \beta)a \quad \text{for any } a \in R^k,$$

which is further equivalent to

$$E[g_a(\hat{\beta}^* - \beta)] \leq E[g_a(\hat{\beta} - \beta)] \quad \text{for any } a \in R^k \tag{8.8}$$

with $g_a(x) = (a'x)^2$. It is easy to see that g_a satisfies

$$K(a, c) \equiv \{x \in R^k \mid g_a(x) \leq c\} \in \mathcal{K}$$

for any $c \in [0, \infty)$ and $a \in R^k$. Thus, Theorem 8.1 applies and the inequality (8.8) is proved.

Anderson's theorem. Next, we state Anderson's theorem (Anderson, 1955), which serves as the main tool for showing (8.4). To do so, two notions on the shape of a probability density function (pdf) $f(x)$ on R^n are introduced: symmetry about the origin and unimodality.

A function f is said to be *symmetric about the origin* if it is an even function in the sense that

$$f(-x) = f(x) \quad \text{for any } x \in R^n.$$

Also, f is called *unimodal* if

$$\{x \in R^n \mid f(x) \geq c\} \quad \text{is convex for each } c > 0. \tag{8.9}$$

Theorem 8.3 (Anderson's theorem) *Let f be a pdf on R^n and suppose that it is symmetric about the origin and unimodal. Let K be a symmetric convex set. Then for each $\theta \in R^n$, the function*

$$\Psi(a) = \int_K f(x - a\theta) \, dx \tag{8.10}$$

defined on R^1 satisfies $\Psi(-a) = \Psi(a)$ and is nonincreasing on $[0, \infty)$.

Proof. Omitted. See Anderson (1955).

Anderson's theorem has a rich potentiality for deriving various results in statistical analysis, which includes not only concentration inequalities we aim at in this section but also the monotonicity of the power functions of invariant tests, the construction of conservative confidence regions, and so on. The theorem, its background and its extensions are fully investigated in Eaton (1982) from the viewpoint of group invariance theory. See also Problems 8.2.4 and 8.2.5.

A typical sufficient condition for f to be unimodal is given in the following proposition (but we shall not use it).

Proposition 8.4 *Suppose that a nonnegative function $f(x)$ on R^n is log-concave, that is,*

$$f(ax + (1 - a)y) \geq f(x)^a f(y)^{1-a} \tag{8.11}$$

holds for any $x, y \in R^n$ and $a \in [0, 1]$. Then f is unimodal.

Proof. For each $c > 0$, let

$$K(c) = \{x \in R^n \mid f(x) \geq c\}.$$

Then for any $x, y \in K(c)$ and any $a \in [0, 1]$, we have

$$f(ax + (1 - a)y) \geq f(x)^a f(y)^{1-a} \geq c^a c^{1-a} = c,$$

where the first inequality is due to the log-concavity of f and the second follows since $x, y \in K(c)$. Hence, $ax + (1 - a)y \in K(c)$, completing the proof.

The term "log-concave" is due to the fact that the condition (8.11) is equivalent to the concavity of $\log f(x)$ when $f(x)$ is a positive real-valued function.

Concentration inequality. Now to establish the main theorems, consider the model (8.1) and suppose $P \equiv \mathcal{L}(\varepsilon)$, the distribution of the error term ε of the model, is in the class $\mathbf{E}_n(0, \Sigma)$ of elliptically symmetric distributions. Suppose further that P has a pdf f_P with respect to the Lebesgue measure on R^n. Hence, for each $P \in \mathbf{E}_n(0, \Sigma)$, the function f_P can be written as

$$f_P(\varepsilon) = |\Sigma|^{-1/2} \tilde{f}_P(\varepsilon' \Sigma^{-1} \varepsilon) \tag{8.12}$$

for some $\tilde{f}_P : [0, \infty) \to [0, \infty)$.

Let f_P be unimodal. A sufficient condition on $\tilde{f}_P$ for which f_P is unimodal is given by the following proposition.

Proposition 8.5 *If the function $\tilde{f}_P$ in (8.12) is nonincreasing, then f_P is unimodal, and the converse is true.*

Proof. The first statement ("if" part) will be proved in the next proposition under a more general setup. Hence, we show the converse.

Suppose that $f_p(\varepsilon)$ is unimodal. Let $0 \le u_1 < u_2$, and choose any $\varepsilon_2 \in R^n$ such that

$$u_2 = \varepsilon_2' \Sigma^{-1} \varepsilon_2.$$

Let $\varepsilon_1 = \sqrt{\frac{u_1}{u_2}} \varepsilon_2$. Then ε_1 satisfies $u_1 = \varepsilon_1' \Sigma^{-1} \varepsilon_1$. Let

$$k_2 \equiv f_P(\varepsilon_2) = |\Sigma|^{-1/2} \tilde{f}_P(u_2).$$

Then, of course, the two vectors 0 and ε_2 are in the set K_2, where

$$K_2 = \{x \in R^k \mid f_P(x) \ge k_2\}.$$

Since K_2 is convex,

$$\varepsilon_1 = \left(1 - \sqrt{\frac{u_1}{u_2}}\right) 0 + \sqrt{\frac{u_1}{u_2}} \varepsilon_2 \in K_2.$$

This means that $f_P(\varepsilon_1) \ge f_P(\varepsilon_2)$, which in turn implies that

$$\tilde{f}_P(u_1) \ge \tilde{f}_P(u_2).$$

This completes the proof.

Proposition 8.6 *Let $Q(x)$ be a real-valued function on R^n. Suppose that $Q(x)$ is convex and symmetric about the origin.*

(1) *If a nonnegative function f on R^1 is nonincreasing, then $f(Q(x))$ is unimodal.*

(2) *For any nonnegative and nondecreasing function L, the set*

$$\{x \in R^n \mid L(Q(x)) \le c\}$$

is convex for any $c > 0$.

Proof. For each $c > 0$, let

$$K(c) = \{x \in R^n \mid f(Q(x)) \ge c\}$$

and let

$$q(c) = \sup\{q \in R^1 \mid f(q) \ge c\}.$$

Then $x \in K(c)$ is equivalent to

$$Q(x) \le q(c).$$

Therefore, for $x, y \in K(c)$ and $a \in [0, 1]$,

$$Q(ax + (1-a)y) \le aQ(x) + (1-a)Q(y) \le q(c)$$

holds and hence,

$$f(Q(ax + (1-a)y)) \ge c,$$

which implies that $ax + (1-a)y \in K(c)$, proving the unimodality of $f(Q(x))$. The statement for L is quite similar and omitted. This completes the proof.

Note that letting $f(x) = |\Sigma|^{-1/2} \tilde{f}_P(x)$ and $Q(\varepsilon) = \varepsilon' \Sigma^{-1} \varepsilon$ in (1) of Proposition 8.6 yields the first part of Proposition 8.5.

Let $\mathbf{C}_2$ be the class of location-equivariant estimators of β. An estimator $\hat{\beta} = \hat{\beta}(y)$ is called a *location-equivariant estimator* if it satisfies

$$\hat{\beta}(y + Xg) = \hat{\beta}(y) + g \quad \text{for any } g \in R^k.$$

See Section 2.3 of Chapter 2. By using Proposition 2.4, the class $\mathbf{C}_2$ is characterized as

$$\mathbf{C}_2 = \{\hat{\beta}(y) = b(I_n) + d(e) \mid d \text{ is a } k \times 1 \text{ vector-valued measurable}$$

$$\text{function on } R^n\}, \tag{8.13}$$

where $b(I_n)$ is the ordinary least squares estimator (OLSE) of β:

$$b(I_n) = (X'X)^{-1} X' y,$$

and e is the ordinary least squares (OLS) residual vector defined by

$$e = Ny \quad \text{with} \quad N = I_n - X(X'X)^{-1}X'. \tag{8.14}$$

We use the notation $\mathbf{C}_2$ in order to make it clear that the class contains estimators without moment. Recall that the class includes as its subclass the class $\mathbf{C}_1$ of GLSEs, where

$$\mathbf{C}_1 = \{\hat{\beta} = C(e)y \mid C(e) \text{ is a } k \times n \text{ matrix-valued measurable}$$

$$\text{function on } R^n \text{ such that } C(e)X = I_k \}. \tag{8.15}$$

The class $\mathbf{C}_1$ also contains GLSEs without moments. Note that since the distribution P has a pdf with respect to the Lebesgue measure, the two classes $\mathbf{C}_1$ and $\mathbf{C}_2$ are essentially the same, that is,

$$\mathbf{C}_1 = \mathbf{C}_2 \quad \text{a.s.}$$

See Proposition 2.5.

The following expression of $\hat{\beta}(y) \in \mathbf{C}_2$ is more convenient for our purpose:

$$\hat{\beta}(y) = b(\Sigma) + h(e), \tag{8.16}$$

where h is a $k \times 1$ vector-valued measurable function. This can be proved easily.

Let Z be any $n \times (n - k)$ matrix such that

$$X'Z = 0, \ Z'Z = I_{n-k} \text{ and } ZZ' = N,$$

and let

$$\overline{X} = \Sigma^{-1/2} X A^{-1/2} : n \times k \quad \text{with} \quad A = X'\Sigma^{-1}X \in \mathcal{S}(k),$$

$$\overline{Z} = \Sigma^{1/2} Z B^{-1/2} : n \times (n - k) \quad \text{with} \quad B = Z'\Sigma Z \in \mathcal{S}(n - k).$$

Then the matrix

$$\overline{\Gamma} = \begin{pmatrix} \overline{X}' \\ \overline{Z}' \end{pmatrix} \tag{8.17}$$

is an $n \times n$ orthogonal matrix. Let

$$\eta = \Sigma^{-1/2} \varepsilon$$

so that $\mathcal{L}(\eta) \in \mathbf{E}_n(0, I_n)$. From (8.12), the pdf of η is given by

$$|\Sigma|^{1/2} f_P(\Sigma^{1/2}\eta) = \tilde{f}_P(\eta'\eta).$$

Hence, the random vector η thus defined satisfies $\mathcal{L}(\Gamma\eta) = \mathcal{L}(\eta)$ for any $\Gamma \in \mathcal{O}(n)$. By choosing $\Gamma = \overline{\Gamma}$ in (8.17) and letting

$$\tilde{\eta} \equiv \overline{\Gamma}\eta = \begin{pmatrix} \overline{X}'\eta \\ \overline{Z}'\eta \end{pmatrix} \equiv \begin{pmatrix} \tilde{\eta}_1 \\ \tilde{\eta}_2 \end{pmatrix}, \tag{8.18}$$

we have

$$\hat{\beta} - \beta = A^{-1/2}\tilde{\eta}_1 + h(ZB^{1/2}\tilde{\eta}_2)$$
$$= A^{-1/2}\tilde{\eta}_1 + \tilde{h}(\tilde{\eta}_2) \quad \text{(say)}, \tag{8.19}$$

since $e = ZB^{1/2}\tilde{\eta}_2$. Here note that

$$\mathcal{L}(\eta) = \mathcal{L}(\tilde{\eta}) \in \mathbf{E}_n(0, I_n), \tag{8.20}$$

and hence the pdf of $\tilde{\eta} = (\tilde{\eta}_1', \tilde{\eta}_2')'$ is expressed as

$$\tilde{f}_P(\tilde{\eta}'\tilde{\eta}) = \tilde{f}_P(\tilde{\eta}_1'\tilde{\eta}_1 + \tilde{\eta}_2'\tilde{\eta}_2). \tag{8.21}$$

We use the following fact on a symmetric convex set.

Lemma 8.7 *If K is a convex set in R^n, then for any nonsingular matrix G, the set*

$$GK = \{Gx \mid x \in K\}$$

is also convex. If in addition K is symmetric, so is GK.

Proof. Since any $x^*, y^* \in GK$ can be written as $x^* = Gx$ and $y^* = Gy$ for some $x, y \in K$, it holds that

$$ax^* + (1 - a)y^* = G[ax + (1 - a)y] \quad \text{for any } a \in [0, 1].$$

Thus, $ax^* + (1 - a)y^* \in GK$ since the convexity of K implies $ax + (1 - a)y \in K$. The rest is clear. This completes the proof.

Theorem 8.8 *Let $\mathcal{K}$ be the set of all symmetric convex sets in R^k. Suppose that $P = \mathcal{L}(\varepsilon) \in \mathbf{E}_n(0, \Sigma)$ and P has a pdf f_P with respect to the Lebesgue measure on R^n. If f_P is symmetric about the origin and unimodal, then the GME $b(\Sigma)$ is most concentrated to β with respect to $\mathcal{K}$ in $\mathbf{C}_2$, that is,*

$$P(b(\Sigma) - \beta \in K) \geq P(\hat{\beta} - \beta \in K) \tag{8.22}$$

for any $\beta \in R^k$, $K \in \mathcal{K}$ and $\hat{\beta} \in \mathbf{C}_2$.

Proof. For any $\hat{\beta} \in \mathbf{C}_2$, the probability in question is written as

$$
\begin{aligned}
P(\hat{\beta} - \beta \in K) &= P(A^{-1/2}\tilde{\eta}_1 + \tilde{h}(\tilde{\eta}_2) \in K) \\
&= E\{P(A^{-1/2}\tilde{\eta}_1 + \tilde{h}(\tilde{\eta}_2) \in K \mid \tilde{\eta}_2)\}.
\end{aligned}
$$

Thus, it is sufficient to show that for almost all $\tilde{\eta}_2$,

$$P(A^{-1/2}\tilde{\eta}_1 \in K \mid \tilde{\eta}_2) \geq P(A^{-1/2}\tilde{\eta}_1 + \tilde{h}(\tilde{\eta}_2) \in K \mid \tilde{\eta}_2).$$

This is equivalent to

$$P(\tilde{\eta}_1 \in A^{1/2}K \mid \tilde{\eta}_2) \geq P(\tilde{\eta}_1 + A^{1/2}\tilde{h}(\tilde{\eta}_2) \in A^{1/2}K \mid \tilde{\eta}_2), \tag{8.23}$$

where $A^{1/2}K$ is also convex (Lemma 8.7).

Next, we apply Anderson's theorem to the conditional pdf, say $g_P(\tilde{\eta}_1|\tilde{\eta}_2)$, of $\tilde{\eta}_1$ given $\tilde{\eta}_2$. To do so, we show that $g_P(\tilde{\eta}_1|\tilde{\eta}_2)$ is symmetric about the origin and unimodal. The conditional pdf g_P is given by

$$g_P(\tilde{\eta}_1|\tilde{\eta}_2) \equiv \tilde{f}_P(\tilde{\eta}_1'\tilde{\eta}_1 + \tilde{\eta}_2'\tilde{\eta}_2) \Big/ \int_{R^k} \tilde{f}_P(\tilde{\eta}_1'\tilde{\eta}_1 + \tilde{\eta}_2'\tilde{\eta}_2) \, d\tilde{\eta}_1.$$

This function is defined for $\tilde{\eta}_2 \in S$, where

$$S = \left\{ \tilde{\eta}_2 \in R^{n-k} \ \Big| \ \int_{R^k} \tilde{f}_P(\tilde{\eta}_1'\tilde{\eta}_1 + \tilde{\eta}_2'\tilde{\eta}_2) d\tilde{\eta}_1 > 0 \right\}.$$

Here, $P(S) = 1$. For each $\tilde{\eta}_2 \in S$, $g_P(\cdot|\tilde{\eta}_2)$ is clearly symmetric about the origin. Furthermore, the unimodality of $g_P(\cdot|\tilde{\eta}_2)$ follows from Propositions 8.5 and 8.6. In fact, for each $\tilde{\eta}_2 \in S$, let $Q(\tilde{\eta}_1) = \tilde{\eta}_1'\tilde{\eta}_1 + \tilde{\eta}_2'\tilde{\eta}_2$. Then Q is convex and symmetric about the origin. Furthermore, by Proposition 8.5, the function $\tilde{f}_P$ is nonincreasing. Hence, from Proposition 8.6, it is shown that the function $\tilde{f}_P(Q(\tilde{\eta}_1))$ is unimodal, from which the unimodality of $g_P(\cdot|\tilde{\eta}_2)$ follows.

Finally, for each $\tilde{\eta}_2 \in S$,

$$
\begin{aligned}
&P(\tilde{\eta}_1 + A^{1/2}\tilde{h}(\tilde{\eta}_2) \in A^{1/2}K \mid \tilde{\eta}_2) \\
&= \int_{R^k} \chi_{\{\tilde{\eta}_1 + A^{1/2}\tilde{h}(\tilde{\eta}_2) \in A^{1/2}K\}} \, g_P(\tilde{\eta}_1|\tilde{\eta}_2) \, d\tilde{\eta}_1
\end{aligned}
$$

$$= \int_{R^k} \chi_{\{\tilde{\eta}_1 \in A^{1/2}K\}} \, g_P(\tilde{\eta}_1 - A^{1/2}\tilde{h}(\tilde{\eta}_2)|\tilde{\eta}_2) \, d\tilde{\eta}_1$$

$$= \int_{A^{1/2}K} g_P(\tilde{\eta}_1 - A^{1/2}\tilde{h}(\tilde{\eta}_2)|\tilde{\eta}_2) \, d\tilde{\eta}_1$$

$$\leq \int_{A^{1/2}K} g_P(\tilde{\eta}_1|\tilde{\eta}_2) \, d\tilde{\eta}_1$$

$$= P(\tilde{\eta}_1 \in A^{1/2}K \mid \tilde{\eta}_2),$$

where χ denotes the indicator function and the inequality in the fifth line is due to Anderson's theorem. This completes the proof.

Theorem 8.9 *Under the assumption of Theorem 8.8, the following inequality*

$$E_P\{L[h(b(\Sigma) - \beta)]\} \leq E_P\{L[h(\hat{\beta} - \beta)]\} \tag{8.24}$$

holds for any $\beta \in R^k$ and $\hat{\beta} \in C_2$, where L is any nonnegative and nondecreasing function, and h is any convex function that is symmetric about the origin.

Proof. By the equivalence theorem, it suffices to show that

$$g(x) \equiv L(h(x))$$

is nonnegative and satisfies the condition (8.6). Since L is nonnegative, so is g. The condition (8.6) readily follows from the latter part of Proposition 8.6. This completes the proof.

Finally, it is noted that the inequality (8.22) remains true even if the set K is replaced by the random set that depends on y only through the OLS residual vector e, since e is a function of $\tilde{\eta}_2$. This yields a further extension of Theorem 8.8. See Eaton (1988).

8.3 Efficiency of GLSEs under Elliptical Symmetry

In this section, some results of Chapter 4 established under normality are extended to the case in which the distribution of the error term is elliptically symmetric.

The SUR model. To state the problem, let a general linear regression model be

$$y = X\beta + \varepsilon \tag{8.25}$$

with

$$E(\varepsilon) = 0 \text{ and } \text{Cov}(\varepsilon) = \Omega,$$

where

$$y : n \times 1, \ X : n \times k, \ rankX = k \text{ and } \varepsilon : n \times 1.$$

The SUR model considered here is the model (8.25) with the following structure:

$$y = \begin{pmatrix} y_1 \\ \vdots \\ y_p \end{pmatrix} : n \times 1, \quad X = \begin{pmatrix} X_1 & & 0 \\ & \ddots & \\ 0 & & X_p \end{pmatrix} : n \times k,$$

$$\beta = \begin{pmatrix} \beta_1 \\ \vdots \\ \beta_p \end{pmatrix} : k \times 1, \quad \varepsilon = \begin{pmatrix} \varepsilon_1 \\ \vdots \\ \varepsilon_p \end{pmatrix} : n \times 1, \tag{8.26}$$

$$\Omega = \Sigma \otimes I_m \quad \text{and} \quad \Sigma = (\sigma_{ij}) \in \mathcal{S}(p), \tag{8.27}$$

where $\mathcal{S}(p)$ denotes the set of $p \times p$ positive definite matrices,

$$y_j : m \times 1, \quad X_j : m \times k_j, \quad rank X_j = k_j,$$

$$n = pm, \quad k = \sum_{j=1}^{p} k_j$$

and $\otimes$ denotes the Kronecker product. Suppose that the distribution of the error term ε is elliptically symmetric with covariance matrix $\Sigma \otimes I_m$:

$$\mathcal{L}(\varepsilon) \in \mathcal{E}_n(0, \Sigma \otimes I_m), \tag{8.28}$$

which includes the normal distribution $N_n(0, \Sigma \otimes I_m)$ as its special element.

To define a GLSE, let $\hat{\Sigma} : \mathcal{S}(p) \to \mathcal{S}(p)$ be any measurable function and let

$$\hat{\Sigma} \equiv \hat{\Sigma}(S) \tag{8.29}$$

be an estimator of the matrix Σ, which depends on the observation vector y only through the random matrix

$$S = Y' N_* Y = E' N_* E : p \times p \tag{8.30}$$

with

$$N_* = I_m - X_*(X_*' X_*)^+ X_*',$$

where

$$Y = (y_1, \ldots, y_p) : m \times p, \quad X_* = (X_1, \ldots, X_p) : m \times k,$$

$$E = (\varepsilon_1, \ldots, \varepsilon_p) : m \times p$$

and A^+ denotes the Moore–Penrose inverse of A. The matrix S can be rewritten as a function of the OLS residual vector e:

$$S = (e_i' N_* e_j) \equiv S_0(e) \quad \text{(say).} \tag{8.31}$$

Here, the OLS residual vector e is given by

$$e = Ny \quad \text{with} \quad N = I_n - X(X'X)^{-1}X',$$

from which it follows that

$$e = \begin{pmatrix} e_1 \\ \vdots \\ e_p \end{pmatrix} \quad \text{with} \quad e_j = N_j y_j : m \times 1,$$

where $N_j = I_m - X_j(X_j'X_j)^{-1}X_j'$. (See Example 2.7).

Let C^* be a class of GLSEs of the form

$$b(\hat{\Sigma} \otimes I_m) = (X'(\hat{\Sigma}^{-1} \otimes I_m)X)^{-1}X'(\hat{\Sigma}^{-1} \otimes I_m)y \tag{8.32}$$

with $\hat{\Sigma} = \hat{\Sigma}(S)$. The class C^* is actually a class of GLSEs, since the estimator $\hat{\Sigma}$ is a function of the OLS residual vector e:

$$\hat{\Sigma} = \hat{\Sigma}(S)$$
$$= \hat{\Sigma}(S_0(e))$$
$$\equiv \hat{\Sigma}_0(e) \text{ (say)}. \tag{8.33}$$

Obviously, the class C^* contains the UZE $b(S \otimes I_m)$ that is obtained by letting $\hat{\Sigma}(S) = S$. The UZE is an unbiased estimator of β under the condition (8.28). More generally,

Proposition 8.10 *Any GLSE $b(\hat{\Sigma} \otimes I_m)$ in class C^* is an unbiased estimator of β as long as the first moment is finite.*

Proof. Since the matrix $S = S_0(e)$ is an even function of e in the sense that $S_0(-e) = S_0(e)$, so is $\hat{\Sigma} = \hat{\Sigma}_0(e)$. Hence, by Proposition 2.6, the unbiasedness follows. This completes the proof.

In Chapter 4, it was observed that if $\mathcal{L}(\varepsilon) = N_n(0, \Sigma \otimes I_m)$, then the matrix S is distributed as the Wishart distribution $W_p(\Sigma, q)$ with mean $q\Sigma$ and degrees of freedom q, where

$$r = rankX_* \quad \text{and} \quad q = m - r. \tag{8.34}$$

Furthermore, any GLSE $b(\hat{\Sigma} \otimes I_m)$ in C^* satisfies

$$\mathcal{L}(b(\hat{\Sigma} \otimes I_m)|\hat{\Sigma}) = N_k(\beta, \ H(\hat{\Sigma} \otimes I_m, \Sigma \otimes I_m)), \tag{8.35}$$

that is, the conditional distribution of $b(\hat{\Sigma} \otimes I_m)$ given $\hat{\Sigma}$ is normal with conditional mean

$$E[b(\hat{\Sigma} \otimes I_m)|\hat{\Sigma}] = \beta \tag{8.36}$$

and conditional covariance matrix

$$\mathrm{Cov}(b(\hat{\Sigma} \otimes I_m)|\hat{\Sigma}) = H(\hat{\Sigma} \otimes I_m, \Sigma \otimes I_m), \tag{8.37}$$

where the function H is defined by

$$H(\hat{\Sigma} \otimes I_m, \Sigma \otimes I_m)$$
$$= (X'(\hat{\Sigma}^{-1} \otimes I_m)X)^{-1}X'(\hat{\Sigma}^{-1}\Sigma\hat{\Sigma}^{-1} \otimes I_m)X(X'(\hat{\Sigma}^{-1} \otimes I_m)X)^{-1} \tag{8.38}$$

and is called a *simple covariance structure*.

The distributional properties (8.35), (8.36) and (8.37) strongly depend on the normality of ε. In this section, the problem of how the assumption of elliptical symmetry influences the conditional mean and covariance matrix is considered. For this purpose, we impose the following additional conditions on the estimator: $\hat{\Sigma} = \hat{\Sigma}(S)$ of Σ:

(1) $\hat{\Sigma} = \hat{\Sigma}(S)$ is a one-to-one continuous function of S;

(2) There exists a function $\gamma : (0, \infty) \to (0, \infty)$ such that

$$\hat{\Sigma}(aS) = \gamma(a)\hat{\Sigma}(S) \quad \text{for any } a > 0. \tag{8.39}$$

The condition in (2) will be used to show the finiteness of the second moment of $b(\hat{\Sigma} \otimes I_m)$. Define a subclass C^{**} of C^* by

$$C^{**} = \{b(\hat{\Sigma} \otimes I_m) \in C^* \mid \hat{\Sigma} \text{ satisfies (1) and (2)}\}. \tag{8.40}$$

The class C^{**} thus defined contains the GLSEs with such $\hat{\Sigma}(S)$'s as

$$\hat{\Sigma}(S) = TDT', \tag{8.41}$$

where T is the lower-triangular matrix with positive diagonal elements such that $S = TT'$ (see Lemma 1.8, Cholesky decomposition) and D is a diagonal matrix with positive elements. In this case, the function γ in condition (2) is given by

$$\gamma(a) = a.$$

Such GLSEs have already appeared in Section 4.4 of Chapter 4: the UZE $b(S \otimes I_m)$ is obtained by letting $D = I_p$; when $p = 2$, the optimal GLSE $b(\hat{\Sigma}_B \otimes I_m)$, which is derived in Theorem 4.14, is also an element of C^{**}, where

$$\hat{\Sigma} = TD_BT' \quad \text{and} \quad D_B = \begin{pmatrix} 1 & 0 \\ 0 & \sqrt{\dfrac{q+1}{q-3}} \end{pmatrix}. \tag{8.42}$$

Proposition 8.11 *Any GLSE $b(\hat{\Sigma} \otimes I_m)$ in C^{**} has a finite second moment.*

Proof. By Proposition 2.6, it is sufficient to see that the function

$$C(e) = (X'(\hat{\Sigma}^{-1} \otimes I_m)X)^{-1}X'(\hat{\Sigma}^{-1} \otimes I_m) \quad \text{with} \quad \hat{\Sigma} = \hat{\Sigma}_0(e)$$

is scale-invariant in the sense that

$$C(ae) = C(e) \quad \text{for any } a > 0. \tag{8.43}$$

To see this, note that

$$S_0(ae) = a^2 S_0(e) \quad \text{for any } a > 0. \tag{8.44}$$

This implies that

$$\hat{\Sigma}_0(ae) = \hat{\Sigma}(a^2 S) = \gamma(a^2)\hat{\Sigma}(S) = \gamma(a^2)\hat{\Sigma}_0(e),$$

where the second equality is due to condition (2). For $\Omega \in \mathcal{S}(n)$, let

$$B(\Omega) = (X'\Omega^{-1}X)^{-1}X'\Omega^{-1}.$$

Then the function B is a scale-invariant function of Ω in the sense that $B(\gamma\Omega) = B(\Omega)$ for any $\gamma > 0$. Thus, we obtain

$$C(ae) = B(\hat{\Sigma}_0(ae) \otimes I_m) = B(\gamma(a^2)\hat{\Sigma}_0(e) \otimes I_m) = B(\hat{\Sigma}_0(e) \otimes I_m) = C(e).$$

This completes the proof.

Conditional covariance structure. Let $\tilde{X}$ and $\tilde{Z}$ be any $m \times r$ and $m \times q$ matrices such that

$$\tilde{X}\tilde{X}' = X_*(X_*'X_*)^+X_*', \quad \tilde{X}'\tilde{X} = I_r \tag{8.45}$$

and

$$\tilde{Z}\tilde{Z}' = I_m - X_*(X_*'X_*)^+X_*', \quad \tilde{Z}'\tilde{Z} = I_q \tag{8.46}$$

Then

$$\Gamma \equiv (\tilde{X}, \tilde{Z}) \in \mathcal{O}(m),$$

where $\mathcal{O}(m)$ denotes the group of $m \times m$ orthogonal matrices. Let

$$\tilde{\varepsilon} = (\Sigma^{-1/2} \otimes I_m)\varepsilon$$

$$= (\Sigma^{-1/2} \otimes I_m)\begin{pmatrix} \varepsilon_1 \\ \vdots \\ \varepsilon_p \end{pmatrix}$$

$$\equiv \begin{pmatrix} \tilde{\varepsilon}_1 \\ \vdots \\ \tilde{\varepsilon}_p \end{pmatrix} : n \times 1 \quad \text{with} \quad \tilde{\varepsilon}_j : m \times 1. \tag{8.47}$$

Then

$$\mathcal{L}(\tilde{\varepsilon}) \in \mathcal{E}_n(0, I_p \otimes I_m) \tag{8.48}$$

and hence $\mathcal{L}(\Psi\tilde{\varepsilon}) = \mathcal{L}(\tilde{\varepsilon})$ holds for any $\Psi \in \mathcal{O}(n)$. By choosing $\Psi = I_p \otimes \Gamma'$, we define

$$\eta \equiv (I_p \otimes \Gamma')\tilde{\varepsilon}$$

$$= \begin{pmatrix} \Gamma' & & 0 \\ & \ddots & \\ 0 & & \Gamma' \end{pmatrix} \begin{pmatrix} \tilde{\varepsilon}_1 \\ \vdots \\ \tilde{\varepsilon}_p \end{pmatrix}$$

$$\equiv \begin{pmatrix} \eta_1 \\ \vdots \\ \eta_p \end{pmatrix} : n \times 1 \tag{8.49}$$

with

$$\eta_j = \Gamma'\tilde{\varepsilon}_j \begin{pmatrix} \tilde{X}'\tilde{\varepsilon}_j \\ \tilde{Z}'\tilde{\varepsilon}_j \end{pmatrix} = \begin{pmatrix} \delta_j \\ \xi_j \end{pmatrix}, \tag{8.50}$$

where $\delta_j : p \times 1$ and $\xi_j : q \times 1$. Since any permutation matrix is in $\mathcal{O}(n)$, we can easily see that

$$\mathcal{L}(\tilde{\varepsilon}) = \mathcal{L}(\eta) = \mathcal{L}\left(\begin{pmatrix} \delta \\ \xi \end{pmatrix} \right) \in \mathcal{E}_n(0, I_n) \tag{8.51}$$

holds, where

$$\delta = \begin{pmatrix} \delta_1 \\ \vdots \\ \delta_p \end{pmatrix} : pr \times 1 \text{ and } \xi = \begin{pmatrix} \xi_1 \\ \vdots \\ \xi_p \end{pmatrix} : pq \times 1. \tag{8.52}$$

By Proposition 1.19, the conditional distribution of δ given ξ, and the marginal distributions of δ and ξ are also elliptically symmetric:

$$\mathcal{L}(\delta|\xi) \in \tilde{\mathcal{E}}_{pr}(0, I_{pr}) = \bigcup_{\gamma > 0} \mathcal{E}_{pr}(0, \gamma I_{pr});$$

$$\mathcal{L}(\delta) \in \mathcal{E}_{pr}(0, I_{pr});$$

and

$$\mathcal{L}(\xi) \in \mathcal{E}_{pq}(0, I_{pq}).$$

Hence, from this,

$$E(\delta|\xi) = 0 \qquad (8.53)$$

and

$$\text{Cov}(\delta|\xi) = E(\delta\delta'|\xi) = \tilde{c}(\|\xi\|^2) \, I_{pr} \quad \text{for some function } \tilde{c}, \qquad (8.54)$$

where $\tilde{c}$ satisfies

$$E\left[\tilde{c}(\|\xi\|^2)\right] = 1, \qquad (8.55)$$

since $\text{Cov}(\delta) = I_{pr}$. Clearly, $\tilde{c}(\|\xi\|^2) = 1$ when ε is normally distributed (see Section 1.3 of Chapter 1).

It is important for later discussion to note that S is a function of ξ only. Say

$$S = S(\xi).$$

In fact,

Lemma 8.12 *The matrix S in (8.30) can be expressed as*

$$S = \Sigma^{1/2} U' U \Sigma^{1/2} \quad \text{with} \quad U = (\xi_1, \ldots, \xi_p) : q \times p. \qquad (8.56)$$

Proof. The proof is straightforward and omitted.

As a function of ξ, $S = S(\xi)$ satisfies

$$S(a\xi) = a^2 S(\xi) \quad \text{for any } a > 0. \qquad (8.57)$$

Theorem 8.13 *Suppose that $\mathcal{L}(\varepsilon) \in \mathcal{E}_n(0, \Sigma \otimes I_m)$.*

(1) *If $b(\hat{\Sigma} \otimes I_m) \in C^*$, then*

$$E[b(\hat{\Sigma} \otimes I_m)|\hat{\Sigma}] = \beta.$$

(2) *If $b(\hat{\Sigma} \otimes I_m) \in C^*$, then*

$$\text{Cov}(b(\hat{\Sigma} \otimes I_m)|\hat{\Sigma}) = c(\hat{\Sigma}) \, H(\hat{\Sigma} \otimes I_m, \Sigma \otimes I_m) \qquad (8.58)$$

for some function c such that $E[c(\hat{\Sigma})] = 1$.

(3) *If $b(\hat{\Sigma} \otimes I_m) \in C^{**}$, then $c(\hat{\Sigma})$ and $H(\hat{\Sigma} \otimes I_m, \Sigma \otimes I_m)$ are independent.*

Proof. By using $X'_j \tilde{Z} = 0$, the following equality is proved:

$$X'(\hat{\Sigma}^{-1} \otimes I_m)\varepsilon = X'(\hat{\Sigma}^{-1}\Sigma^{1/2} \otimes \tilde{X})\delta, \qquad (8.59)$$

from which it follows that

$$E[b(\hat{\Sigma} \otimes I_m)|\hat{\Sigma}] = (X'(\hat{\Sigma}^{-1} \otimes I_m)X)^{-1} X'(\hat{\Sigma}^{-1}\Sigma^{1/2} \otimes \tilde{X}) \, E(\delta|\hat{\Sigma}) + \beta.$$

Here, the first term of the right-hand side of this equality vanishes. In fact, since $\hat{\Sigma}$ is a function of ξ, it follows from (8.53) that

$$E(\delta|\hat{\Sigma}) = E[E(\delta|\xi)\,|\hat{\Sigma}] = 0.$$

This proves (1).

Similarly, by using (8.54),

$$E(\delta\delta'|\hat{\Sigma}) = E[E(\delta\delta'|\xi)\,|\hat{\Sigma}]$$
$$= E[\tilde{c}(\|\xi\|^2)|\hat{\Sigma}]\,I_{pr}$$

holds for some function $\tilde{c}$. Hence, letting

$$c(\hat{\Sigma}) = E[\tilde{c}(\|\xi\|^2)|\hat{\Sigma}] \tag{8.60}$$

yields

$$E(\delta\delta'|\hat{\Sigma}) = c(\hat{\Sigma})I_{pr}. \tag{8.61}$$

The function c satisfies $E\{c(\hat{\Sigma})\} = 1$, since $\mathrm{Cov}(\delta) = I_{pr}$. Therefore, from (8.59) and (8.61), we obtain

$$\mathrm{Cov}(\hat{\beta}(\hat{\Sigma}\otimes I_m)|\hat{\Sigma})$$
$$= (X'(\hat{\Sigma}^{-1}\otimes I_m)X)^{-1}X'(\hat{\Sigma}^{-1}\Sigma^{1/2}\otimes\tilde{X})$$
$$\times E(\delta\delta'|\hat{\Sigma})\,(\Sigma^{1/2}\hat{\Sigma}^{-1}\otimes\tilde{X}')X(X'(\hat{\Sigma}^{-1}\otimes I_m)X)^{-1}$$
$$= c(\hat{\Sigma})\,(X'(\hat{\Sigma}^{-1}\otimes I_m)X)^{-1}X'(\hat{\Sigma}^{-1}\Sigma\hat{\Sigma}^{-1}\otimes\tilde{X}\tilde{X}')X(X'(\hat{\Sigma}^{-1}\otimes I_m)X)^{-1}.$$

It is easily proved by direct calculation that

$$X'(\hat{\Sigma}^{-1}\Sigma\hat{\Sigma}^{-1}\otimes\tilde{X}\tilde{X}')X = X'(\hat{\Sigma}^{-1}\Sigma\hat{\Sigma}^{-1}\otimes I_m)X.$$

Thus, (2) is obtained.

To prove (3), we assume that $b(\hat{\Sigma}\otimes I_m)\in C^{**}$. As a function of ξ, let

$$\tilde{H}(\xi) \equiv H(\hat{\Sigma}\,(S(\xi))\otimes I_m,\ \Sigma\otimes I_m). \tag{8.62}$$

The function $\tilde{H}(\xi)$ depends on ξ only through $\xi/\|\xi\|$, since for any $a > 0$,

$$\tilde{H}(a\xi) = H(\hat{\Sigma}(S(a\xi))\otimes I_m,\ \Sigma\otimes I_m)$$
$$= H(\hat{\Sigma}(a^2 S(\xi))\otimes I_m,\ \Sigma\otimes I_m)$$
$$= H(\gamma(a^2)\hat{\Sigma}(S(\xi))\otimes I_m,\ \Sigma\otimes I_m)$$
$$= H(\hat{\Sigma}(S(\xi))\otimes I_m,\ \Sigma\otimes I_m)$$
$$= \tilde{H}(\xi),$$

where the second equality follows from (8.57), the third follows from the definition of C^{**} and the fourth follows because

$$H(a\hat{\Sigma} \otimes I_m, \Sigma \otimes I_m) = H(\hat{\Sigma} \otimes I_m, \Sigma \otimes I_m) \text{ holds for any } a > 0.$$

(See (8.38)). Hence, we can write

$$H(\hat{\Sigma} \otimes I_m, \Sigma \otimes I_m) = \tilde{H}(\xi/\|\xi\|).$$

On the other hand, the function c in (8.60) is a function of $\|\xi\|^2$ because

$$\begin{aligned}
c(\hat{\Sigma}) &= E[\tilde{c}(\|\xi\|^2)|\hat{\Sigma}] \\
&= E[\tilde{c}(tr(S\Sigma^{-1}))|\hat{\Sigma}] \\
&= E[\tilde{c}(tr(S\Sigma^{-1}))|S] \\
&= \tilde{c}(tr(S\Sigma^{-1})) \\
&= \tilde{c}(\|\xi\|^2),
\end{aligned}$$

where the third equality follows since $\hat{\Sigma} = \hat{\Sigma}(S)$ is a one-to-one function of S and we also used

$$\|\xi\|^2 = tr(S\Sigma^{-1}),$$

which follows from (8.56). Since the two quantities $\xi/\|\xi\|$ and $\|\xi\|$ are independent (see Proposition 1.13), statement (3) is proved. This completes the proof.

Covariance matrix of a GLSE. Theorem 8.13 implies that deviation from the normality does not affect the magnitude of the covariance matrix of a GLSE in C^{**}. In fact, by independence of the functions c and H, the covariance matrix is evaluated as

$$\begin{aligned}
\text{Cov}(b(\hat{\Sigma} \otimes I_m)) &= E[c(\hat{\Sigma})] \times E[H(\hat{\Sigma} \otimes I_m, \Sigma \otimes I_m)] \\
&= E[H(\hat{\Sigma} \otimes I_m, \Sigma \otimes I_m)] \\
&= E[\tilde{H}(\xi/\|\xi\|)],
\end{aligned}$$

where $E[c(\hat{\Sigma})] = 1$ is used in the second line. The quantity $\xi/\|\xi\|$ is distributed as the (unique) uniform distribution on the unit sphere

$$\mathcal{U}(pq) = \{u \in R^{pq} \mid \|u\| = 1\}$$

(see Proposition 1.13 and Corollary 1.14). Hence, we obtain the following.

Theorem 8.14 *For any GLSE* $b(\hat{\Sigma} \otimes I_m) \in C^{**}$, *the covariance matrix* $\text{Cov}(b(\hat{\Sigma} \otimes I_m))$ *remains the same as long as* $\mathcal{L}(\varepsilon) \in \mathcal{E}_n(0, \Sigma \otimes I_m)$, *that is, let* $P_0 = N_n(0, \Sigma \otimes I_m)$. *Then*

$$\text{Cov}_P(b(\hat{\Sigma} \otimes I_m)) = \text{Cov}_{P_0}(b(\hat{\Sigma} \otimes I_m)).$$

This result is not surprising, since this property is shared by all the linear unbiased estimators, and since the structure of the conditional covariance matrices of the GLSEs considered here are similar to those of linear unbiased estimators. Here, the covariance matrices of the GME

$$b(\Sigma \otimes I_m) = (X'(\Sigma^{-1} \otimes I_m)X)^{-1}X'(\Sigma^{-1} \otimes I_m)y$$

and the OLSE

$$b(I_p \otimes I_m) = (X'X)^{-1}X'y$$

are given by

$$\text{Cov}(b(\Sigma \otimes I_m)) = (X'(\Sigma^{-1} \otimes I_m)X)^{-1} \tag{8.63}$$

and

$$\text{Cov}(b(I_p \otimes I_m)) = (X'X)^{-1}X'(\Sigma \otimes I_m)X(X'X)^{-1} \tag{8.64}$$

respectively, as long as

$$E(\varepsilon) = 0 \quad \text{and} \quad \text{Cov}(\varepsilon) = \Sigma \otimes I_m.$$

Therefore, as for the relative efficiency in terms of the covariance matrices, the results derived under normality are still valid as long as $\mathcal{L}(\varepsilon) \in \mathcal{E}_n(0, \Sigma \otimes I_m)$. Hence, for example, when $p = 2$, the following inequality for the covariance matrix of the UZE $b(S \otimes I_m)$ in C^{**} remains true:

$$\text{Cov}(b(\Sigma \otimes I_m)) \leq \text{Cov}(b(S \otimes I_m))$$

$$\leq \left[1 + \frac{2}{q-3}\right] \text{Cov}(b(\Sigma \otimes I_m)). \tag{8.65}$$

More generally, we obtain the following proposition: let

$$\alpha_0(b(\hat{\Sigma} \otimes I_m)) = E[L_0(\hat{\Sigma}, \Sigma)], \tag{8.66}$$

$$L_0(\hat{\Sigma}, \Sigma) = \frac{(\pi_1 + \pi_p)^2}{4\pi_1\pi_p},$$

where $\pi_1 \leq \cdots \leq \pi_p$ are the latent roots of $\Sigma^{-1/2}\hat{\Sigma}(S)\Sigma^{-1/2}$.

Proposition 8.15 *For a GLSE $b(\hat{\Sigma} \otimes I_m)$ in C^{**}, $\alpha_0(b(\hat{\Sigma} \otimes I_m))$ remains the same as long as $\mathcal{L}(\varepsilon) \in \mathcal{E}_n(0, \Sigma \otimes I_m)$, that is, let $P_0 = N_n(0, \Sigma \otimes I_m)$. Then*

$$E_P[L_0(\hat{\Sigma}, \Sigma)] = E_{P_0}[L_0(\hat{\Sigma}, \Sigma)] \quad \text{for any } P \in \mathcal{E}_n(0, \Sigma \otimes I_m). \tag{8.67}$$

Proof. It is sufficient to see that L_0 in (8.66) depends on ξ only through $\xi/\|\xi\|$. Let

$$\tilde{L}_0(\xi) = L_0(\hat{\Sigma}(S(\xi)), \Sigma).$$

For any $a > 0$,

$$\begin{aligned}
\tilde{L}_0(a\xi) &= L_0(\hat{\Sigma}(S(a\xi)), \Sigma) \\
&= L_0(\hat{\Sigma}(a^2 S(\xi)), \Sigma) \\
&= L_0(\gamma(a^2)\hat{\Sigma}(S(\xi)), \Sigma) \\
&= L_0(\hat{\Sigma}(S(\xi)), \Sigma) \\
&= \tilde{L}_0(\xi),
\end{aligned}$$

where the last equality is due to

$$L_0(\gamma\hat{\Sigma}, \Sigma) = L_0(\hat{\Sigma}, \Sigma) \quad \text{for any } \gamma > 0,$$

see (4.89). This completes the proof.

This yields the following extension of Theorem 4.14.

Corollary 8.16 *Let $p = 2$. The GLSE $b(\hat{\Sigma}_B \otimes I_m)$ given in (8.42) minimizes the upper bound $\alpha_0(b(\hat{\Sigma} \otimes I_m))$ among the GLSEs of the form (8.41).*

8.4   Degeneracy of the Distributions of GLSEs

In this section, it is proved that the distribution of the difference between a GLSE and the GME is degenerate when the covariance matrix of the error term has a certain kind of simple structure.

The model.   Let us consider the following general linear regression model

$$y = X\beta + \varepsilon \tag{8.68}$$

with

$$P \equiv \mathcal{L}(\varepsilon) \in \mathcal{P}_n(0, \sigma^2\Sigma) \quad \text{and} \quad \sigma^2\Sigma \in \mathcal{S}(n),$$

where

$$y : n \times 1, \quad X : n \times k, \quad \text{rank} X = k,$$

$\mathcal{P}_n(0, \sigma^2\Sigma)$ denotes the set of distributions on R^n with mean 0 and covariance matrix $\sigma^2\Sigma$, and $\mathcal{S}(n)$ the set of $n \times n$ positive definite matrices.

Suppose for a moment that the matrix Σ is of the structure

$$\Sigma^{-1} = \Sigma(\theta)^{-1} = I_n + \theta C \quad \text{with} \quad \theta \in \Theta, \tag{8.69}$$

where Θ is a subset of R^1 on which $\Sigma(\theta)$ is positive definite, and C is an $n \times n$ symmetric matrix satisfying

$$C1_n = 0 \quad \text{with} \quad 1_n = (1, \ldots, 1)' : n \times 1. \tag{8.70}$$

Note that the set Θ contains 0, since

$$\Sigma(0) = I_n \in S(n).$$

The model includes some serial correlation models such as the Anderson model and the model with circularly distributed error.

The Anderson model, which was defined in Example 2.1 and has been repeatedly treated in the previous chapters, is the model (8.68) with covariance structure (8.69), where $\Theta = (-1/4, \infty)$ and the matrix C is given by

$$C = \begin{pmatrix} 1 & -1 & & & & 0 \\ -1 & 2 & -1 & & & \\ & \ddots & \ddots & \ddots & & \\ & & \ddots & \ddots & \ddots & \\ & & & \ddots & 2 & -1 \\ 0 & & & & -1 & 1 \end{pmatrix} \equiv C_A \text{ (say).} \tag{8.71}$$

The model is an approximation of the AR(1) error model whose error term satisfies

$$\varepsilon_j = \theta_* \varepsilon_{j-1} + \xi_j \quad \text{with} \quad |\theta_*| < 1, \tag{8.72}$$

where $E(\xi_j) = 0$, $\text{Var}(\xi_j) = \sigma_*^2$ and $\text{Cov}(\xi_i, \xi_j) = 0$ $(i \neq j)$. (Here, the two parameters σ^2 and θ in (8.68) and (8.69) are written in terms of σ_*^2 and θ_* in (8.72) as $\sigma^2 = \sigma_*^2/(1 - \theta_*)^2$ and $\theta = \theta_*/(1 - \theta_*)^2$, respectively.)

On the other hand, the model with circularly distributed errors is defined by imposing the additional assumption $\varepsilon_n = \varepsilon_0$ to the equation (8.72), that is,

$$\varepsilon_j = \theta_* \varepsilon_{j-1} + \xi_j \quad \text{with} \quad |\theta_*| < 1 \text{ and } \varepsilon_0 = \varepsilon_n. \tag{8.73}$$

This model is also an example of the model (8.68) that satisfies (8.69) and (8.70). In fact, the quantities σ^2, θ, Θ and C are given respectively by

$$\sigma^2 = \frac{\sigma^2}{(1 - \theta_*)^2}, \quad \theta = \frac{\theta_*}{(1 - \theta_*)^2},$$
$$\Theta = (-1/4, \infty)$$

and

$$
C = \begin{pmatrix}
2 & -1 & & & & & -1 \\
-1 & 2 & -1 & & & & \\
& \ddots & \ddots & \ddots & & & \\
& & & \ddots & \ddots & \ddots & \\
& & & & \ddots & 2 & -1 \\
-1 & & & & & -1 & 2
\end{pmatrix} \equiv C_C \text{ (say).} \qquad (8.74)
$$

Let a GLSE of β in the model (8.68) be

$$
b(\hat{\Sigma}) = (X'\hat{\Sigma}^{-1}X)^{-1}X'\hat{\Sigma}^{-1}y \quad \text{with} \quad \hat{\Sigma} = \Sigma(\hat{\theta}), \qquad (8.75)
$$

where $\hat{\theta} = \hat{\theta}(y)$ is an estimator of θ such that

$$
\hat{\theta}(y) \in \Theta \quad \text{a.s.}
$$

In this section, no assumption is imposed on the functional form of the estimator $\hat{\theta}$. Hence, $\hat{\theta}$ is not necessarily a function of the OLS residual vector

$$
e = \{I_n - X(X'X)^{-1}X'\}y.
$$

Clearly, the GLSE with $\hat{\theta}(y) \equiv \theta$ is the GME

$$
b(\Sigma) = b(\Sigma(\theta)) = (X'\Sigma^{-1}X)^{-1}X'\Sigma^{-1}y
$$

and the GLSE with $\hat{\theta}(y) \equiv 0$ is the OLSE

$$
b(I_n) = b(\Sigma(0)) = (X'X)^{-1}X'y.
$$

The GME is not feasible in our setup.

Simple linear regression model. We begin with the following interesting result established by Usami and Toyooka (1997a).

Theorem 8.17 *Let $k = 2$ and*

$$
X = (x_1, x_2) : n \times 2 \quad \text{with} \quad x_1 = 1_n.
$$

Suppose that $\mathcal{L}(\varepsilon) \in \mathcal{P}_n(0, \sigma^2\Sigma)$, and that Σ satisfies (8.69) with

$$
C = C_A \text{ or } C_C.
$$

Then for any GLSE $b(\hat{\Sigma})$ of the form (8.75), the distribution of the quantity $b(\hat{\Sigma}) - b(\Sigma)$ degenerates into the one-dimensional linear subspace of R^2. More specifically, any GLSE $b(\hat{\Sigma})$ satisfies

$$
b_1(\hat{\Sigma}) + \bar{x}_2 b_2(\hat{\Sigma}) = \bar{y}, \qquad (8.76)
$$

where

$$b(\hat{\Sigma}) = \begin{pmatrix} b_1(\hat{\Sigma}) \\ b_2(\hat{\Sigma}) \end{pmatrix} : 2 \times 1, \quad \bar{x}_2 = \frac{1'_n x_2}{n} \text{ and } \bar{y} = \frac{1'_n y}{n}.$$

And therefore

$$b_1(\hat{\Sigma}) - b_1(\Sigma) = -\bar{x}_2[b_2(\hat{\Sigma}) - b_2(\Sigma)]. \tag{8.77}$$

Note that (8.77) follows since the right-hand side of (8.76) does not depend on $\hat{\theta}$.
   The theorem states that

$$b(\hat{\Sigma}) - b(\Sigma) \in L^{\perp}(v) \text{ a.s. with } v = \begin{pmatrix} 1 \\ \bar{x}_2 \end{pmatrix} : 2 \times 1,$$

where $L(A)$ denotes the linear subspace spanned by the column vectors of matrix
A, and $L^{\perp}(A)$ denotes the orthogonally complementary subspace of $L(A)$.
   The equation (8.76) holds without distinction of the functional form of $\hat{\theta} = \hat{\theta}(y)$, and hence the equation (8.77) remains true if the GME $b(\Sigma)$ is replaced by
another GLSE $b(\hat{\Sigma}^*)$ with $\hat{\Sigma}^* = \Sigma(\hat{\theta}^*)$, which includes the OLSE $b(\Sigma(0))$. Note
also that no distributional assumption (such as elliptical symmetry) is imposed on
the error term except that $\mathcal{L}(\varepsilon) \in \mathcal{P}_n(0, \sigma^2\Sigma)$.
   The proof given by Usami and Toyooka (1997a) contains element-wise calculation of $b(\hat{\Sigma})$ and is therefore somewhat complicated. In the following theorem,
we extend their result to the case in which the matrix C in (8.69) is not necessarily
of the form C_A in (8.71) or C_C in (8.74). Furthermore, the proof given below is a
much simpler one.

Theorem 8.18 *Let $k = 2$ and let $x_1 = 1_n$ in $X = (x_1, x_2)$. Suppose that $\mathcal{L}(\varepsilon) \in$
$\mathcal{P}_n(0, \sigma^2\Sigma)$ and that the matrix Σ is of the structure (8.69) and C satisfies the
condition (8.70). Then any GLSE $b(\hat{\Sigma})$ with $\hat{\Sigma} = \Sigma(\hat{\theta})$ satisfies*

$$1'_n X b(\hat{\Sigma}) = 1'_n y, \tag{8.78}$$

and hence

$$1'_n X[b(\hat{\Sigma}) - b(\Sigma)] = 0. \tag{8.79}$$

   Proof. Note first that the condition (8.70) implies that

$$\Sigma(\theta)^{-1} 1_n = 1_n \text{ for any } \theta \in \Theta, \tag{8.80}$$

which in turn implies that for any $\hat{\theta} = \hat{\theta}(y)$,

$$\Sigma(\hat{\theta})^{-1} 1_n = 1_n \text{ a.s.} \tag{8.81}$$

Fix any $\hat{\theta}$, and let

$$P_{\hat{\Sigma}} = \hat{\Sigma}^{-1/2} X (X' \hat{\Sigma}^{-1} X)^{-1} X' \hat{\Sigma}^{-1/2} \text{ with } \hat{\Sigma} = \Sigma(\hat{\theta}).$$

Then $P_{\hat{\Sigma}}$ is the orthogonal projection matrix onto the subspace $L(\hat{\Sigma}^{-1/2}X)$. Since $1_n \in L(X)$ implies $\hat{\Sigma}^{-1/2}1_n \in L(\hat{\Sigma}^{-1/2}X)$, the following equality is clear:

$$1'_n \hat{\Sigma}^{-1/2} P_{\hat{\Sigma}} = 1'_n \hat{\Sigma}^{-1/2},$$

which is written in the original notation as

$$1'_n \hat{\Sigma}^{-1} X (X'\hat{\Sigma}^{-1}X)^{-1} X'\hat{\Sigma}^{-1/2} = 1'_n \hat{\Sigma}^{-1/2}.$$

Postmultiplying by $\hat{\Sigma}^{-1/2}$ and using (8.81) yields

$$1'_n X (X'\hat{\Sigma}^{-1}X)^{-1} X'\hat{\Sigma}^{-1} = 1'_n. \tag{8.82}$$

This completes the proof.

Clearly, the statements (8.78) and (8.79) are equivalent to (8.76) and (8.77) respectively.

Corollary 8.19 *If in addition* $x_2 = (1, 2, \ldots, n)'$, *then*

$$b_1(\hat{\Sigma}) + \frac{n+1}{2} b_2(\hat{\Sigma}) = \bar{y},$$

and hence

$$[b_1(\hat{\Sigma}) - b_1(\Sigma)] = -\frac{n+1}{2}[b_2(\hat{\Sigma}) - b_2(\Sigma)].$$

The two equalities in the above corollary hold without distinction of C as long as C satisfies the condition (8.70).

A numerical example. As an illustration of the theorem above, we give a simple simulation result here. Consider the following simple linear regression model:

$$y_j = \beta_1 + \beta_2 x_{2j} + \varepsilon_j \quad (j = 1, \ldots, n),$$

where $n = 27$, $\beta_1 = 4.5$, $\beta_2 = 0.4$ and $x_{2j} = \log(\text{GNP}_j)$. Here, GNP_j's are Japanese GNP data treated in Section 2.5 of Chapter 2. Let the error term $\varepsilon = (\varepsilon_1, \ldots, \varepsilon_n)'$ be distributed as the normal distribution with the covariance structure of the Anderson model:

$$\mathcal{L}(\varepsilon) = N_n(0, \sigma^2 \Sigma(\theta))$$

with

$$\Sigma(\theta)^{-1} = I_n + \lambda(\theta) C_A,$$

where $\sigma^2 = 1$, $\lambda(\theta) = \theta/(1-\theta)^2$, $\theta = 0.7$ and the matrix C_A is given in (8.71).

In this case, for any estimator $\hat{\theta} = \hat{\theta}(y)$, the difference d between the GLSE $b(\hat{\Sigma})$ with $\hat{\Sigma} = \Sigma(\hat{\theta})$ and the GME $b(\Sigma)$

$$d = \begin{pmatrix} d_1 \\ d_2 \end{pmatrix} = b(\hat{\Sigma}) - b(\Sigma) = \begin{pmatrix} b_1(\hat{\Sigma}) - b_1(\Sigma) \\ b_2(\hat{\Sigma}) - b_2(\Sigma) \end{pmatrix} : 2 \times 1$$

lies in the straight line

$$d_2 = -\frac{1}{\bar{x}_2} d_1 \qquad (8.83)$$

in $R^2 = \{(d_1, d_2) \mid -\infty < d_1, d_2 < \infty\}$, where

$$\bar{x}_2 = 5.763 \quad \text{and hence} \quad \frac{1}{\bar{x}_2} = 0.1735.$$

Figure 8.1 shows the scatter plot of the realized values of the difference between the GME and the OLSE obtained by 100 replication, where the OLSE is a GLSE with $\hat{\theta}(y) \equiv 0$. It is observed that all the quantities d's are on the line in (8.83). The mean vector $\bar{d}$ and the covariance matrix S_d of d's are calculated by

$$\bar{d} = \begin{pmatrix} -0.0038 \\ 0.0218 \end{pmatrix} \quad \text{and} \quad S_d = \begin{pmatrix} 0.0136 & -0.0786 \\ -0.0786 & 0.4532 \end{pmatrix}.$$

The correlation coefficient of d_1 and d_2 is clearly -1. The readers may try other cases and observe such degeneracy phenomena.

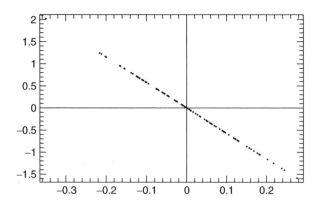

Figure 8.1   Scatter plot of the realized values of the difference between the GME and the OLSE.

An extension to multiple linear regression models. By arguing in the same way as in the proof of Theorem 8.18, we obtain the following generalization:

Theorem 8.20 *Suppose that the matrix Σ is of the structure (8.69). If there exists an $n \times k_0$ matrix X_0 such that*

$$CX_0 = 0 \quad \text{and} \quad L(X_0) \subset L(X), \tag{8.84}$$

then any GLSE $b(\hat{\Sigma})$ with $\hat{\Sigma} = \Sigma(\hat{\theta})$ satisfies

$$X_0'Xb(\hat{\Sigma}) = X_0'y \tag{8.85}$$

and hence

$$X_0'X[b(\hat{\Sigma}) - b(\Sigma)] = 0. \tag{8.86}$$

Proof. See Problem 8.4.1.

Clearly, the equation (8.86) can be restated as

$$b(\hat{\Sigma}) - b(\Sigma) \in L^{\perp}(X'X_0) \quad \text{a.s.} \tag{8.87}$$

In our context, the dimension of the linear subspace $L^{\perp}(X'X_0)$ is of interest. Needless to say, the smaller $\dim L^{\perp}(X'X_0)$ is, the more informative the result in (8.87) will be, where dim denotes the dimension of linear subspace. The dimension in question depends on the relation between the two linear subspaces L and $L(X)$, where

$$L = \{x \in R^n \mid Cx = 0\}. \tag{8.88}$$

Here, L is the linear subspace spanned by the latent vectors corresponding to zero latent roots of C. To see this more precisely, suppose without loss of generality that C has zero latent roots with multiplicity k_0. Then clearly $\dim L = k_0$. Let X_0 be any $n \times k_0$ matrix such that $L(X_0) = L$, or equivalently,

$$CX_0 = 0 \quad \text{and} \quad rank X_0 = k_0.$$

If, in addition, the matrix X_0 satisfies $L(X_0) \subset L(X)$, then (8.87) holds with

$$\dim L^{\perp}(X'X_0) = k - k_0. \tag{8.89}$$

Further extension. The proofs of Theorems 8.18 and 8.20 do not use the full force of the assumption (8.69) imposed on the structure of $\Sigma(\theta)$. This suggests a further extension of the above results to a more general covariance structure. To this end, consider the general linear regression model

$$y = X\beta + \varepsilon \quad \text{with} \quad \mathcal{L}(\varepsilon) \in \mathcal{P}_n(0, \sigma^2\Sigma)$$

and suppose that the matrix Σ is a function of an unknown but estimable parameter θ:

$$\Sigma = \Sigma(\theta) \quad \text{with} \quad \theta \in \Theta \subset R^m, \tag{8.90}$$

where Θ is a subset of R^m and $\Sigma(\theta)$ is assumed to be positive definite on Θ. Let $\hat{\theta} = \hat{\theta}(y)$ be an estimator of θ such that $\hat{\theta}(y) \in \Theta$ a.s., and consider the GLSE

$$b(\hat{\Sigma}) = (X'\hat{\Sigma}^{-1}X)^{-1}X'\hat{\Sigma}^{-1}y \quad \text{with } \hat{\Sigma} = \Sigma(\hat{\theta}). \tag{8.91}$$

In what follows, it is convenient to use a spectral decomposition of $\Sigma(\theta)$:

$$\Sigma(\theta) = \Psi(\theta)\, \Lambda(\theta)\, \Psi(\theta)' \tag{8.92}$$

with

$$\Psi = \Psi(\theta) \in \mathcal{O}(n),$$

where $\mathcal{O}(n)$ is the group of $n \times n$ orthogonal matrices, and

$$\Lambda = \Lambda(\theta) = \begin{pmatrix} \lambda_1(\theta) & & 0 \\ & \ddots & \\ 0 & & \lambda_n(\theta) \end{pmatrix} : n \times n,$$

is a diagonal matrix with $\lambda_i = \lambda_i(\theta)$ $(i = 1, \ldots, n)$ being the latent roots of $\Sigma = \Sigma(\theta)$. The inverse matrix Σ^{-1} is expressed as

$$\Sigma(\theta)^{-1} = \Psi(\theta)\, \Lambda(\theta)^{-1}\, \Psi(\theta)'. \tag{8.93}$$

Let

$$\psi_i = \psi_i(\theta)$$

be the ith column vector of Ψ $(i = 1, \ldots, n)$, which is of course a latent vector of Σ corresponding to the latent root $\lambda_i = \lambda_i(\theta)$. By interchanging the order of ψ_i's in Ψ, let

$$\psi_1, \ldots, \psi_{k_0}$$

be the latent vectors of Σ such that

(1) $\psi_1, \ldots, \psi_{k_0}$ are free from θ;

(2) $\psi_1, \ldots, \psi_{k_0} \in L(X)$.

Note that this assumption does not lose any generality. In fact, if there is no ψ_i satisfying the conditions (1) and (2), we set $k_0 = 0$. Partition Ψ and Λ as

$$\Psi = (\Psi_1, \Psi_2)$$

with

$$\Psi_1 = (\psi_1, \ldots, \psi_{k_0}) : n \times k_0,$$

$$\Psi_2 = \Psi_2(\theta) = (\psi_{k_0+1}(\theta), \ldots, \psi_n(\theta)) : n \times (n - k_0),$$

and

$$\Lambda = \begin{pmatrix} \Lambda_1 & 0 \\ 0 & \Lambda_2 \end{pmatrix}$$

with

$$\Lambda_1 = \Lambda_1(\theta) : k_0 \times k_0,$$

$$\Lambda_2 = \Lambda_2(\theta) : (n - k_0) \times (n - k_0),$$

respectively. Ψ_1 is free from θ.

Theorem 8.21 *Under assumptions (1) and (2), any GLSE $b(\hat{\Sigma})$ in (8.91) satisfies*

$$\Psi_1' X b(\hat{\Sigma}) = \Psi_1' y, \tag{8.94}$$

and hence

$$\Psi_1' X[b(\hat{\Sigma}) - b(\Sigma)] = 0. \tag{8.95}$$

Proof. The proof is essentially the same as that of Theorem 8.18. But we produce it here. Since $L(\Psi_1) \subset L(X)$, it holds that $\Psi_1' \Sigma^{-1/2} P_\Sigma = \Psi_1' \Sigma^{-1/2}$, from which we obtain

$$\Psi_1' \Sigma(\theta)^{-1} X (X' \Sigma(\theta)^{-1} X)^{-1} X' \Sigma(\theta)^{-1} = \Psi_1' \Sigma(\theta)^{-1}$$

for any $\theta \in \Theta$. Here, noting that

$$\Sigma(\theta)^{-1} \Psi_1 = \Psi_1 \Lambda_1(\theta)^{-1} \quad (\theta \in \Theta) \tag{8.96}$$

yields

$$\Lambda_1(\theta)^{-1} \Psi_1' X (X' \Sigma(\theta)^{-1} X)^{-1} X' \Sigma(\theta)^{-1} = \Lambda_1(\theta)^{-1} \Psi_1' \quad (\theta \in \Theta), \tag{8.97}$$

from which (8.94) follows. This completes the proof.

8.5 Problems

8.2.1 For two random variables A and B, A is said to be stochastically no greater than B, and is written as $A \leq_{st} B$, if for any $x \in R^1$

$$P(A \geq x) \leq P(B \geq x).$$

In particular, if $P(A \geq x) = P(B \geq x)$ holds for any $x \in R^1$, then $\mathcal{L}(A) = \mathcal{L}(B)$. Let us write $A =_{st} B$ (or $A \neq_{st} B$), when A and B have identical (or nonidentical) distribution. And define $A <_{st} B$ by

$$A \leq_{st} B \quad \text{and} \quad A \neq_{st} B.$$

Show that if $A \leq_{st} B$, then there exist two nondecreasing functions a and b, and a random variable X such that

$$a(x) \leq b(x) \quad \text{for any } x \in R^1, \quad a(X) =_{st} A \quad \text{and} \quad b(X) =_{st} B.$$

The answer will be found in Lemma 1 (page 84) of Lehmann (1986).

8.2.2 Prove the following two statements.

(1) If $A \leq_{st} B$, then

$$E(A) \leq E(B)$$

holds.

(2) If $|E(A)| < \infty$, then $A <_{st} B$ if and only if $E(A) < E(B)$.

The answers will be found in Lemma 2.4 of Hwang (1985)

8.2.3 (The equivalence theorem by Hwang (1985)) Let $y : n \times 1$ be a random vector, and let $\hat{\theta} \equiv \hat{\theta}(y)$ be an estimator of a parameter $\theta : p \times 1$. For a fixed nonnegative definite matrix D, let

$$\|\hat{\theta} - \theta\|_D = \{(\hat{\theta} - \theta)'D(\hat{\theta} - \theta)\}^{1/2},$$

and consider a loss function

$$L(\|\hat{\theta} - \theta\|_D),$$

where L is nondecreasing. Show that the following two statements are equivalent:

(1) An estimator $\hat{\theta}_1$ universally dominates $\hat{\theta}_2$ with respect to $\| \cdot \|_D$. (Here $\hat{\theta}_1$ universally dominates $\hat{\theta}_2$ with respect to $\| \cdot \|_D$, if for every $\theta \in R^p$ and every nondecreasing loss function L,

$$E_\theta\{L(\|\hat{\theta}_1 - \theta\|_D)\} \leq E_\theta\{L(\|\hat{\theta}_2 - \theta\|_D)\},$$

holds, and for a particular L, the inequality is strict.)

(2) An estimator $\hat{\theta}_1$ stochastically dominates $\hat{\theta}_2$ with respect to $\| \cdot \|_D$. (Here $\hat{\theta}_1$ stochastically dominates $\hat{\theta}_2$ with respect to $\| \cdot \|_D$, if for any $\theta \in R^p$,

$$\|\hat{\theta}_1 - \theta\|_D \leq_{st} \|\hat{\theta}_2 - \theta\|_D,$$

and for some θ, the inequality is strict.)

The proof is given in Theorem 2.3 of Hwang (1985).

8.2.4 Suppose that the two $n \times 1$ random vectors y and z satisfy

$$\mathcal{L}(y) = N_n(0, \Sigma) \quad \text{and} \quad \mathcal{L}(z) = N_n(0, \Sigma + \theta\theta'),$$

where $\Sigma \in \mathcal{S}(n)$ and $\theta \in R^n$. Show that for any symmetric convex set K, the vector y is more concentrated to 0 than is z:

$$P(y \in K) \geq P(z \in K).$$

Hint: Let w be a random variable, which is independent of y and satisfies $\mathcal{L}(w) = N(0, 1)$. Then $\mathcal{L}(z) = \mathcal{L}(y + w\theta)$. Hence, Anderson's theorem is applicable.

8.2.5 Suppose that

$$\mathcal{L}(y) = N_n(0, \Sigma) \quad \text{and} \quad \mathcal{L}(z) = N_n(0, \Psi),$$

(1) Show that y is more concentrated to 0 than z is when

$$\Sigma \leq \Psi. \tag{8.98}$$

(2) Prove the converse, that is, if y is more concentrated to 0 than z is, then the inequality (8.98) holds.

The answers of (1) and (2) will be found in Eaton (1982) and Theorem 2 of Liski and Zaigraev (2001) respectively.

8.3.1 Show that the GLSEs $b(\hat{\Sigma} \otimes I_m)$ with $\hat{\Sigma} = TDT'$ in (8.41) are in the class $\mathcal{C}^{**}$. Verify that the function γ in condition (2) is given by $\gamma(a) = a$.

8.3.2 Establish Lemma 8.12.

8.3.3 Derive similar results to those in Section 8.3 under p-equation heteroscedastic model. The answer will be found in Kurata (1999).

8.4.1 Establish Theorem 8.20.

9

Growth Curve Model and GLSEs

9.1 Overview

In this last chapter, we treat generalized least squares estimators (GLSEs) in a growth curve model as a multivariate case.

As has been observed, whether univariate or multivariate, a general linear regression model is formally expressed as

$$y = X\beta + \varepsilon \quad \text{with } \mathcal{L}(\varepsilon) \in \mathcal{P}_{n_0}(0, \Phi), \tag{9.1}$$

where

$$y : n_0 \times 1, \quad X : n_0 \times k_0, \quad rankX = k_0, \text{ and } \Phi \in \mathcal{S}(n_0).$$

Here, $\mathcal{S}(n_0)$ denotes the set of $n_0 \times n_0$ positive definite matrices. This includes the multivariate model expressed by

$$Y = X_1 B X_2 + E \quad \text{with } \mathcal{L}(E) \in \mathcal{P}_{n \times p}(0, I_n \otimes \Omega), \tag{9.2}$$

where

$$Y : n \times p, \quad X_1 : n \times k \text{ with } rankX_1 = k,$$

$$B : k \times q, \quad X_2 : q \times p \text{ with } rankX_2 = q,$$

$$\Omega \in \mathcal{S}(p).$$

Here, $\mathcal{L}(E)$ is understood as $\mathcal{L}(vec(E'))$, where $vec(E') : np \times 1$. See (9.9) for detail. This model is often called the *growth curve model* . In biometric applications,

Generalized Least Squares   Takeaki Kariya and Hiroshi Kurata
© 2004 John Wiley & Sons, Ltd   ISBN: 0-470-86697-7 (PPC)

the model is used to describe growth patterns for both groups with and without treatments. This chapter is devoted to the estimation problem in the growth curve model (9.2).

The model in (9.2) describes the structure of growth patterns more directly than the model in (9.1) although they are equivalent. In other words, a model expressed as in (9.1) contains many specific models according to the structure of X and Φ, and the estimation problem in each model is differentiated according to the structure of the model. Some cases were observed in Chapters 4 and 5.

GLSEs. However, a common feature we enjoyed when writing a model in the form of (9.1) is that the generalized least squares (GLS) estimation procedure for estimating the unknown parameter β is commonly applied to any model when it is expressed as (9.1), and the Gauss–Markov estimator (GME) and GLSEs are derived by

$$b(\Phi) = (X'\Phi^{-1}X)^{-1}X'\Phi^{-1}y \qquad (9.3)$$

and

$$b(\hat{\Phi}) = (X'\hat{\Phi}^{-1}X)^{-1}X'\hat{\Phi}^{-1}y \qquad (9.4)$$

respectively, where $\hat{\Phi} = \hat{\Phi}(e)$ is an estimator of Φ based on the ordinary least squares (OLS) residual vector e:

$$e = [I_{n_0} - X(X'X)^{-1}X']y.$$

The linear and nonlinear Gauss–Markov theorems and upper bound problems can be discussed for the risk matrix

$$R(b(\hat{\Phi}), \beta) = E\{(b(\hat{\Phi}) - \beta)(b(\hat{\Phi}) - \beta)'\} : k_0 \times k_0 \qquad (9.5)$$

as long as the second moment of $b(\hat{\Phi})$ is finite. We will keep this approach for the estimation problem of the coefficient matrix B in the growth curve model (9.2).

Special forms of the growth curve model have been considered in various forms and in various problems in association with growth curves. But it is Potthoff and Roy (1964) that systematically formulated the models in the form of (9.2). An excellent review of this model will be found in the survey papers of von Rosen (1991) and Kanda (1994). In Kanda (1994), the growth curve models with several specific covariance structures are treated in detail. See also the recent textbook by Pan and Fang (2002). The problem of testing the general linear hypothesis on B

$$R_1 B R_2 = R_0 \quad \text{with} \quad R_i\text{'s known} \qquad (9.6)$$

is often called the *general multivariate analysis of variance (GMANOVA)* problem and the hypothesis in (9.6) is often called a GMANOVA hypothesis. In this line, the model in (9.2) is often called a GMANOVA model. Note that the model with

$X_2 = I_p$ is a multivariate linear regression model, sometimes referred to as the MANOVA model, in which the GME and the OLS estimator (OLSE) are identically equal. The problem of testing

$$R_1 B = R_0$$

is called a MANOVA problem.

Notation and some basic facts. We assume that the error term matrix E in (9.2) is distributed with mean 0 and covariance matrix $I_n \otimes \Omega$, which is denoted by

$$\mathcal{L}(E) \in \mathcal{P}_{n \times p}(0, I_n \otimes \Omega), \tag{9.7}$$

where, as defined earlier, $\mathcal{L}(E) = \mathcal{L}(vec(E'))$. Hence, the covariance matrix $\text{Cov}(E)$ of E is defined by

$$\text{Cov}(E) = \text{Cov}(\varepsilon) \quad \text{with } \varepsilon = vec(E') : np \times 1. \tag{9.8}$$

Here, the vectorization $vec(A)$ of a matrix $A : n \times m$ is defined by the vector stacking column vectors of A:

$$a = vec(A) \quad \text{if and only if} \quad a = \begin{pmatrix} a_1 \\ \vdots \\ a_m \end{pmatrix}, \tag{9.9}$$

where

$$A = \begin{pmatrix} a_1, \ldots, a_m \end{pmatrix} \quad \text{and} \quad a_i : n \times 1.$$

This notation is often used in this chapter. By this definition, the n rows of E are uncorrelatedly distributed with mean 0 and covariance matrix $\Omega \in \mathcal{S}(p)$.

Lemma 9.1 *When $\mathcal{L}(E) \in \mathcal{P}_{n \times p}(0, I_n \otimes \Omega)$ holds,*

$$\mathcal{L}(CED) \in \mathcal{P}_{m \times r}(0, CC' \otimes D'\Omega D) \tag{9.10}$$

for any $C : m \times n$ with $rank C = m$ and $D : p \times r$ with $rank D = r$.

Proof. By Problem 2.2.6,

$$vec((CED)') = vec(D'E'C') = (C \otimes D')vec(E').$$

From this, (9.10) follows. This completes the proof.

Now let us express the model (9.2) as a form $y = X\beta + \varepsilon$ in (9.1) by using the vectorization scheme in (9.9):

$$y = vec(Y') : np \times 1, \quad \beta = vec(B') : kq \times 1,$$
$$\varepsilon = vec(E') : np \times 1, \tag{9.11}$$

which naturally yields

$$X = X_1 \otimes X_2' : np \times kq \quad \text{with } rankX = kq. \tag{9.12}$$

Then the GME in (9.3) is expressed as

$$b(I_n \otimes \Omega) = [(X_1'X_1)^{-1}X_1' \otimes (X_2\Omega^{-1}X_2')^{-1}X_2\Omega^{-1}]y, \tag{9.13}$$

which we write as

$$\hat{B}(\Omega) = (X_1'X_1)^{-1}X_1'Y\Omega^{-1}X_2'(X_2\Omega^{-1}X_2')^{-1} \tag{9.14}$$

by the definition of (9.9). It is easy to see that the GME is unbiased. The risk matrix, or equivalently, the covariance matrix of the GME is given by

$$\begin{aligned}
R(\hat{B}(\Omega), B) &\equiv R(b(I_n \otimes \Omega), \beta) \\
&= E\{(b(I_n \otimes \Omega) - \beta)(b(I_n \otimes \Omega) - \beta)'\} \\
&= \text{Cov}(b(I_n \otimes \Omega)) \\
&= (X_1'X_1)^{-1} \otimes (X_2\Omega^{-1}X_2')^{-1}. \tag{9.15}
\end{aligned}$$

Also, the risk matrix of the OLSE is

$$\begin{aligned}
R(\hat{B}(I_p), B) &\equiv R(b(I_n \otimes I_p), \beta) \\
&= \text{Cov}(b(I_n \otimes I_p)) \\
&= (X_1'X_1)^{-1} \otimes (X_2X_2')^{-1}X_2\Omega X_2'(X_2X_2')^{-1}, \tag{9.16}
\end{aligned}$$

where the OLSE is clearly expressed as

$$b(I_n \otimes I_p) = [(X_1'X_1)^{-1}X_1' \otimes (X_2X_2')^{-1}X_2]y$$

and

$$\hat{B}(I_p) = (X_1'X_1)^{-1}X_1'YX_2'(X_2X_2')^{-1}$$

according to (9.13) and (9.14) respectively.

When the matrix Ω is of the form

$$\Omega = \sigma^2\Sigma,$$

the GME clearly satisfies

$$\hat{B}(\Omega) = \hat{B}(\Sigma) \quad \text{and} \quad b(I_n \otimes \Omega) = b(I_n \otimes \Sigma).$$

Furthermore, if the matrix Σ is known, the GME $\hat{B}(\Sigma)$ is the best linear unbiased estimator by the Gauss–Markov theorem. Hence,

$$R(\hat{B}(\Sigma), B) \leq R(\hat{B}(I_p), B).$$

In fact, we obtain

$$R(\hat{B}(I_p), B) - R(\hat{B}(\Sigma), B)$$
$$= (X_1'X_1)^{-1} \otimes \sigma^2[(X_2X_2')^{-1}X_2\Sigma X_2'(X_2X_2')^{-1} - (X_2\Sigma^{-1}X_2')^{-1}]$$
$$= (X_1'X_1)^{-1} \otimes \sigma^2[(X_2X_2')^{-1}X_2\Sigma Z_2'(Z_2\Sigma Z_2')^{-1}Z_2\Sigma X_2'(X_2X_2')^{-1}],$$

$$(9.17)$$

which is nonnegative definite. The last equality follows by using the following matrix identity:

$$\Sigma = X_2'(X_2\Sigma^{-1}X_2')^{-1}X_2 + \Sigma Z_2'(Z_2\Sigma Z_2')^{-1}Z_2\Sigma$$

(see Problem 3.3.1), where Z_2 is a $(p - q) \times p$ matrix satisfying

$$Z_2'Z_2 = N_2 \quad \text{and} \quad Z_2Z_2' = I_{p-q} \qquad (9.18)$$

with

$$N_2 = I_p - M_2 \quad \text{and} \quad M_2 = X_2'(X_2X_2')^{-1}X_2.$$

Note that the difference (9.17) of the two risk matrices is equal to

$$E\{(b(I_n \otimes \Sigma) - b(I_n \otimes I_p))(b(I_n \otimes \Sigma) - b(I_n \otimes I_p))'\}, \qquad (9.19)$$

the proof of which is clear from (2.40). Hence, $R(\hat{B}(I_p), B) = R(\hat{B}(\Sigma), B)$ is equivalent to the identical equality $\hat{B}(\Sigma) \equiv \hat{B}(I_p)$, which is further equivalent to

$$X_2\Sigma Z_2' = 0 \qquad (9.20)$$

(see the last line of (9.17)). This fact will be used in the next section.

This chapter develops the arguments in the following order:

9.2 Condition for the Identical Equality between the GME and the OLSE

9.3 GLSEs and Nonlinear Version of the Gauss–Markov Theorem

9.4 Analysis Based on a Canonical Form

9.5 Efficiency of GLSEs.

In Section 9.2, we establish a necessary and sufficient condition on the structure of Σ as well as Ω for which the GME $\hat{B}(\Sigma)$ is identically equal to the OLSE $\hat{B}(I_p)$. In Section 9.3, a GLSE is defined in a growth curve model and a nonlinear version of the Gauss–Markov theorem is established. In Sections 9.4 and 9.5, the model is reduced to a canonical form and the efficiency of GLSEs is discussed.

9.2 Condition for the Identical Equality between the GME and the OLSE

In this section, a necessary and sufficient condition for the GME to be identically equal to the OLSE is derived.

Rao's covariance structure. Consider the growth curve model in (9.2) with

$$\Omega = \sigma^2 \Sigma.$$

When the GME $\hat{B}(\Sigma)$ is identically equal to the OLSE $\hat{B}(I_p)$, no GLS estimation problem is involved in a model. As is discussed in Chapter 7, Rao's covariance structure is a covariance structure under which the identical equality between the GME and the OLSE holds. In the case of a growth curve model, the structure of Σ for

$$\hat{B}(\Sigma) \equiv \hat{B}(I_p) \tag{9.21}$$

becomes

$$\Sigma = X_2' \Upsilon X_2 + Z_2' \Delta Z_2 \tag{9.22}$$

with $\Upsilon \in \mathcal{S}(q)$ and $\Delta \in \mathcal{S}(p-q)$, where $Z_2 : (p-q) \times p$ is a fixed matrix satisfying (9.18). Thus, we obtain

Proposition 9.2 *The GME $\hat{B}(\Sigma)$ is identically equal to the OLSE $\hat{B}(I_p)$ if and only if Σ is of the structure in (9.22).*

A comprehensive description of Rao's covariance structure is given in Section 7.2 of Chapter 7.

Example 9.1 (Equi-correlated model) In the growth curve model (9.2), assume that Ω is of the structure

$$\Omega = \sigma^2 \Sigma(\theta) \quad \text{with} \quad \Sigma(\theta) = (1-\theta)I_p + \theta 1_p 1_p', \tag{9.23}$$

where $1_p = (1, \dots, 1)' : p \times 1$ and $-1/(p-1) < \theta < 1$.
    Then $\hat{B}(\Sigma) = \hat{B}(I_p)$ holds if and only if

$$X_2 \Sigma(\theta) Z_2' = 0 \quad \text{for any } \theta \in (-1/(p-1), 1),$$

or equivalently

$$X_2 1_p 1_p' Z_2' = 0,$$

which equals either (1) $X_2 1_p = 0$ or (2) $Z_2 1_p = 0$.

In case (1), the vector 1_p is of the form $1_p = Z_2'c$ for some $c \in R^{p-q}$, and the matrix $\Sigma(\theta)$ is rewritten as

$$
\begin{aligned}
\Sigma(\theta) &= (1 - \theta)I_p + \theta 1_p 1_p' \\
&= (1 - \theta)[X_2'(X_2 X_2')^{-1}X_2 + Z_2'Z_2] + \theta Z_2'cc'Z_2 \\
&= X_2'[(1 - \theta)(X_2 X_2')^{-1}]X_2 + Z_2'[(1 - \theta)I_{p-q} + \theta cc']Z_2,
\end{aligned}
$$

where the matrix identity $I_p = X_2'(X_2 X_2')^{-1}X_2 + Z_2'Z_2$ is used in the second line. In case (2), $1_p = X_2'd$ for some $d \in R^q$ and

$$
\Sigma(\theta) = X_2'[(1 - \theta)(X_2 X_2')^{-1} + \theta dd']X_2 + Z_2'[(1 - \theta)I_{p-q}]Z_2.
$$

In Kariya (1985b), a necessary and sufficient condition for which a Gauss–Markov-type estimator $\hat{\sigma}_{GM}^2$ of σ^2 (in $\Omega = \sigma^2 \Sigma$) is identically equal to an OLS-type estimator $\hat{\sigma}_{OLS}^2$ is also given. Combining the condition with the one in Proposition 9.2 leads to a necessary and sufficient condition for which the two equalities $\hat{B}(\Sigma) \equiv \hat{B}(I_p)$ and $\hat{\sigma}_{GM}^2 \equiv \hat{\sigma}_{OLS}^2$ simultaneously hold.

9.3   GLSEs and Nonlinear Version of the Gauss–Markov Theorem

In this section, we consider GLSEs and develop a nonlinear version of the Gauss–Markov theorem in line with Chapter 3.

A nonlinear version of the Gauss–Markov theorem. In the growth curve model (9.1), or equivalently (9.2), the GME is given by (9.13) or equivalently by (9.14). Hence, when Ω is unknown, a GLSE is of the form

$$
\hat{B}(\hat{\Omega}) = (X_1'X_1)^{-1}X_1'Y\hat{\Omega}^{-1}X_2'(X_2\hat{\Omega}^{-1}X_2')^{-1}, \tag{9.24}
$$

which is equivalent to

$$
b(I_n \otimes \hat{\Omega}) = [(X_1'X_1)^{-1}X_1' \otimes (X_2\hat{\Omega}^{-1}X_2')^{-1}X_2\hat{\Omega}^{-1}]y.
$$

Here, $\hat{\Omega} = \hat{\Omega}(\hat{E})$ is an estimator of Ω based on the OLS residual matrix

$$
\hat{E} = Y - X_1\hat{B}(I_p)X_2 = E - M_1 E M_2, \tag{9.25}
$$

where $M_1 = X_1(X_1'X_1)^{-1}X_1'$ and $M_2 = X_2'(X_2 X_2')^{-1}X_2$. This is equivalent to

$$
e = vec(\hat{E}') = [I_{np} - X(X'X)^{-1}X']y \quad \text{with } X = X_1 \otimes X_2'. \tag{9.26}
$$

The risk matrix of a GLSE is given by

$$R(\hat{B}(\hat{\Omega}), \beta) \equiv R(b(I_n \otimes \hat{\Omega}), \beta)$$
$$= E\{(b(I_n \otimes \hat{\Omega}) - \beta)(b(I_n \otimes \hat{\Omega}) - \beta)'\} \qquad (9.27)$$

as long as it is finite.

To show that this risk matrix is bounded below by that of the GME, we take the same approach taken in Chapter 3. The result that we obtain by this approach is applicable to the case in which Ω is a function of a parameter vector θ:

$$\Omega = \Omega(\theta),$$

that is, Ω is of a certain structure.

In a similar manner as in (3.37), let

$$C = \begin{pmatrix} (X_1' X_1)^{-1/2} X_1' \\ Z_1' \end{pmatrix} \equiv \begin{pmatrix} C_1' \\ C_2' \end{pmatrix} : n \times n \qquad (9.28)$$

with $C_1 : n \times k$ and $C_2 : n \times (n - k)$, and

$$D = \left(\Omega^{-1} X_2', Z_2' \right) \equiv (D_1, D_2) : p \times p \qquad (9.29)$$

with $D_1 : p \times q$ and $D_2 : p \times (p - q)$, where Z_1 and Z_2 are $n \times (n - k)$ and $(p - q) \times p$ matrices such that

$$Z_1' Z_1 = I_{n-k} \quad \text{and} \quad Z_1 Z_1' = N_1$$

and

$$Z_2 Z_2' = I_{p-q} \quad \text{and} \quad Z_2' Z_2 = N_2,$$

respectively. Here,

$$N_1 = I_n - M_1 \quad \text{with} \quad M_1 = X_1 (X_1' X_1)^{-1} X_1'$$

and

$$N_2 = I_p - M_2 \quad \text{with} \quad M_2 = X_2' (X_2 X_2')^{-1} X_2.$$

The matrices C and D are $n \times n$ orthogonal and $p \times p$ nonsingular matrices respectively. Let

$$U = CED$$
$$= \begin{pmatrix} C_1' E D_1 & C_1' E D_2 \\ C_2' E D_1 & C_2' E D_2 \end{pmatrix}$$
$$= \begin{pmatrix} (X_1' X_1)^{-1/2} X_1' E \Omega^{-1} X_2' & (X_1' X_1)^{-1/2} X_1' E Z_2' \\ Z_1' E \Omega^{-1} X_2' & Z_1' E Z_2' \end{pmatrix}$$
$$\equiv \begin{pmatrix} U_{11} & U_{12} \\ U_{21} & U_{22} \end{pmatrix} \qquad (9.30)$$

and let

$$\Upsilon_1 = X_1' X_1 \in \mathcal{S}(k),$$
$$\Upsilon_2 = X_2 \Omega^{-1} X_2' \in \mathcal{S}(q),$$
$$\Delta_1 = Z_1' Z_1 = I_{n-k},$$
$$\Delta_2 = Z_2 \Omega Z_2' \in \mathcal{S}(p-q). \tag{9.31}$$

Then it follows from Lemma 9.1 that

$$\mathcal{L}(U) = \mathcal{L}\left(\begin{pmatrix} U_{11} & U_{12} \\ U_{21} & U_{22} \end{pmatrix}\right)$$
$$\in \mathcal{P}_{n \times p}\left(\begin{pmatrix} 0 & 0 \\ 0 & 0 \end{pmatrix}, I_n \otimes \begin{pmatrix} \Upsilon_2 & 0 \\ 0 & \Delta_2 \end{pmatrix}\right). \tag{9.32}$$

In fact, since

$$vec(U') = vec(D'E'C') = (C \otimes D')vec(E'),$$

using $CC' = I_n$ and $X_2 Z_2' = 0$ yields (9.32).

Let

$$K(\hat{\Omega}, \Omega) = \Delta_2^{-1} Z_2 \Omega \hat{\Omega}^{-1} X_2' (X_2 \hat{\Omega}^{-1} X_2')^{-1} : (p-q) \times q.$$

Note that when $\hat{\Omega} = \Omega$, the function K is zero:

$$K(\Omega, \Omega) = 0.$$

Note also that the OLS residual $\hat{E}$ in (9.25) is a function of U_{12}, U_{21} and U_{22}. To see this, by using

$$C^{-1} = C' = (X_1 \Upsilon_1^{-1/2}, Z_1)$$

and

$$D^{-1} = (\Omega^{-1} X_2', Z_2')^{-1} = \begin{pmatrix} \Upsilon_2^{-1} X_2 \\ \Delta_2^{-1} Z_2 \Omega \end{pmatrix}, \tag{9.33}$$

the matrix $\hat{E}$ is rewritten as

$$\hat{E} = C'UD^{-1} - M_1 C'UD^{-1} M_2$$
$$= X_1 \Upsilon_1^{-1/2} U_{12} \Delta_2^{-1} Z_2 \Omega N_2 + Z_1 U_{21} \Upsilon_2^{-1} X_2 + Z_1 U_{22} \Delta_2^{-1} Z_2 \Omega, \tag{9.34}$$

where $E = C'UD^{-1}$ is used.

Theorem 9.3   (1) *For a GLSE $\hat{B}(\hat{\Omega})$ of the form (9.24), it holds that*

$$\hat{B}(\hat{\Omega}) - B = [\hat{B}(\Omega) - B] + [\hat{B}(\hat{\Omega}) - \hat{B}(\Omega)]$$
$$= \Upsilon_1^{-1/2} U_{11} \Upsilon_2^{-1} + \Upsilon_1^{-1/2} U_{12} K(\hat{\Omega}, \Omega). \qquad (9.35)$$

Hence, for the GME $\hat{B}(\Omega)$,

$$\hat{B}(\Omega) - B = \Upsilon_1^{-1/2} U_{11} \Upsilon_2^{-1}. \qquad (9.36)$$

(2) *If the distribution $\mathcal{L}(E)$ of the error term E satisfies*

$$E(U_{11}|U_{12}, U_{21}, U_{22}) = 0 \text{ a.s.}, \qquad (9.37)$$

then for a GLSE $\hat{B}(\hat{\Omega})$ such that $\hat{\Omega}$ is a function of $\hat{E}$ only, the risk matrix is bounded below by that of the GME:

$$R(\hat{B}(\hat{\Omega}), B) \geq R(\hat{B}(\Omega), B) = (X_1' X_1)^{-1} \otimes (X_2 \Omega^{-1} X_2')^{-1}. \qquad (9.38)$$

Proof. Since

$$(X_1' X_1)^{-1/2} X_1' E = U_{11} \Upsilon_2^{-1} X_2 + U_{12} \Delta_2^{-1} Z_2 \Omega,$$

the GLSE $\hat{B}(\hat{\Omega})$ and the GME $\hat{B}(\Omega)$ are respectively rewritten by

$$\hat{B}(\hat{\Omega}) = B + (X_1' X_1)^{-1} X_1' E \hat{\Omega}^{-1} X_2' (X_2 \hat{\Omega}^{-1} X_2')^{-1}$$
$$= B + (X_1' X_1)^{-1/2} [(X_1' X_1)^{-1/2} X_1' E] \hat{\Omega}^{-1} X_2' (X_2 \hat{\Omega}^{-1} X_2')^{-1}$$
$$= B + \Upsilon_1^{-1/2} [U_{11} \Upsilon_2^{-1} X_2 + U_{12} \Delta_2^{-1} Z_2 \Omega] \hat{\Omega}^{-1} X_2' (X_2 \hat{\Omega}^{-1} X_2')^{-1}$$
$$= B + \Upsilon_1^{-1/2} U_{11} \Upsilon_2^{-1} + \Upsilon_1^{-1/2} U_{12} \Delta_2^{-1} Z_2 \Omega \hat{\Omega}^{-1} X_2' (X_2 \hat{\Omega}^{-1} X_2')^{-1}$$
$$= B + \Upsilon_1^{-1/2} U_{11} \Upsilon_2^{-1} + \Upsilon_1^{-1/2} U_{12} K(\hat{\Omega}, \Omega), \qquad (9.39)$$

and

$$\hat{B}(\Omega) = B + \Upsilon_1^{-1/2} U_{11} \Upsilon_2^{-1},$$

since $K(\Omega, \Omega) = 0$. Hence (1) follows.

For (2), we use the expression $b(I_n \otimes \hat{\Omega})$ and $b(I_n \otimes \Omega)$. It suffices to show that

$$E\{(b(I_n \otimes \hat{\Omega}) - b(I_n \otimes \Omega))(b(I_n \otimes \Omega) - \beta)'\} = 0,$$

which is equivalent to

$$E\{[\Upsilon_1^{-1/2} \otimes K(\hat{\Omega}, \Omega)'] u_{12} u_{11}' [\Upsilon_1^{-1/2} \otimes \Upsilon_2^{-1}]\} = 0. \qquad (9.40)$$

Here, u_{ij}'s are defined by

$$u_{ij} = vec(U_{ij}') \quad (i, j = 1, 2).$$

The condition $E(U_{11}|U_{12}, U_{21}, U_{22}) = 0$ is clearly equivalent to

$$E(u_{11}|u_{12}, u_{21}, u_{22}) = 0$$

because of the one-to-one correspondence between U_{ij}'s and u_{ij}'s. Since K is a function of (U_{12}, U_{21}, U_{22}), the left-hand side of (9.40) is

$$E\{[\Upsilon_1^{-1/2} \otimes K(\hat{\Omega}, \Omega)']u_{12}E(u'_{11}|u_{12}, u_{21}, u_{22})[\Upsilon_1^{-1/2} \otimes \Upsilon_2^{-1}]\},$$

which is zero. This completes the proof.

GLSE. It should be mentioned here that if $\mathcal{L}(E) = \mathcal{L}(-E)$ and if $\hat{\Omega}$ is an even function of

$$U_2 \equiv (U_{21}, U_{22}) : (n - k) \times p,$$

then the GLSE $\hat{B}(\hat{\Omega})$ is unbiased. Here, $\hat{\Omega}$ is even in U_2 means that

$$\hat{\Omega}(-U_2) = \hat{\Omega}(U_2),$$

when we write $\hat{\Omega} = \hat{\Omega}(U_2)$. A typical example is a GLSE $\hat{B}(\hat{\Omega})$ with

$$\begin{aligned}
\hat{\Omega} &= Y'N_1Y = E'N_1E \\
&= \hat{E}'N_1\hat{E} \equiv W \quad \text{(say).}
\end{aligned} \tag{9.41}$$

In fact, by using (9.34), the matrix $N_1\hat{E}$ is expressed as

$$\begin{aligned}
N_1\hat{E} &= N_1\{X_1\Upsilon_1^{-1/2}U_{12}\Delta_2^{-1}Z_2\Omega N_2 + Z_1U_{21}\Upsilon_2^{-1}X_2 + Z_1U_{22}\Delta_2^{-1}Z_2\Omega\} \\
&= Z_1U_{21}\Upsilon_2^{-1}X_2 + Z_1U_{22}\Delta_2^{-1}Z_2\Omega \\
&= Z_1U_2D^{-1},
\end{aligned} \tag{9.42}$$

and the matrix W is rewritten as

$$W = D'^{-1}U'_2N_1U_2D^{-1}, \tag{9.43}$$

which is an even function of U_2. Hence, the GLSE $\hat{B}(W)$ is an unbiased estimator of B.

Elliptically symmetric distributions. As has been discussed in Section 3.3, the condition $E(U_{11}|U_{12}, U_{21}, U_{22}) = 0$ is satisfied by a class of elliptically symmetric distributions, that is, when $\mathcal{L}(E) \in \mathcal{E}_{n \times p}(0, I_n \otimes \Omega)$, or equivalently when

$$\mathcal{L}(vec(E')) \in \mathcal{E}_{np}(0, I_n \otimes \Omega), \tag{9.44}$$

then, as in (9.32), we obtain

$$\mathcal{L}(U) = \mathcal{L}\left(\begin{pmatrix} U_{11} & U_{12} \\ U_{21} & U_{22} \end{pmatrix} \right)$$

$$\in \mathcal{E}_{n \times p}\left(\begin{pmatrix} 0 & 0 \\ 0 & 0 \end{pmatrix}, I_n \otimes \begin{pmatrix} \Upsilon_2 & 0 \\ 0 & \Delta_2 \end{pmatrix} \right) \tag{9.45}$$

This implies that

$$\mathcal{L}\left(\begin{pmatrix} u_{11} \\ u_{12} \\ u_{21} \\ u_{22} \end{pmatrix} \right)$$

$$\in \mathcal{E}_{np}\left(\begin{pmatrix} 0 \\ 0 \\ 0 \\ 0 \end{pmatrix}, \begin{pmatrix} I_k \otimes \Upsilon_2 & 0 & 0 & 0 \\ 0 & I_k \otimes \Delta_2 & 0 & 0 \\ 0 & 0 & I_{n-k} \otimes \Upsilon_2 & 0 \\ 0 & 0 & 0 & I_{n-k} \otimes \Delta_2 \end{pmatrix} \right),$$

from which it readily follows that $E(u_{11}|u_{12}, u_{21}, u_{22}) = 0$ (see Proposition 1.19 in Section 1.3).

9.4 Analysis Based on a Canonical Form

In this section, the original model and its parameters are transformed in order that the structure of the estimation problem itself is more revealing in the transformed model. The transformed model is called a *canonical form*. Once a solution is obtained in the canonical form, it is inversely transformed to the original problem.

Canonical form. Let us consider the growth curve model (9.2). To transform our estimation problem to a canonical form, we use the following fact on matrices:

Lemma 9.4 *For any $n \times m$ matrix A with $rank A = m$, there exists an $m \times m$ non-singular matrix $G \in \mathcal{Gl}(m)$ and an $n \times n$ orthogonal matrix $P \in \mathcal{O}(n)$ such that*

$$A = P \begin{pmatrix} I_m \\ 0 \end{pmatrix} G,$$

where $\mathcal{Gl}(m)$ and $\mathcal{O}(n)$ denote the groups of $m \times m$ nonsingular matrices and $n \times n$ orthogonal matrices respectively.

Proof. Let B be an $n \times (n - m)$ matrix such that

$$A'B = 0 \quad \text{and} \quad B'B = I_{n-m}.$$

Then

$$P \equiv (A(A'A)^{-1/2}, B) \in \mathcal{O}(n).$$

Let $G = (A'A)^{1/2} \in \mathcal{G}\ell(m)$. Then

$$A = (A(A'A)^{-1/2}, B) \begin{pmatrix} I_m \\ 0 \end{pmatrix} (A'A)^{1/2} = P \begin{pmatrix} I_m \\ 0 \end{pmatrix} G$$

follows. (Of course, there are other choices for P and G.)

By this lemma, write X_1 and X_2 as

$$X_1 = P_1 \begin{pmatrix} I_k \\ 0 \end{pmatrix} G_1 \quad \text{with} \quad P_1 \in \mathcal{O}(n), \ G_1 \in \mathcal{G}\ell(k),$$

$$X_2 = G_2(I_q, 0)P_2 \quad \text{with} \quad P_2 \in \mathcal{O}(p), \ G_2 \in \mathcal{G}\ell(q). \tag{9.46}$$

Using this, let

$$Y^* = P_1'Y P_2' : n \times p, \qquad \Theta = G_1 B G_2 : k \times q$$

$$\Omega^* = P_2 \Omega P_2' : p \times p, \qquad E^* = P_1' E P_2' : n \times p. \tag{9.47}$$

Then the model becomes

$$Y^* = \Pi + E^* \quad \text{with} \quad \mathcal{L}(E^*) \in \mathcal{P}_{n \times p}(0, I_n \otimes \Omega^*), \tag{9.48}$$

where the matrix Π is of the form

$$\Pi = \begin{pmatrix} \Pi_1 \\ \Pi_2 \end{pmatrix} = \overset{\displaystyle q \quad p-q}{\begin{pmatrix} \Theta & 0 \\ 0 & 0 \end{pmatrix}} \begin{matrix} k \\ n-k \end{matrix} : n \times p,$$

and Y^* and E^* are also partitioned according to Π:

$$Y^* = \begin{pmatrix} Y_1^* \\ Y_2^* \end{pmatrix} = \begin{pmatrix} Y_{11}^* & Y_{12}^* \\ Y_{21}^* & Y_{22}^* \end{pmatrix},$$

$$E^* = \begin{pmatrix} E_1^* \\ E_2^* \end{pmatrix} = \begin{pmatrix} E_{11}^* & E_{12}^* \\ E_{21}^* & E_{22}^* \end{pmatrix}. \tag{9.49}$$

In this canonical form, the problem is to estimate Θ on the basis of Y^*. The original parameter matrix B is estimated by

$$\hat{B} = G_1^{-1} \hat{\Theta} G_2^{-1} \tag{9.50}$$

once an estimator $\hat{\Theta}$ of Θ is obtained. Understanding the relation in (9.46), we omit the asterisk $*$ in the notation for simplicity in the sequel.

The GME. To find the GME for Θ, let

$$\Theta = \begin{pmatrix} \theta_1' \\ \vdots \\ \theta_k' \end{pmatrix} : k \times q \quad \text{with} \quad \theta_i : q \times 1$$

and

$$\theta = vec(\Theta') = \begin{pmatrix} \theta_1 \\ \vdots \\ \theta_k \end{pmatrix} : kq \times 1$$

and rewrite the model as

$$\begin{pmatrix} vec(Y_{11}') \\ vec(Y_{12}') \\ vec(Y_{21}') \\ vec(Y_{22}') \end{pmatrix} = \begin{pmatrix} \theta \\ 0 \\ 0 \\ 0 \end{pmatrix} + \begin{pmatrix} vec(E_{11}') \\ vec(E_{12}') \\ vec(E_{21}') \\ vec(E_{22}') \end{pmatrix}.$$

This model is denoted by a univariate linear regression model of the form

$$y = X\theta + \varepsilon, \tag{9.51}$$

where

$$y = \begin{pmatrix} vec(Y_{11}') \\ vec(Y_{12}') \\ vec(Y_{21}') \\ vec(Y_{22}') \end{pmatrix} : np \times 1,$$

$$\varepsilon = \begin{pmatrix} vec(E_{11}') \\ vec(E_{12}') \\ vec(E_{21}') \\ vec(E_{22}') \end{pmatrix} \equiv \begin{pmatrix} \varepsilon_{11} \\ \varepsilon_{12} \\ \varepsilon_{21} \\ \varepsilon_{22} \end{pmatrix} : np \times 1$$

and

$$X = \begin{pmatrix} I_{kq} \\ 0 \\ 0 \\ 0 \end{pmatrix} \begin{matrix} kq \\ k(p-q) \\ (n-k)q \\ (n-k)(p-q) \end{matrix} : np \times kq,$$

and the distribution of ε is obtained as

$$\mathcal{L}(\varepsilon) \in \mathcal{P}_{np}(0, \Phi) \tag{9.52}$$

with

$$\Phi = \begin{pmatrix} I_k \otimes \Omega_{11} & I_k \otimes \Omega_{12} & 0 & 0 \\ I_k \otimes \Omega_{21} & I_k \otimes \Omega_{22} & 0 & 0 \\ 0 & 0 & I_{n-k} \otimes \Omega_{11} & I_{n-k} \otimes \Omega_{12} \\ 0 & 0 & I_{n-k} \otimes \Omega_{21} & I_{n-k} \otimes \Omega_{22} \end{pmatrix} \qquad (9.53)$$

$$= \begin{pmatrix} \Phi_1 & 0 \\ 0 & \Phi_2 \end{pmatrix} \quad \text{(say)}$$

The inverse matrix Φ^{-1} of Φ is given by

$$\Phi^{-1} = \begin{pmatrix} \Phi_1^{-1} & 0 \\ 0 & \Phi_2^{-1} \end{pmatrix}$$

with

$$\Phi_1^{-1} = \begin{pmatrix} I_k \otimes \Omega_{11.2}^{-1} & I_k \otimes (-\Omega_{11.2}^{-1}\Omega_{12}\Omega_{22}^{-1}) \\ I_k \otimes (-\Omega_{22}^{-1}\Omega_{21}\Omega_{11.2}^{-1}) & I_k \otimes \Omega_{22.1}^{-1} \end{pmatrix}$$

and

$$\Phi_1^{-1} = \begin{pmatrix} I_{n-k} \otimes \Omega_{11.2}^{-1} & I_{n-k} \otimes (-\Omega_{11.2}^{-1}\Omega_{12}\Omega_{22}^{-1}) \\ I_{n-k} \otimes (-\Omega_{22}^{-1}\Omega_{21}\Omega_{11.2}^{-1}) & I_{n-k} \otimes \Omega_{22.1}^{-1} \end{pmatrix}.$$

Recall that $\Omega_{11.2} = \Omega_{11} - \Omega_{12}\Omega_{22}^{-1}\Omega_{21}$ and $\Omega_{22.1} = \Omega_{22} - \Omega_{21}\Omega_{11}^{-1}\Omega_{12}$. Hence, the GME of θ is given by

$$\hat{\theta}(\Omega) = (X'\Phi^{-1}X)^{-1}X'\Phi^{-1}y$$

$$= (I_k \otimes \Omega_{11.2}^{-1})^{-1}[I_k \otimes \Omega_{11.2}^{-1}, \ I_k \otimes (-\Omega_{11.2}^{-1}\Omega_{12}\Omega_{22}^{-1}), \ 0, \ 0]$$

$$\times \begin{pmatrix} vec(Y_{11}') \\ vec(Y_{12}') \\ vec(Y_{21}') \\ vec(Y_{22}') \end{pmatrix}$$

$$= [I_k \otimes I_q, \ I_k \otimes (-\Omega_{12}\Omega_{22}^{-1}), \ 0, \ 0] \begin{pmatrix} vec(Y_{11}') \\ vec(Y_{12}') \\ vec(Y_{21}') \\ vec(Y_{22}') \end{pmatrix}$$

$$= vec(Y_{11}') - [I_k \otimes (\Omega_{12}\Omega_{22}^{-1})]vec(Y_{12}'). \qquad (9.54)$$

The OLSE is given by letting $\Omega = I_p$:

$$\hat{\theta}(I_p) = vec(Y_{11}').$$

In matrix notation, the GME and the OLSE are respectively expressed as

$$\hat{\Theta}(\Omega) = Y_{11} - Y_{12}\Omega_{22}^{-1}\Omega_{21}, \tag{9.55}$$

$$\hat{\Theta}(I_p) = Y_{11}. \tag{9.56}$$

The following result is a natural consequence of this.

Proposition 9.5 *The covariance matrices of the GME and the OLSE are given by*

$$\text{Cov}(\hat{\Theta}(\Omega)) = \text{Cov}(\hat{\theta}(\Omega)) = I_k \otimes \Omega_{11.2}, \tag{9.57}$$

$$\text{Cov}(\hat{\Theta}(I_p)) = \text{Cov}(\hat{\theta}(I_p)) = I_k \otimes \Omega_{11}, \tag{9.58}$$

respectively.

It is clear that $\text{Cov}(\hat{\Theta}(I_p)) \geq \text{Cov}(\hat{\Theta}(\Omega))$, since $\Omega_{11} \geq \Omega_{11.2}$.

GLSEs. The discussion above leads to the following GLSE

$$\hat{\theta}(\hat{\Omega}) = vec(Y_{11}') - [I_k \otimes (\hat{\Omega}_{12}\hat{\Omega}_{22}^{-1})]vec(Y_{12}'), \tag{9.59}$$

whose matrix notation is given by

$$\hat{\Theta}(\hat{\Omega}) = Y_{11} - Y_{12}\hat{\Omega}_{22}^{-1}\hat{\Omega}_{21}, \tag{9.60}$$

where

$$\hat{\Omega} = \begin{pmatrix} \hat{\Omega}_{11} & \hat{\Omega}_{12} \\ \hat{\Omega}_{21} & \hat{\Omega}_{22} \end{pmatrix}$$

is an estimator of Ω. Consider a GLSE $\hat{\Theta}(\hat{\Omega})$ with such $\hat{\Omega}$'s being a function of Y_{12}, Y_{21} and Y_{22}, or equivalently, E_{12}, E_{21} and E_{22}. Typical examples are the GLSE $\hat{\Theta}(\hat{\Omega})$ with

$$\hat{\Omega} = W, \tag{9.61}$$

where the matrix W is defined by

$$\begin{aligned} W &= \begin{pmatrix} W_{11} & W_{12} \\ W_{21} & W_{22} \end{pmatrix} \\ &= Y_2'Y_2 \\ &= \begin{pmatrix} Y_{21}' \\ Y_{22}' \end{pmatrix}(Y_{21}, Y_{22}) \\ &= \begin{pmatrix} Y_{21}'Y_{21} & Y_{21}'Y_{22} \\ Y_{22}'Y_{21} & Y_{22}'Y_{22} \end{pmatrix} \\ &= \begin{pmatrix} E_{21}'E_{21} & E_{21}'E_{22} \\ E_{22}'E_{21} & E_{22}'E_{22} \end{pmatrix} \end{aligned}$$

$$= \begin{pmatrix} E'_{21} \\ E'_{22} \end{pmatrix} (E_{21}, E_{22})$$

$$= E'_2 E_2 : p \times p \tag{9.62}$$

and the GLSE $\hat{\Theta}(\hat{\Omega})$ with

$$\hat{\Omega} = S, \tag{9.63}$$

where

$$S = W + \begin{pmatrix} 0_{q \times q} & 0_{q \times (p-q)} \\ 0_{(p-q) \times q} & V_{22} \end{pmatrix} \tag{9.64}$$

and

$$V_{22} \equiv Y'_{12} Y_{12}$$

$$= E'_{12} E_{12} : (p - q) \times (p - q). \tag{9.65}$$

See Problem 9.4.2. Here, recall that a GLSE has a scale-invariance property: $\hat{\Theta}(c\Psi) = \hat{\Theta}(\Psi)$ for any $c > 0$. Hence, when we estimate Ω by $\hat{\Omega} = S/n$, the corresponding GLSE $\hat{\Theta}(S/n)$ reduces to $\hat{\Theta}(S)$. While the matrix S in $\hat{\Theta}(S)$ depends on E_{12}, E_{21} and E_{22}, the matrix W in $\hat{\Theta}(W)$ is a function of E_{21} and E_{22} only.

Let

$$\Xi = \Omega_{12}\Omega_{22}^{-1} : q \times (p - q) \quad \text{and} \quad \hat{\Xi} = \hat{\Omega}_{12}\hat{\Omega}_{22}^{-1}, \tag{9.66}$$

and let

$$z = [I_k \otimes (\hat{\Xi} - \Xi)]vec(Y'_{12})$$

$$= [I_k \otimes (\hat{\Xi} - \Xi)]vec(E'_{12}). \tag{9.67}$$

Note that $\hat{\Xi}$ is, in general, a function of Y_{12} and $Y_2 = (Y_{21}, Y_{22})$ (or equivalently, E_{12} and $E_2 = (E_{21}, E_{22})$).

Theorem 9.6 *When the distribution $\mathcal{L}(E)$ of E satisfies*

$$E[E_{11}|E_{12}, E_{21}, E_{22}] = E_{12}\Xi' \quad a.s., \tag{9.68}$$

the risk matrix of $\hat{\Theta}(\hat{\Omega})$ is expressed as

$$R(\hat{\Theta}(\hat{\Omega}), \Theta) = R(\hat{\Theta}(\Omega), \Theta) + E(zz'), \tag{9.69}$$

and hence it is bounded below by that of the GME $\hat{\Theta}(\Omega)$:

$$R(\hat{\Theta}(\hat{\Omega}), \Theta) \geq R(\hat{\Theta}(\Omega), \Theta). \tag{9.70}$$

Proof. Since the GLSE $\hat{\Theta}(\hat{\Omega})$ is written as

$$\hat{\Theta}(\hat{\Omega}) = Y_{11} - Y_{12}\hat{\Xi}'$$
$$= (\Theta + E_{11}) - E_{12}\hat{\Xi}',$$

it holds that

$$\hat{\Theta}(\hat{\Omega}) - \Theta = [\hat{\Theta}(\Omega) - \Theta] + [\hat{\Theta}(\hat{\Omega}) - \hat{\Theta}(\Omega)]$$
$$= (E_{11} - E_{12}\Xi') - E_{12}(\hat{\Xi} - \Xi)'. \tag{9.71}$$

Hence, in vector notation,

$$\hat{\theta}(\hat{\Omega}) - \theta = [\hat{\theta}(\Omega) - \theta] - z,$$

from which the risk matrix in question is evaluated as

$$R(\hat{\Theta}(\hat{\Omega}), \Theta) = R(\hat{\theta}(\hat{\Omega}), \theta)$$
$$= E\{[\hat{\theta}(\Omega) - \theta][\hat{\theta}(\Omega) - \theta]'\} + E(zz')$$
$$- E\{[\hat{\theta}(\Omega) - \theta]z'\} - E\{z[\hat{\theta}(\Omega) - \theta]'\}.$$

Since z depends on E only through E_{12}, E_{21} and E_{22}, the last two terms vanish. In fact, by noting that (9.68) is equivalent to

$$E(\varepsilon_{11}|\varepsilon_{12}, \varepsilon_{21}, \varepsilon_{22}) = (I_k \otimes \Xi)\varepsilon_{12} \quad \text{a.s.,} \tag{9.72}$$

we have

$$E\{[\hat{\theta}(\Omega) - \theta]z'\} = E\{[\varepsilon_{11} - (I_k \otimes \Xi)\varepsilon_{12}]z'\}$$
$$= E\{E[[\varepsilon_{11} - (I_k \otimes \Xi)\varepsilon_{12}]z' \mid \varepsilon_{12}, \varepsilon_{21}, \varepsilon_{22}]\}$$
$$= E\{E[\varepsilon_{11}|\varepsilon_{12}, \varepsilon_{21}, \varepsilon_{22}]z' - (I_k \otimes \Xi)\varepsilon_{12}z'\}$$
$$= E[(I_k \otimes \Xi)\varepsilon_{12}z' - (I_k \otimes \Xi)\varepsilon_{12}z']$$
$$= 0.$$

This completes the proof.

Elliptically symmetric distributions. When the distribution of the error term E is elliptically symmetric:

$$\mathcal{L}(E) \in \mathcal{E}_{n \times p}(0, \Omega),$$

the condition (9.68) is satisfied. In fact, since noting that

$$\mathcal{L}(E_{12}|E_{12}, E_{21}, E_{22}) = \mathcal{L}(\varepsilon_{11}|\varepsilon_{12}, \varepsilon_{21}, \varepsilon_{22}),$$

by using (9.53) and Proposition 1.19 of Section 1.3 of Chapter 1, we can see that the conditional distribution is also elliptically symmetric with mean

$$E\left(\varepsilon_{11}|\varepsilon_{12}, \varepsilon_{21}, \varepsilon_{22}\right)$$

$$= (I_k \otimes \Omega_{12}, \ 0, \ 0) \begin{pmatrix} I_k \otimes \Omega_{22} & 0 & 0 \\ 0 & I_{n-k} \otimes \Omega_{11} & I_{n-k} \otimes \Omega_{12} \\ 0 & I_{n-k} \otimes \Omega_{21} & I_{n-k} \otimes \Omega_{22} \end{pmatrix}^{-1}$$

$$\times \begin{pmatrix} \varepsilon_{12} \\ \varepsilon_{21} \\ \varepsilon_{22} \end{pmatrix}$$

$$= (I_k \otimes \Xi)\varepsilon_{12}.$$

Hence, in such a class of distributions, the risk matrices of the GLSEs $\hat{\Theta}(\hat{\Omega})$ are bounded below by the covariance matrix of the GME $\hat{\Theta}(\Omega)$.

9.5   Efficiency of GLSEs

By Theorem 9.6, it turns out that the problem of evaluating the efficiency of a GLSE of the form $\hat{\Theta}(\hat{\Omega})$ is reduced to the problem of evaluating the following quantity

$$R_1(\hat{\Xi}, \Xi) \equiv E(zz')$$

$$= E\{[I_k \otimes (\hat{\Xi} - \Xi)]\varepsilon_{12} \ \varepsilon'_{12}[I_k \otimes (\hat{\Xi} - \Xi)']\}. \tag{9.73}$$

In this section, we evaluate this quantity when the distribution of E is elliptically symmetric, and Ω is estimated by W.

Preliminaries. Consider the model (9.48). As before, we omit the asterisk $*$, for simplicity in the notation. The model can also be expressed as (9.51).

To evaluate the $R_1(\hat{\Xi}, \Xi)$, we assume that the distribution $P \equiv \mathcal{L}(E)$ is elliptically symmetric:

$$\mathcal{L}(E) \in \mathcal{E}_{n \times p}(0, I_n \otimes \Omega), \tag{9.74}$$

and P has a probability density function (pdf) with respect to the Lebesgue measure on R^{np}. (Here R^{np} is regarded as the set of all $n \times p$ matrices.) As is clear from (9.52), the distribution of $\varepsilon : np \times 1$ in (9.51) is elliptically symmetric:

$$\mathcal{L}(\varepsilon) \in \mathcal{E}_{np}(0, \Phi),$$

where Φ is in (9.53). Hence, by Proposition 1.17, the marginal distribution of $(\varepsilon'_{12}, \varepsilon'_{21}, \varepsilon'_{22})'$ is given by

$$\mathcal{L}\left(\begin{pmatrix} \varepsilon_{12} \\ \varepsilon_{21} \\ \varepsilon_{22} \end{pmatrix}\right) \in \mathcal{E}_{n_0}\left(\begin{pmatrix} 0 \\ 0 \\ 0 \end{pmatrix}, \begin{pmatrix} I_k \otimes \Omega_{22} & 0 & 0 \\ 0 & I_{n-k} \otimes \Omega_{11} & I_{n-k} \otimes \Omega_{12} \\ 0 & I_{n-k} \otimes \Omega_{21} & I_{n-k} \otimes \Omega_{22} \end{pmatrix}\right) \tag{9.75}$$

with $n_0 = k(p - q) + (n - k)p$. Furthermore, the following distributional results are given by the assumption (9.74):

Lemma 9.7 (1) *For W in (9.62), let*

$$F = W_{12}W_{22}^{-1} : q \times (p - q).$$

Then

$$\mathcal{L}(\varepsilon_{12}|F, \varepsilon_{22}) \in \mathcal{E}_{k(p-q)}(0, \ c_1(h)I_k \otimes \Omega_{22}), \tag{9.76}$$

where

$$h = tr[(F - \Xi)W_{22}(F - \Xi)'\Omega_{11.2}^{-1}] + tr(W_{22}\Omega_{22}^{-1}). \tag{9.77}$$

Here $c_i(\cdot)$'s in (1) and (2) below are functions of $\cdot$ and their functional forms depend on the distribution $P = \mathcal{L}(E)$.

(2) *The distributions of F and E_{22} are given by*

$$\mathcal{L}(F|E_{22}) \in \mathcal{E}_{q \times (p-q)}(\Xi, \ c_2(tr[W_{22}\Omega_{22}^{-1}])\Omega_{11.2} \otimes W_{22}^{-1}),$$
$$\mathcal{L}(E_{22}) \in \mathcal{E}_{(n-k) \times (p-q)}(0, \ I_{n-k} \otimes \Omega_{22}) \tag{9.78}$$

respectively.

Proof. The marginal distribution of E_{22} can be directly obtained by applying Proposition 1.17 to the joint distribution of $(\varepsilon_{11}, \varepsilon_{12}, \varepsilon_{21}, \varepsilon_{22})$.

Next, from (9.75), a joint pdf of $(\varepsilon_{12}, \varepsilon_{21}, \varepsilon_{22})$ is written by

$$f(\varepsilon_{12}, \varepsilon_{21}, \varepsilon_{22})$$

$$= d(\Omega) \times g\left(\varepsilon_{12}'(I_k \otimes \Omega_{22}^{-1})\varepsilon_{12}\right.$$

$$+ (\varepsilon_{21}', \varepsilon_{22}') \begin{pmatrix} I_{n-k} \otimes \Omega_{11} & I_{n-k} \otimes \Omega_{12} \\ I_{n-k} \otimes \Omega_{21} & I_{n-k} \otimes \Omega_{22} \end{pmatrix}^{-1} \left.\begin{pmatrix} \varepsilon_{21} \\ \varepsilon_{22} \end{pmatrix}\right)$$

$$= d(\Omega) \times g(\varepsilon_{12}'(I_k \otimes \Omega_{22}^{-1})\varepsilon_{12}$$

$$+ [\varepsilon_{21} - (I_{n-k} \otimes \Xi)\varepsilon_{22}]'(I_{n-k} \otimes \Omega_{11.2}^{-1})[\varepsilon_{21} - (I_{n-k} \otimes \Xi)\varepsilon_{22}]$$

$$+ \varepsilon_{22}'(I_{n-k} \otimes \Omega_{22}^{-1})\varepsilon_{22}) \tag{9.79}$$

for some function $g : [0, \infty) \to [0, \infty)$. Here, the second line of the above equalities follows from the matrix identity:

$$\begin{pmatrix} I \otimes \Omega_{11} & I \otimes \Omega_{12} \\ I \otimes \Omega_{21} & I \otimes \Omega_{22} \end{pmatrix}^{-1}$$

$$= \begin{pmatrix} I \otimes I & 0 \\ -I \otimes \Xi' & I \otimes I \end{pmatrix} \begin{pmatrix} I \otimes \Omega_{11.2}^{-1} & 0 \\ 0 & I \otimes \Omega_{22}^{-1} \end{pmatrix} \begin{pmatrix} I \otimes I & -I \otimes \Xi \\ 0 & I \otimes I \end{pmatrix},$$

which is due to

$$\begin{pmatrix} I \otimes \Omega_{11} & I \otimes \Omega_{12} \\ I \otimes \Omega_{21} & I \otimes \Omega_{22} \end{pmatrix}$$

$$= \begin{pmatrix} I \otimes I & I \otimes \Xi \\ 0 & I \otimes I \end{pmatrix} \begin{pmatrix} I \otimes \Omega_{11.2} & 0 \\ 0 & I \otimes \Omega_{22} \end{pmatrix} \begin{pmatrix} I \otimes I & 0 \\ I \otimes \Xi' & I \otimes I \end{pmatrix}.$$

Recall that $\Xi = \Omega_{12}\Omega_{22}^{-1}$. Here, the function $d(\Omega)$ is given by

$$d(\Omega) = \begin{vmatrix} I_k \otimes \Omega_{22} & 0 & 0 \\ 0 & I_{n-k} \otimes \Omega_{11} & I_{n-k} \otimes \Omega_{12} \\ 0 & I_{n-k} \otimes \Omega_{21} & I_{n-k} \otimes \Omega_{22} \end{vmatrix}^{-1/2}$$

$$= |I_k \otimes \Omega_{22}|^{-1/2} \begin{vmatrix} I_{n-k} \otimes \Omega_{11} & I_{n-k} \otimes \Omega_{12} \\ I_{n-k} \otimes \Omega_{21} & I_{n-k} \otimes \Omega_{22} \end{vmatrix}^{-1/2}$$

$$= |I_k \otimes \Omega_{22}|^{-1/2}|I_{n-k} \otimes \Omega_{11.2}|^{-1/2}|I_{n-k} \otimes \Omega_{22}|^{-1/2}$$

$$= |\Omega_{11.2}|^{-(n-k)/2}|\Omega_{22}|^{-n/2},$$

where the formulas $|I_m \otimes A| = |A|^m$ and $|A| = |A_{11.2}||A_{22}|$ are used. In the rest of the proof, the factor $d(\Omega)$ will be absorbed into the function g, since the factor is not essential.

In matrix form, the three quadratic forms in g are written by

$$Q_{12} \equiv \varepsilon'_{12}(I_k \otimes \Omega_{22}^{-1})\varepsilon_{12}$$

$$= tr(\Omega_{22}^{-1}E'_{12}E_{12}), \tag{9.80}$$

$$Q_{21} \equiv [\varepsilon_{21} - (I_{n-k} \otimes \Xi)\varepsilon_{22}]'(I_{n-k} \otimes \Omega_{11.2}^{-1})[\varepsilon_{21} - (I_{n-k} \otimes \Xi)\varepsilon_{22}]$$

$$= tr[\Omega_{11.2}^{-1}(E_{21} - E_{22}\Xi')'(E_{21} - E_{22}\Xi')] \tag{9.81}$$

and

$$Q_{22} \equiv \varepsilon'_{22}(I_{n-k} \otimes \Omega_{22}^{-1})\varepsilon_{22}$$

$$= tr(\Omega_{22}^{-1}E'_{22}E_{22}) \tag{9.82}$$

respectively.

Let

$$N_{22} = I_{n-k} - E_{22}(E'_{22}E_{22})^{-1}E'_{22} : (n-k) \times (n-k),$$

and let Z_{22} be an $(n-k) \times [(n-k) - (p-q)]$ random matrix such that

$$N_{22} = Z_{22}Z'_{22} \quad \text{and} \quad Z'_{22}Z_{22} = I_{(n-k)-(p-q)}.$$

By using Z_{22}, define a matrix F_{21} as

$$F_{21} = \begin{pmatrix} F' \\ G' \end{pmatrix} : (n-k) \times q$$

with

$$F = E'_{21} E_{22} (E'_{22} E_{22})^{-1} : q \times (p - q)$$

and

$$G = E'_{21} Z_{22} : q \times [(n - k) - (p - q)].$$

Then

$$E_{21} = E_{22} F' + Z_{22} G'$$
$$= (E_{22}, Z_{22}) F_{21}.$$

Here, since $E'_{22} Z_{22} = 0$, it holds that

$$(E_{21} - E_{22} \Xi')'(E_{21} - E_{22} \Xi')$$
$$= [E_{22}(F' - \Xi') + Z_{22} G']'[E_{22}(F' - \Xi') + Z_{22} G']$$
$$= (F - \Xi) E'_{22} E_{22} (F' - \Xi') + GG', \tag{9.83}$$

from which Q_{21} is decomposed into two parts as

$$Q_{21} = tr[\Omega_{11.2}^{-1}(F - \Xi) E'_{22} E_{22} (F' - \Xi')] + tr(\Omega_{11.2}^{-1} GG')$$
$$= Q_{21}^{(1)} + Q_{21}^{(2)} \quad \text{(say)}. \tag{9.84}$$

By viewing f as a pdf of (E_{12}, E_{21}, E_{22}) and transforming (E_{12}, E_{21}, E_{22}) to (E_{12}, F, G, E_{22}), the Jacobian becomes $|E'_{22} E_{22}|^{q/2}$. Hence, the joint pdf of (E_{12}, F, G, E_{22}) is of the form

$$|E'_{22} E_{22}|^{q/2} g(Q_{12} + Q_{21} + Q_{22}), \tag{9.85}$$

where Q_{12}, Q_{21} and Q_{22} are given in (9.80), (9.81) and (9.82) respectively. (Recall that $d(\Omega)$ is absorbed into g.) Integrating f with respect to G yields the pdf of (E_{12}, F, E_{22}) of the form

$$f_2(E_{12}, F, E_{22}) \equiv |E'_{22} E_{22}|^{q/2} \int g(Q_{12} + Q_{21}^{(1)} + Q_{21}^{(2)} + Q_{22}) dG$$
$$\equiv |E'_{22} E_{22}|^{q/2} g_2(Q_{12} + Q_{21}^{(1)} + Q_{22}) \quad \text{(say)}.$$

Integrating f_2 with respect to E_{12} yields the marginal pdf f_3 of (F, E_{22}):

$$f_3(F, E_{22}) \equiv \int f_2(E_{12}, F, E_{22}) dE_{12}$$
$$= |E'_{22} E_{22}|^{q/2} \int g_2(Q_{12} + Q_{21}^{(1)} + Q_{22}) dE_{12}$$
$$\equiv |E'_{22} E_{22}|^{q/2} g_3(Q_{21}^{(1)} + Q_{22}) \quad \text{(say)}. \tag{9.86}$$

Hence, the conditional pdf f_4 of E_{12} given (F, E_{22}) is obtained as

$$
\begin{aligned}
f_4(E_{12}|F, E_{22}) &\equiv f_2(E_{12}, F, E_{22})/f_3(F, E_{22}) \\
&= g_2(Q_{12} + Q_{21}^{(1)} + Q_{22})/g_3(Q_{21}^{(1)} + Q_{22}) \\
&\equiv g_4(Q_{12}; \ Q_{21}^{(1)} + Q_{22}) \quad \text{(say)} \\
&= g_4(\varepsilon_{12}'(I_k \otimes \Omega_{22}^{-1})\varepsilon_{12}; \ Q_{21}^{(1)} + Q_{22}).
\end{aligned}
\tag{9.87}
$$

This shows that

$$
\mathcal{L}(E_{12}|F, E_{22}) \in \mathcal{E}_{k\times(p-q)}(0, \ c_1(Q_{21}^{(1)} + Q_{22})I_k \otimes \Omega_{22}), \tag{9.88}
$$

which is equivalent to (1).

The marginal pdf f_5 of E_{22} is obtained by integrating f_3 in (9.86) with respect to F:

$$
\begin{aligned}
f_5(E_{22}) &\equiv \int f_3(F, E_{22})\mathrm{d}F \\
&= |E_{22}'E_{22}|^{q/2} \int g_3(Q_{21}^{(1)} + Q_{22})\mathrm{d}F \\
&\equiv |E_{22}'E_{22}|^{q/2}g_5(Q_{22}) \quad \text{(say)}.
\end{aligned}
\tag{9.89}
$$

This yields the conditional pdf f_6 of F given E_{22} as

$$
\begin{aligned}
f_6(F|E_{22}) &\equiv f_3(F, E_{22})/f_5(E_{22}) \\
&= g_3(Q_{21}^{(1)} + Q_{22})/g_5(Q_{22}) \\
&= g_6(Q_{21}^{(1)}; Q_{22}),
\end{aligned}
\tag{9.90}
$$

where

$$
Q_{21}^{(1)} = [vec(F') - vec(\Xi')]'(\Omega_{11.2}^{-1} \otimes E_{22}'E_{22})[vec(F') - vec(\Xi')].
$$

Thus,

$$
\mathcal{L}(F|E_{22}) \in \mathcal{E}_{q\times(p-q)}(\Xi, \ c_2(Q_{22}) \ \Omega_{11.2} \otimes (E_{22}'E_{22})^{-1})
$$

follows, which is equivalent to the first statement of (2). This completes the proof.

Note that when the distribution of E is normal: $\mathcal{L}(E) = N_{n\times p}(0, I_n \otimes \Omega)$, the functions c_1 and c_2 are

$$
c_1(\cdot) \equiv c_2(\cdot) \equiv 1,
$$

and E_{12} and $E_2 = (E_{21}, E_{22})$ are independent.

Evaluation of the risk matrix. Now using this lemma, the matrix in (9.73) is evaluated as

$$R_1(\hat{\Xi}, \Xi)$$

$$= E\{[I_k \otimes (\hat{\Xi} - \Xi)]\varepsilon_{12}\,\varepsilon'_{12}[I_k \otimes (\hat{\Xi} - \Xi)']\}$$

$$= E\{E\{[I_k \otimes (\hat{\Xi} - \Xi)]\varepsilon_{12}\,\varepsilon'_{12}[I_k \otimes (\hat{\Xi} - \Xi)']|F, E_{22}\}\}$$

$$= E\{c_1(h)I_k \otimes [(\hat{\Xi} - \Xi)\Omega_{22}(\hat{\Xi} - \Xi)']\}$$

$$= I_k \otimes R_0(\hat{\Xi}, \Xi), \tag{9.91}$$

where

$$R_0(\hat{\Xi}, \Xi) = E\{c_1(h)\,(\hat{\Xi} - \Xi)\Omega_{22}(\hat{\Xi} - \Xi)'\}.$$

Hence, to evaluate $R_1(\hat{\Xi}, \Xi)$, it suffices to evaluate $R_0(\hat{\Xi}, \Xi)$.

Theorem 9.8 *Suppose that the assumptions (9.74) holds. Then for the GLSE $\hat{\Theta}(W)$, the matrix $R_0(\hat{\Xi}, \Xi)$ with*

$$\hat{\Xi} = F = W_{12}W_{22}^{-1}$$

is evaluated as

$$R_0(\hat{\Xi}, \Xi) = E\{d(tr(\tilde{W}_{22}))\,tr(\tilde{W}_{22})\}\,\Omega_{11.2} \tag{9.92}$$

for some function $d : [0, \infty) \to [0, \infty)$, where the functional form of d depends on $P = \mathcal{L}(E)$, and the matrix $\tilde{W}_{22}$ is defined by

$$\tilde{W}_{22} = \tilde{E}'_{22}\tilde{E}_{22} \quad with \quad \tilde{E}_{22} = E_{22}\Omega_{22}^{-1/2}.$$

Proof. Note first that

$$\tilde{W}_{22} = \Omega_{22}^{-1/2}W_{22}\Omega_{22}^{-1/2}. \tag{9.93}$$

So $\tilde{W}_{22}$ and W_{22} are in one-to-one correspondence.

Using $\hat{\Xi} = F$ and (2) of Lemma 9.7, the conditional distribution of F given E_{22} is

$$\mathcal{L}(F|E_{22}) \in \mathcal{E}_{q \times (p-q)}(\Xi,\ c_2(tr[W_{22}\Omega_{22}^{-1}])\Omega_{11.2} \otimes W_{22}^{-1}). \tag{9.94}$$

Let

$$\tilde{F} = \Omega_{11.2}^{-1/2}\,F\,\Omega_{22}^{1/2} : q \times (p - q),$$

$$\Lambda = \Omega_{11.2}^{-1/2}\,\Xi\,\Omega_{22}^{1/2} : q \times (p - q),$$

and

$$\tilde{U} = (\tilde{F} - \Lambda)\tilde{W}_{22}^{1/2} : q \times (p - q).$$

Then

$$\mathcal{L}(\tilde{U}|\tilde{E}_{22}) \in \mathcal{E}_{q\times(p-q)}(0, \ c_2(tr(\tilde{W}_{22}))I_q \otimes I_{p-q}),$$

$$\mathcal{L}(\tilde{E}_{22}) \in \mathcal{E}_{(n-k)\times(p-q)}(0, \ I_{n-k} \otimes I_{p-q}), \qquad (9.95)$$

and

$$R_0(F, \Xi) = \Omega_{11.2}^{1/2} \ E[c_1(tr(\tilde{U}\tilde{U}') + tr(\tilde{W}_{22})) \ \tilde{U}\tilde{W}_{22}^{-1}\tilde{U}'] \ \Omega_{11.2}^{1/2}. \qquad (9.96)$$

Let the ith row vector of $\tilde{U}$ be $u_i' : 1 \times (p - q)$ $(i = 1, \ldots, q)$. Since, conditional on $\tilde{E}_{22}$, the distribution of

$$\mathbf{u} = vec(\tilde{U}') = \begin{pmatrix} u_1 \\ \vdots \\ u_q \end{pmatrix} : q(p - q) \times 1$$

is spherically symmetric, u_i's are uncorrelated and

$$\mathrm{Cov}(\mathbf{u}|\tilde{E}_{22}) = c_2(tr(\tilde{W}_{22})) \ I_{q(p-q)}. \qquad (9.97)$$

Note here that the conditional distribution of $\mathbf{u}$ given $\tilde{E}_{22}$ depends on $\tilde{E}_{22}$ only through $\tilde{W}_{22}$. We often use this in the discussion below. It also holds that

$$\mathcal{L}(\mathbf{u} \ |\tilde{E}_{22}) = \mathcal{L}(\mathbf{u}_{(j)} \ |\tilde{E}_{22}) \qquad (9.98)$$

for any $j = 1, \ldots, q$, where $\mathbf{u}_{(j)}$ is $\mathbf{u}$ with u_j replaced by $-u_j$:

$$\mathbf{u}_{(j)} = \begin{pmatrix} u_1 \\ \vdots \\ u_{j-1} \\ -u_j \\ u_{j+1} \\ \vdots \\ u_q \end{pmatrix} : q(p - q) \times 1.$$

This is because there exists an orthogonal matrix Γ such that $\Gamma\mathbf{u} = \mathbf{u}_{(j)}$.

On the other hand, since

$$\tilde{U}\tilde{W}_{22}^{-1}\tilde{U}' = \begin{pmatrix} u_1'\tilde{W}_{22}^{-1}u_1 & \cdots & u_1'\tilde{W}_{22}^{-1}u_q \\ \vdots & & \vdots \\ u_q'\tilde{W}_{22}^{-1}u_1 & \cdots & u_q'\tilde{W}_{22}^{-1}u_q \end{pmatrix} : q \times q, \qquad (9.99)$$

and since

$$tr(\tilde{U}\tilde{U}') = \|\mathbf{u}\|^2 = \mathbf{u}'\mathbf{u} = \sum_{i=1}^{q} u_i'u_i,$$

the (i, j)th element Δ_{ij} of the matrix

$$\Delta \equiv E[c_1(tr(\tilde{U}\tilde{U}') + tr(\tilde{W}_{22})) \, \tilde{U}\tilde{W}_{22}^{-1}\tilde{U}'|\tilde{E}_{22}] \equiv (\Delta_{ij}) \qquad (9.100)$$

is given by

$$\Delta_{ij} = E\left[c_1 \left(\sum_{i=1}^{q} u_i'u_i + tr(\tilde{W}_{22}) \right) u_i'\tilde{W}_{22}^{-1}u_j \bigg| \tilde{E}_{22} \right]. \qquad (9.101)$$

For $i \neq j$, this quantity is equal to the one in which u_j is replaced by $-u_j$ (or equivalently, $\mathbf{u}$ is replaced by $\mathbf{u}_{(j)}$):

$$\Delta_{ij} = E\left[c_1 \left(\sum_{i=1}^{q} u_i'u_i + tr(\tilde{W}_{22}) \right) u_i'\tilde{W}_{22}^{-1}(-u_j) \bigg| \tilde{E}_{22} \right]$$

$$= -E\left[c_1 \left(\sum_{i=1}^{q} u_i'u_i + tr(\tilde{W}_{22}) \right) u_i'\tilde{W}_{22}^{-1}u_j \bigg| \tilde{E}_{22} \right]$$

$$= -\Delta_{ij}. \qquad (9.102)$$

This implies that

$$\Delta_{ij} = 0 \quad (i \neq j), \qquad (9.103)$$

since the expectation is clearly finite.

Furthermore, for Δ_{ii}'s, we have

$$\Delta_{11} = \cdots = \Delta_{qq}. \qquad (9.104)$$

To see this, let

$$\mathbf{v} = \mathbf{u}/\|\mathbf{u}\|,$$

and recall that $\mathcal{L}(\mathbf{v}|\tilde{E}_{22})$ is the uniform distribution on the unit sphere on $\mathcal{U}(q(p-q))$ (see Proposition 1.13), and hence

$$E\{\mathbf{v}\mathbf{v}'|\tilde{E}_{22}\} = \frac{1}{q(p-q)} I_{q(p-q)} \qquad (9.105)$$

holds by Corollaries 1.14 and 1.15. Let

$$\mathbf{c} = \|\mathbf{u}\|,$$

and let $\mathbf{W_{22}}$ be the $q(p-q) \times q(p-q)$ block diagonal matrix with ith diagonal block being $\tilde{W}_{22}^{-1}$ and the others 0:

$$\mathbf{W_{22}} = \begin{pmatrix} 0 & & & & & & \\ & \ddots & & & & & \\ & & 0 & & & & \\ & & & \tilde{W}_{22}^{-1} & & & \\ & & & & 0 & & \\ & & & & & \ddots & \\ & & & & & & 0 \end{pmatrix}.$$

Then we have

$$\Delta_{ii} = E\left[c_1 \left(\sum_{i=1}^{q} u_i' u_i + tr(\tilde{W}_{22}) \right) u_i' \tilde{W}_{22}^{-1} u_i \,\Big|\, \tilde{E}_{22} \right]$$

$$= E[c_1(\mathbf{c}^2 + tr(\tilde{W}_{22}))\mathbf{u}'\mathbf{W_{22}}\mathbf{u}|\tilde{E}_{22}]$$

$$= E\left\{ c_1(\mathbf{c}^2 + tr(\tilde{W}_{22}))\mathbf{c}^2 E[\mathbf{v}'\mathbf{W_{22}}\mathbf{v}|\mathbf{c}, \tilde{E}_{22}]|\tilde{E}_{22} \right\}$$

$$= E\{ c_1(\mathbf{c}^2 + tr(\tilde{W}_{22}))\mathbf{c}^2[tr(\tilde{W}_{22}^{-1})/q(p-q)]|\tilde{E}_{22} \}$$

$$\text{(by (9.105))}$$

$$= tr(\tilde{W}_{22}^{-1}) \times \frac{1}{q(p-q)} E[c_1(\mathbf{c}^2 + tr(\tilde{W}_{22}))\mathbf{c}^2|\tilde{E}_{22}]$$

$$\equiv tr(\tilde{W}_{22}^{-1})\, d(tr(\tilde{W}_{22})), \tag{9.106}$$

where

$$d(tr(\tilde{W}_{22})) = \frac{1}{q(p-q)} E[c_1(\mathbf{c}^2 + tr(\tilde{W}_{22}))\mathbf{c}^2|\tilde{E}_{22}] \tag{9.107}$$

and the last line of (9.106) is due to the fact that the conditional distribution $\mathcal{L}(\mathbf{u}|\tilde{E}_{22})$ depends on $\tilde{E}_{22}$ only through $\tilde{W}_{22}$. The quantity Δ_{ii} thus derived is clearly independent of fixed i. Hence, (9.104) is proved with $\Delta_{ii} = tr(\tilde{W}_{22}^{-1})d(tr(\tilde{W}_{22}))$. Thus, we have

$$\Delta = tr(\tilde{W}_{22}^{-1})\, d(tr(\tilde{W}_{22}))\, I_q.$$

By combining this with (9.96), it follows that

$$R_0(F, \Xi) = \Omega_{11.2}^{1/2}\, E(\Delta)\, \Omega_{11.2}^{1/2}$$

$$= tr(\tilde{W}_{22}^{-1})\, d(tr(\tilde{W}_{22}))\, \Omega_{11.2}. \tag{9.108}$$

This completes the proof.

As a particular case, when E is normal, we obtain the following result, which was first established by Sugiura and Kubokawa (1988).

Corollary 9.9 *When the error term E is normally distributed:* $\mathcal{L}(E) = N_{n \times p}(0, I_n \otimes \Omega)$,

$$R_0(F, \Xi) = d_n \, \Omega_{11.2} \;\; with \;\; d_n = \frac{p-q}{(n-k)-(p-q)-1}. \tag{9.109}$$

Proof. Under the normal distribution, the function d in (9.107) is

$$d(\cdot) \equiv 1.$$

See Problem 9.5.2. (Since in this case, the function c_1 in (9.107) is $c_1(\cdot) \equiv 1$. Hence, the function d is rewritten as

$$d(tr(\tilde{W}_{22})) = \frac{1}{q(p-q)} E(\mathbf{c}^2 | \tilde{E}_{22}).$$

Here, $\mathcal{L}(\mathbf{u}) = N_{q(p-q)}(0, I_q \otimes I_{p-q}).$) Further, the matrix $\tilde{W}_{22}$ is distributed as the Wishart distribution with mean $(n-k)I_{p-q}$ and degrees of freedom $n-k$:

$$\mathcal{L}(\tilde{W}_{22}) = W_{p-q}(I_{p-q}, \; n-k).$$

Since

$$E(\tilde{W}_{22}^{-1}) = \frac{1}{(n-k)-(p-q)-1} \, I_{p-q},$$

(see Problem 1.2.4), the result follows.

By these results, the risk matrix of the GLSE $\hat{\Theta}(W)$ is given by

$$R(\hat{\Theta}(W), \Theta) = (1 + \alpha_0) \, (I_k \otimes \Omega_{11.2})$$
$$= (1 + \alpha_0) \, R(\hat{\Theta}(\Omega), \Theta),$$

where

$$\alpha_0 = E\{d(tr(\tilde{W}_{22})) \, tr(\tilde{W}_{22}^{-1})\}.$$

Furthermore, when $\mathcal{L}(E) = N_{n \times p}(0, I_n \otimes \Omega)$, it holds that $\alpha_0 = d_n$, and hence, when n or $(n-k)-(p-q)-1$ is large, the quantity d_n is small.

9.6 Problems

9.1.1 Prove that $b(I_n \otimes \Omega)$ in (9.13) and $\hat{B}(\Omega)$ in (9.14) are equivalent.

9.1.2 Under model (9.2), show that $\hat{B}(\Omega) = \hat{B}(I_p)$ for any $Y : n \times p$, when $X_2 = I_p$. Consider model (9.2) with $X_2 = I_p$. Express the model as $y = X\beta + \varepsilon$ in (9.1), and let Z be an $np \times (np - kq)$ matrix such that $Z'X = 0$ and $Z'Z = I_{np-kq}$, where $X : np \times kq$ is given by (9.12). Show that the matrix $\mathrm{Cov}(\varepsilon) = \Phi = I_n \otimes \Omega$ is of Rao's covariance structure as shown in Section 7.2 of Chapter 7.

9.2.1 In Example 9.1, express the GME $\hat{B}(\Omega)$ explicitly.

9.2.2 When

$$\Omega = \begin{pmatrix} \theta_1 I_{p_1} & 0 \\ 0 & \theta_2 I_{p_2} \end{pmatrix} \in \mathcal{S}(p)$$

with $p_1 + p_2 = p$, express the GME $\hat{B}(\Omega)$ explicitly. Find a condition for $\hat{B}(\Omega) \equiv \hat{B}(I_p)$.

9.3.1 Consider the maximum likelihood estimator (MLE) $\tilde{B}$ of B in the model treated in Section 9.3.

(1) When Ω is known, show that the MLE $\tilde{B}$ of B is equivalent to the GME: $\tilde{B} = \hat{B}(\Omega)$.

(2) When Ω is unknown, show that the MLE $\tilde{B}$ is equivalent to the GLSE $\hat{B}(\hat{\Omega}) = \hat{B}(W)$ in (9.41).

See Khatri (1966).

9.3.2 Consider the following two growth curve models with common coefficient matrix:

$$Y_1 = X_{11} B X_{12} + E_1,$$
$$Y_2 = X_{21} B X_{22} + E_2,$$

where E_1 and E_2 are independent, all X_{ij}'s are of full rank,

$$Y_i : n_i \times p_i, \quad X_{i1} : n_i \times k,$$
$$X_{i2} : q \times p_i, \quad \mathcal{L}(E_i) = N_{n_i \times p_i}(0, I_{n_i} \otimes \Omega_i),$$
$$\Omega_i \in \mathcal{S}(p_i) \quad (i = 1, 2).$$

(1) Derive the MLE of B under the assumption that Ω_i's are known.

(2) Show that the MLE is unbiased.

(3) Evaluate the covariance matrix of the MLE.

(4) Discuss an estimation procedure for this model when Ω_i's are unknown.

See Sugiura and Kubokawa (1988).

9.4.1 Show that the GME $\hat{\Theta}(\Omega)$ in (9.55) and the OLSE $\hat{\Theta}(I_p)$ in (9.56) are equivalent to those in Section 9.2.

9.4.2 Show that the following two GLSEs $\hat{B}(\hat{\Omega})$ with

$$\hat{\Omega} = Y'N_1Y = \hat{E}'N_1\hat{E}$$

in (9.41) and

$$\hat{\Omega} = \hat{E}'\hat{E}$$

are respectively equivalent to $\hat{\Theta}(W)$ in (9.61) and $\hat{\Theta}(S)$ in (9.63).

9.5.1 In the proof of Lemma 9.7, show that the Jacobian of transforming (E_{12}, E_{21}, E_{22}) to (E_{12}, F, G, E_{22}) is given by $|E'_{22}E_{22}|^{q/2}$.

9.5.2 Show that when $\mathcal{L}(E) = N_{n \times p}(0, I_n \otimes \Omega)$, the function d in (9.107) is

$$d(\cdot) \equiv 1.$$

A

Appendix

A.1 Asymptotic Equivalence of the Estimators of θ in the AR(1) Error Model and Anderson Model

Notation. Recall that the AR(1) model for the error term $\varepsilon \equiv \varepsilon_{AR}$ is $\mathcal{L}(\varepsilon_{AR}) = N_n(0, \Omega_{AR})$, where

$$\Omega_{AR} = \tau^2 \Phi_{AR} \quad \text{with} \quad \Phi_{AR} = \frac{1}{1-\theta^2}(\theta^{|i-j|}).$$

Note that

$$\Phi_{AR} = \frac{1}{(1-\theta)^2}(I_n + \lambda C + \psi B)^{-1}$$

$$= \frac{1}{(1-\theta)^2}\Sigma_{AR},$$

where λ and ψ are given by

$$\lambda = \lambda(\theta) = \theta/(1-\theta)^2 \quad \text{and} \quad \psi = \psi(\theta) = \theta/(1-\theta).$$

A typical choice of an estimator of θ in this model is

$$\hat{\theta}_{AR} = e'_{AR} K e_{AR}/e'_{AR} e_{AR}$$

with

$$e_{AR} = N y_{AR} \quad \text{and} \quad K = 2I_n - C - B,$$

where $y_{AR} = X\beta + \varepsilon_{AR}$.

Generalized Least Squares   Takeaki Kariya and Hiroshi Kurata
© 2004 John Wiley & Sons, Ltd   ISBN: 0-470-86697-7 (PPC)

On the other hand, the Anderson model for the error term $\varepsilon \equiv \varepsilon_{AN}$ is $\mathcal{L}(\varepsilon_{AN}) = N_n(0, \Omega_{AN})$, where

$$\Omega_{AN} = \tau^2 \Phi_{AN}$$

$$= \frac{\tau^2}{(1-\theta)^2}(I_n + \lambda C)^{-1}$$

$$= \frac{\tau^2}{(1-\theta)^2}\Sigma_{AN}.$$

Here, it is often the case that θ is estimated by

$$\hat{\theta}_{AN} = e'_{AN} K e_{AN}/e'_{AN} e_{AN} \quad \text{with} \quad e_{AN} = N y_{AN},$$

where $y_{AN} = X\beta + \varepsilon_{AN}$. In this appendix, we shall show that

$$\lim_{n\to\infty} \mathcal{L}[\sqrt{n}(\hat{\theta}_{AN} - \theta)] = \lim_{n\to\infty} \mathcal{L}[\sqrt{n}(\hat{\theta}_{AR} - \theta)]$$

$$= N(0, 1 - \theta^2).$$

AR(1) error model. Consider the AR(1) error model. Drop the suffix AR from all the symbols since we treat the AR model only in this section. When $\beta = 0$ in $y = X\beta + \varepsilon$, it follows that for $\hat{\theta}(e) = e'Ke/e'e$ with $e = y = \varepsilon$, $\sqrt{n}(\hat{\theta}(\varepsilon) - \theta) \to_d N(0, 1 - \theta^2)$ (see, for example, Chapter 5 of Brockwell and Davis, 2002). Also, when $\beta \neq 0$, it holds that $\sqrt{n}(\hat{\theta} - \theta) \to N(0, 1 - \theta^2)$, when

$$\lim_{n\to\infty} \frac{X'AX}{n} \equiv V_A \in \mathcal{S}(k) \quad \text{and} \quad \lim_{n\to\infty} \frac{X'F_j AX}{n} \equiv V_{Aj} \tag{A.1}$$

are assumed to be finite and bounded in j, where

$$F_j = (f_{ik}^j) \quad \text{with} \quad f_{ik}^j = 1 \quad \text{if} \quad |i - k| = j \quad \text{and} \quad f_{ik}^j = 0 \quad \text{otherwise}.$$

To see this, first observe that for $A = I$ and K

$$g \equiv \lim_{n\to\infty} E[(\varepsilon'A\epsilon - e'Ae)^2]$$

is finite. If this holds,

$$\text{plim}_{n\to\infty} \frac{1}{n}e'e = \text{plim}_{n\to\infty} \frac{1}{n}\varepsilon'\varepsilon$$

$$= \sigma^2/(1 - \theta^2)$$

and

$$\text{plim}_{n\to\infty} \frac{1}{\sqrt{n}}e'Ke = \text{plim}_{n\to\infty} \frac{1}{\sqrt{n}}\varepsilon'K\varepsilon$$

and hence

$$\sqrt{n}(\hat\theta(\varepsilon) - \theta) - \sqrt{n}(\hat\theta(e) - \theta) = \sqrt{n}(\hat\theta(\varepsilon) - \hat\theta(e))$$

$$= \frac{\frac{1}{\sqrt{n}}\varepsilon' K\varepsilon}{\frac{1}{n}\varepsilon'\varepsilon} - \frac{\frac{1}{\sqrt{n}}e' K e}{\frac{1}{n}e'e}$$

$$\to_p 0.$$

Consequently, it follows that $\sqrt{n}(\hat\theta(e) - \theta) \to_d N(0, 1 - \theta^2)$. Now to show that g is finite, write

$$\varepsilon = \Omega^{1/2}\eta \quad \text{with} \quad \Omega^{1/2} \in S(n) \quad \text{and} \quad \mathcal{L}(\eta) = N_n(0, I_n).$$

Then

$$g = \lim_{n\to\infty} E\{[\eta'\Omega^{1/2}(A - NAN)\Omega^{1/2}\eta]^2\}$$

$$= \lim_{n\to\infty} E[(\eta' Q\eta)^2]$$

$$= \lim_{n\to\infty} [2\, tr(Q^2) + (tr Q)^2],$$

where with $M = X(X'X)^{-1}X'$

$$Q = \Omega^{1/2}AM\Omega^{1/2} + \Omega^{1/2}MA\Omega^{1/2} - \Omega^{1/2}MAM\Omega^{1/2}. \tag{A.2}$$

The limits of each term in $tr(Q)$ and $tr(Q^2)$ exist. For example,

$$tr(\Omega^{1/2}AM\Omega^{1/2}) = tr\left[\left(\frac{X'\Omega AX}{n}\right)\left(\frac{X'X}{n}\right)^{-1}\right],$$

which has a finite limit by (A.1) when $n \to \infty$. In fact, since $\Omega = \sigma^2 \sum_{j=0}^{n-1}\theta^j F_j$ with $\sigma^2 = \tau^2/(1 - \theta^2)$,

$$\lim_{n\to\infty} tr(\Omega AM) = \sigma^2 \sum_{j=0}^{\infty}\theta^j tr(V_{Aj}V_I^{-1}) \quad \text{for} \quad A = I, K. \tag{A.3}$$

Note that the elements of V_{Aj} are linear combinations of $w_{ik} = \lim_{n\to\infty} x_i'x_k/n$ with x_i' being the ith row of X where $V_I = (v_{Iik})$. Since $tr V A_j V_I^{-1}$ in the right side is bounded, (A.3) converges and equals the left side. Thus, the limit exists. Similarly the other terms are shown to have their limits.

Anderson model. Consider the Anderson model. Here, we shall show that $\hat\theta_{AN}(e)$ is asymptotically equivalent to $\hat\theta_{AN}(\varepsilon)$. We use the following expressions for the covariance matrix of ε_{AN} and ε_{AR} below:

$$\Omega_{AN} = \frac{\tau^2}{(1 - \theta)^2}\Sigma_{AN}$$

and with $\Sigma_{AN} = (I_n + \lambda C)^{-1}$,

$$\Sigma_{AR} = \frac{\tau^2}{(1-\theta)^2}[\Sigma_{AN}^{-1} + \varphi B]^{-1}.$$

Since $\hat{\theta}^{AN}$ and $\hat{\theta}^{AN}$ are scale-invariant, assume for the proof below without loss of generality $\tau^2/(1-\theta)^2 = 1$ so that $\Phi_{AR} = \Sigma_{AR}$ and $\Phi_{AN} = \Sigma_{AN}$. Further, note that

$$B = u_1 u_1' + u_n u_n',$$

where $u_j = (u_{j1}, \ldots, u_{jn})' \in R^n$ is the unit vector having $u_{jj} = 1$ and $u_{jk} = 0$ $(j \neq k)$. Note also that for $a, b \in R^n$

$$(I_n + ab')^{-1} = I_n - ab'/(1 + a'b).$$

Use this equation twice to obtain

$$\Phi_{AN} = \left[I_n + \frac{a_1 b_1'}{1 + a_1' b_1} + \frac{a_2 b_2'}{1 + a_2' b_2} + \frac{a_2 b_1' b_2' a_1}{(1 + a_1' b_1)(1 + a_2' b_2)} \right] \Phi_{AR} \quad \text{(A.4)}$$

with

$$a_1 = \Phi_{AR} u_n, \quad b_1 = \varphi u_n, \quad a_2 = \Phi_{AR} u_1 \quad \text{and} \quad b_2 = \varphi u_1.$$

Then, as in the AR case,

$$\begin{aligned} g_{AN} &= \lim_{n \to \infty} E[(\varepsilon' A \varepsilon - e' A e)^2] \\ &= \lim_{n \to \infty} E[(\eta' Q \eta)^2] \\ &= \lim_{n \to \infty} [2tr(Q^2) + (tr Q)^2] \end{aligned}$$

is finite, where Q here is given by (A.2) with $\Omega = \Phi_{AN}$. In fact, the limits of each term in $tr(Q)$ and $tr(Q^2)$ also exist. For example, for

$$tr(\Phi_{AN}^{1/2} A M \Phi_{AN}^{1/2}) = tr\left[\frac{(X' \Phi_{AN} A X)}{n} \left(\frac{X'X}{n} \right)^{-1} \right],$$

substituting Φ_{AN} above into this, terms such as

$$\frac{1}{n} X' \Phi_{AR} A X, \quad \frac{1}{n} X' \Phi_{AR} u_1 u_1' \Phi_{AR} A X \quad \text{etc}$$

are involved. Since $a_1' b_1 = \varphi$, $a_2' b_2 = \varphi$ and $b_2' a_1 = \varphi \theta^{n-1}$, and $\Phi_{AR} = \nu \sum_{j=0}^{n-1} \theta^j F_j$ with $\nu = 1 - \theta^2$,

$$X' \Phi_{AR} u_1 = \nu \sum_{j=0}^{n-1} \theta^j X' F_j u_1 = \nu \sum_{j=0}^{n-1} \theta^j X' u_j.$$

Here, it holds that $b'_2 a_1 \to 0$ and

$$\lim_{n\to\infty} X' u_j / \sqrt{n} = \lim_{n\to\infty} \text{(the } i\text{th row of X)}' / \sqrt{n} = 0,$$

since

$$\frac{1}{n} \sum_{j=1}^{n} X' u_j u'_j X = \frac{1}{n} X' X \to V_I \quad (n \to \infty).$$

And

$$\lim_{n\to\infty} [X' \Phi_{AN} A X / n] = \lim_{n\to\infty} [X' \Phi_{AR} A X / n]$$

exists. Therefore, it is shown that $g_{AN} = g_{AR}$, and $e'_{AN} A e_{AN}$ is asymptotically equivalent to $\varepsilon'_{AN} A \varepsilon_{AN}$ for $A = I_n$ and K. Note that under $\tau^2 / (1 - \theta)^2 = 1$

$$\frac{1}{n} \varepsilon'_{AN} \varepsilon_{AN} = \frac{1}{n} tr(\Phi_{AN} \eta \eta') \to 1 \quad \text{in the mean,}$$

since

$$\frac{1}{n^2} \{ E[(\eta' \Phi_{AN} \eta)^2] - (E[\eta' \Phi_{AN} \eta])^2 \} = \frac{1}{n^2} 2 \, tr(\Phi_{AN}^2) \to 0$$

and $tr(\Phi_{AN}^2) = \sum_{i=1}^{n} (1 + \lambda d_i)^{-2}$, where d_i's are the latent roots of C and bounded. In fact,

$$E \left[\frac{1}{n} tr(\Phi_{AN} \eta \eta') \right] = \frac{1}{n} tr(\Phi_{AN}).$$

Hence, using (A.4), for example,

$$tr(a_1 b'_1 \Phi_{AR}) = \varphi u'_n \Phi_{AR}^2 u_n = \varphi \sum_{j=0}^{n-1} \theta^{2j} \to \varphi \frac{1}{1 - \theta^2},$$

meaning $\lim_{n\to\infty} \frac{1}{n} tr(\Phi_{AN}) = \lim_{n\to\infty} \frac{1}{n} tr(\Phi_{AR}) = 1$.

Equivalence of the asymptotic distributions. As has been shown, $\sqrt{n}(\hat{\theta}_J - \theta)$ is respectively asymptotically equivalent to

$$\sqrt{n} \left(\frac{\varepsilon'_J K \varepsilon_J}{n} - \theta \right) \quad (J = AR, AN).$$

The characteristic function of this quantity is given by

$$\phi_J(t) = exp \left(-it \sqrt{n} \theta \right) \left| I_n - \frac{it}{\sqrt{n}} \Phi_J K \right|^{-1/2} \quad (J = AR, AN).$$

We will show $\lim_{n\to\infty} \phi_{AN}(t) = \lim_{n\to\infty} \phi_{AR} = \exp(-t^2/2(1-\theta^2))$. Let

$$\psi(t) = \left[\frac{\phi_{AN}(t)}{\phi_{AR}(t)}\right]^2 = |\Xi_{AR}|/|\Xi_{AN}|,$$

where

$$\Xi_J = I_n - \frac{it}{\sqrt{n}}\Phi_J K \quad (J = AR, AN).$$

Then

$$|\Xi_{AR}| = |\Phi_{AR}|\left|\Phi_{AR}^{-1} - \frac{it}{\sqrt{n}}K\right|$$

$$= \delta\left|I_n - \frac{it}{\sqrt{n}}\Phi_{AN}^{1/2}K\Phi_{AN}^{1/2} + \varphi\Phi_{AN}^{1/2}B\Phi_{AN}^{1/2}\right|$$

$$= \delta|\Xi_{AN}|\left|I_n + \varphi\Xi_{AN}^{-1}\Phi_{AN}^{1/2}B\Phi_{AN}^{1/2}\right|,$$

where

$$\delta = |\Phi_{AR}|/|\Phi_{AN}|.$$

Therefore, with $\hat{u}_j = \Phi_{AN}^{1/2}u_j$,

$$\psi(t) = \delta_1\delta_2,$$

where with $\tilde{u}_j = \Phi_{AN}^{1/2}u_j$,

$$\delta_2 = \left|I_n + \varphi\Xi_{AN}^{-1}\Phi_{AN}^{1/2}B\Phi_{AN}^{1/2}\right|$$

$$= \left|I_n + \varphi\Xi_{AN}^{-1}(\tilde{u}_1\tilde{u}_1' + \tilde{u}_n\tilde{u}_n')\right|$$

and

$$\delta_1^{-1} = |I_n + \varphi(\tilde{u}_1\tilde{u}_1' + \tilde{u}_n\tilde{u}_n')|.$$

Here using $|I_n + ab'| = 1 + a'b$,

$$\delta_1^{-1} = |I_n + \varphi\tilde{u}_1\tilde{u}_1'||I_n + (I_n + \varphi\tilde{u}_1\tilde{u}_1')^{-1}\varphi\tilde{u}_n\tilde{u}_n'|$$

$$= (1 + \varphi\tilde{u}_1'\tilde{u}_1)\left[1 + \varphi\tilde{u}_n'\tilde{u}_n - \frac{(\varphi\tilde{u}_n'\tilde{u}_1)^2}{1 + \varphi\tilde{u}_1'\tilde{u}_1}\right].$$

Similarly,

$$\delta_2 = (1 + \varphi\tilde{u}_1'\Xi_{AN}^{-1}\tilde{u}_1)\left[1 + \varphi\tilde{u}_n'\Xi_{AN}^{-1}\tilde{u}_n - \frac{(\varphi\tilde{u}_n'\Xi_{AN}^{-1}\tilde{u}_1)^2}{1 + \varphi\tilde{u}_1'\Xi_{AN}^{-1}\tilde{u}_1}\right].$$

By taking $\triangle \in O(n)$ such that

$$\triangle C \triangle' = D = \begin{pmatrix} d_1 & & \\ & \ddots & \\ & & d_n \end{pmatrix} \quad \text{with} \quad 0 = d_1 < \cdots < d_n < 4,$$

and using $-\frac{1}{4} < \lambda < \infty$,

$$\tilde{u}_i{}' \tilde{u}_i = u_i' \Phi_{AN} u_i$$
$$= u_i' \triangle (I_n + \lambda D)^{-1} \triangle' u_i$$
$$\leq \max\{1, 1/(1 + 4\lambda)\} \quad (i = 1, n).$$

Further, since $K = \frac{1}{2}[2I_n - C - B]$,

$$\Phi_{AN}^{1/2} K \Phi_{AN}^{1/2} = \triangle (I_n + \lambda D)^{-1/2} \frac{1}{2} \left(2I_n - D - \triangle' B \triangle\right) (I_n + \lambda D)^{-1/2} \triangle',$$

and hence the elements of this matrix are bounded. Consequently, when n is given but large, Ξ_N^{-1} is expanded as an infinite series:

$$\tilde{u}_1' \Xi_{AN}^{-1} \tilde{u}_1 = \tilde{u}_1' \left[\sum_{j=0}^{\infty} \left(-\frac{it}{\sqrt{n}}\right)^j (\Phi_{AN}^{1/2} K \Phi_{AN}^{1/2})^j \right] \tilde{u}_1$$
$$= \sum_{j=0}^{\infty} \left(-\frac{it}{\sqrt{n}}\right)^j \tilde{u}_1' (\Phi_{AN}^{1/2} K \Phi_{AN}^{1/2})^j \tilde{u}_1,$$

where each term of the series is bounded and converges to 0 except for $j = 0$. Therefore,

$$\lim_{n \to \infty} \tilde{u}_1' \Xi_{AN}^{-1} \tilde{u}_1 = \lim_{n \to \infty} \tilde{u}_1' \tilde{u}_1.$$

Similarly,

$$\lim_{n \to \infty} \tilde{u}_n' \Xi_{AN}^{-1} \tilde{u}_n = \lim_{n \to \infty} \tilde{u}_n' \tilde{u}_n,$$

and

$$\lim_{n \to \infty} \tilde{u}_n' \Xi_{AN}^{-1} \tilde{u}_1 = \lim_{n \to \infty} \tilde{u}_n' \tilde{u}_1.$$

This implies $\lim_{n \to \infty} \delta_2 = \lim_{n \to \infty} \delta_1^{-1}$ and hence $\lim_{n \to \infty} \psi(t) = 1$. Therefore, the condition (5) in Section 5.3 of Chapter 5 is verified. The condition (6) follows from the fact that the characteristic function of $\sqrt{n}(\hat{\theta}_{AN} - \theta)$ satisfies $\lim_{n \to \infty} \log \Phi_{AN}(t) = -\frac{1}{2}(1 - \theta^2)t^2 \equiv \xi(t)$ and hence $\log \Phi_{AN}(t) = \xi(t) + O(1/n)$. This implies that $E[\hat{\theta}_{AN} - \theta] = O(1/n)$.

Bibliography

Abramowitz M and Stegun IA 1972 *Handbook of Mathematical Functions*. Dover Publications.

Ali MM and Ponnapalli R 1990 An optimality of the Gauss-Markov estimator. *Journal of Multivariate Analysis* **32**, 171–176.

Anderson TW 1948 On the theory of testing serial correlation. *Skandinavisk Aktuarietidskrift* **31**, 88–116.

Anderson TW 1955 The integral of a symmetric unimodal function over a symmetric convex set and some probability inequalities. *Proceedings of American Mathematical Society* **6**, 170–176.

Anderson TW 1971 *The Statistical Analysis of Time Series*. John Wiley & Sons.

Anderson TW 1984 *An Introduction to Multivariate Statistical Analysis*, 2nd edn. John Wiley & Sons.

Andrews DWK 1986 A note on the unbiasedness of feasible GLS, quasi-maximum likelihood, robust, adaptive, and spectral estimators of the linear model. *Econometrica* **54**, 687–698.

Andrews DWK and Phillips PCB 1987 Best median-unbiased estimation in linear regression with bounded asymmetric loss functions. *Journal of American Statistical Association* **82**, 886–893.

Beach CM and MacKinnon JG 1978 A maximum likelihood procedure for regression with autocorrelated errors. *Econometrica* **46**, 51–58.

Berk RH 1986 Sphericity and the normal law. *Annals of Probability* **14**, 696–701.

Berk RH and Hwang JT 1989 Optimality of the least squares estimator. *Journal of Multivariate Analysis* **30**, 245–254.

Bilodeau M 1990 On the choice of the loss function in covariance estimation. *Statistics & Decisions* **8**, 131–139.

Bilodeau M and Brenner D 1999 *Theory of Multivariate Statistics*. Springer-Verlag.

Bischoff W 2000 The structure of a linear model: sufficiency, ancillarity, invariance, and the normal distribution. *Journal of Multivariate Analysis* **73**, 180–198.

Bischoff W, Cremers H and Fieger W 1991 Normal distribution assumption and least squares estimation function in the model of polynomial regression. *Journal of Multivariate Analysis* **36**, 1–17.

Bloomfield P and Watson GS 1975 The inefficiency of least squares. *Biometrika* **62**, 121–128.

Generalized Least Squares Takeaki Kariya and Hiroshi Kurata

© 2004 John Wiley & Sons, Ltd ISBN: 0-470-86697-7 (PPC)

Breusch TS and Pagan AR 1980 The Lagrange multiplier test and its applications to model specification in econometrics. *Review of Economic Studies* **47**, 239–253.

Brockwell PJ and Davis RA 2002 *Introduction to Time Series and Forecasting*, 2nd edn. Springer-Verlag.

Cambanis S, Huang S and Simons G 1981 On the theory of elliptically contoured distributions. *Journal of Multivariate Analysis* **11**, 368–385.

Davis AW 1989 Distribution theory for some tests of independence of seemingly unrelated regressions. *Journal of Multivariate Analysis* **31**, 69–82.

Eaton ML 1981 On the projections of isotropic distributions. *Annals of Statistics* **9**, 391–400.

Eaton ML 1982 A review of selected topics in multivariate probability inequalities. *Annals of Statistics* **10**, 11–43.

Eaton ML 1983 *Multivariate Statistics: A Vector Space Approach*. John Wiley & Sons.

Eaton ML 1985 The Gauss-Markov theorem in multivariate analysis. In *Multivariate Analysis VI* (ed. Krishnaiah PR), pp. 177–201. Elsevier Science Publishers B.V. (North Holland).

Eaton ML 1986 A characterization of spherical distributions. *Journal of Multivariate Analysis* **20**, 272–276.

Eaton ML 1987 Group induced orderings with some applications in statistics. *CWI Newsletter* **16**, 3–31.

Eaton ML 1988 Concentration inequarities for Gauss-Markov estimators. *Journal of Multivariate Analysis* **25**, 119–138.

Eaton ML 1989 *Group Invariance Applications in Statistics* (Regional Conference Series in Probability and Statistics 1). Institute of Mathematical Statistics.

Ezekiel M and Fox FA 1959 *Methods of Correlation and Regression Analysis*. John Wiley & Sons.

Fang KT, Kotz S and Ng KW 1990 *Symmetric Multivariate and Related Distributions*. Chapman & Hall.

Feller W 1966 *An Introduction to Probability Theory and its Applications*. John Wiley & Sons.

Ferguson TS 1967 *Mathematical Statistics*. Academic Press.

Geisser S 1970 Baysian analysis of growth curve. *Sankhya A* **32**, 53–64.

Goldberger AS 1962 Best linear unbiased prediction in the generalized linear regression model. *Journal of American Statistical Association* **57**, 369–375.

Harville DA and Jeske DR 1992 Mean squared error of estimation or prediction under a general linear model. *Journal of American Statistical Association* **87**, 724–731.

Hasegawa H 1995 On small sample properties of Zellner's estimator for the case of two SUR equations with compound normal disturbances. *Communications in Statistics: Simulation and Computation* **24**, 45–59.

Hillier GH and King ML 1987 Linear regression with correlated errors: bounds on coefficient estimates and t-values. In *Specification Analysis in the Linear Model* (In Honour of Donald Cochrane) (eds. King ML and Giles DEA), pp. 74–80. Routledge & Kegan Paul.

Hwang JT 1985 Universal domination and stochastic domination: Estimation simultaneously under a broad class of loss functions. *Annals of Statistics* **13**, 295–314.

Jensen DR 1996 Symmetry and unimodality in linear inference. *Journal of Multivariate Analysis* **60**, 188–202.

Judge GG, Griffiths WE, Hill RC, Lütkepohl M and Lee TC 1985 *The Theory and Practice of Econometrics*, 2nd edn. John Wiley & Sons.

Kanda T 1994 Growth curve model with covariance structures. *Hiroshima Mathematical Journal* **24**, 135–176.

Kariya T 1980 Note on a condition for equality of sample variances in a linear model. *Journal of American Statistical Association* **75**, 701–703.

Kariya T 1981a Bounds for the covariance matrices of Zellner's estimator in the SUR model and the 2SAE in a heteroscedastic model. *Journal of American Statistical Association* **76**, 975–979.

Kariya T 1981b Tests for the independence between two seemingly unrelated regression equations. *Annals of Statistics* **9**, 381–390.

Kariya T 1985a A nonlinear version of the Gauss-Markov theorem. *Journal of American Statistical Association* **80**, 476–477.

Kariya T 1985b *Testing in the Multivariate General Linear Model*. Kinokuniya, Tokyo.

Kariya T 1988 An identity in regression with missing observations. In *Statistical Theory and Data Analysis II* (ed. Matusita K), pp. 365–368. Elsevier Science Publishers B.V. (North Holland).

Kariya T 1993 *Quantitative Methods for Portfolio Analysis*. Kluwer Academic Publishers.

Kariya T, Fujikoshi Y and Krishnaiah PR 1984 Tests for independence of two multivariate regression equations with different design matrices. *Journal of Multivariate Analysis* **15**, 383–407.

Kariya T, Fujikoshi Y and Krishnaiah PR 1987 On tests for selection of variables and independence under multivariate regression models. *Journal of Multivariate Analysis* **21**, 207–237.

Kariya T and Kurata H 2002 A maximal extension of the Gauss-Markov theorem and its nonlinear version. *Journal of Multivariate Analysis* **83**, 37–55.

Kariya T and Sinha BK 1989 *Robustness of Statistical Tests*. Academic Press.

Kariya T and Toyooka Y 1985 Nonlinear versions of the Gauss-Markov theorem and GLSE. In *Multivariate Analysis VI* (ed. Krishnaiah PR), pp. 345–354. Elsevier Science Publishers B.V. (North Holland).

Kariya. T and Toyooka Y 1992 Bounds for normal approximations to the distributions of generalized least squares predictors and estimators. *Journal of Statistical Planning and Inference* **30**, 213–221.

Kariya T and Tsuda H 1994 New bond pricing models with applications to Japanese data. *Financial Engineering and the Japanese Markets* **1**, 1–20.

Kelker D 1970 Distribution theory of spherical distributions and a location-scale parameter. *Sankhya A* **32**, 419–430.

Khatri CG 1966 A note on MANOVA model applied to problems in growth curves. *Annals of Institute of Statistical Mathematics* **18**, 75–86.

Khatri CG and Shah KR 1974 Estimation of location parameters from two linear models under normality. *Communications in Statistics: Theory and Methods* **3**, 647–663.

Khatri CG and Srivastava MS 1971 On exact non-null distributions of likelihood ration criteria for sphericity test and equality of two covariance matrices. *Sankhya A* **33**, 201–206.

Knott M 1975 On the minimum efficiency of least squares. *Biometrika* **62**, 129–132.

Kobayashi M 1985 Comparison of efficiencies of several estimators for linear regressions with autocorrelated errors. *Journal of American Statistical Association* **80**, 951–953.

Kruskal W 1968 When are Gauss-Markov and least squares estimators identical? A coordinate-free approach. *Annals of Mathematical Statistics* **39**, 70–75.

Kubokawa T 1998 Double shrinkage estimation of common coefficients in two regression equations with heteroscedasticity. *Journal of Multivariate Analysis* **67**, 169–189.

Kunitomo N 1977 A note on the efficiency of Zellner's estimator for the case of two seemingly unrelated regression equations. *Economic Studies Quarterly* **28**, 73–77.

Kurata H 1998 A generalization of Rao's covariance structure with applications to several linear models. *Journal of Multivariate Analysis* **67**, 297–305.

Kurata H 1999 On the efficiencies of several generalized least squares estimators in a seemingly unrelated regression model and a heteroscedastic model. *Journal of Multivariate Analysis* **70**, 86–94.

Kurata H 2001 A note on an upper bound for the covariance matrix of a generalized least squares estimator in a heteroscedastic model. *Scientiae Mathematicae Japonicae* **53**, 65–74.

Kurata H and Kariya T 1996 Least upper bound for the covariance matrix of a generalized least squares estimator in regression with applications to a seemingly unrelated regression model and a heteroscedastic model. *Annals of Statistics* **24**, 1547–1559.

Kuritsyn YG 1986 On the least-squares method for elliptically contoured distributions. *Theory of Probability and its Applications* **31**, 738–740.

Lehmann EL 1983 *Theory of Point Estimation*. John Wiley & Sons.

Lehmann EL 1986 *Testing Statistical Hypotheses*, 2nd edn. Springer-Verlag.

Liski EP and Zaigraev A 2001 A stochastic characterization of Loewner optimality design criterion in linear models. *Metrika* **53**, 207–222.

Lu CY and Shi NZ 2000 Universal inadmissibility of least squares estimator. *Journal of Multivariate Analysis* **72**, 22–29.

Magee L 1985 Efficiency of iterative estimators in the regression model with AR(1) disturbances. *Journal of Econometrics* **29**, 275–287.

Magee L, Ullah A and Srivastava VK 1987 Efficiency of estimators in the regression model with first-order autoregressive errors. In *Specification Analysis in the Linear Model* (In Honour of Donald Cochrane) (eds. King ML and Giles DEA), pp. 81–98. Routledge & Kegan Paul.

Magnus JR 1978 Maximum likelihood estimation of the GLS model with unknown parameters in the disturbance covariance matrix. *Journal of Econometrics* **7**, 281–312.

Mathew T 1983 Linear estimation with an incorrect dispersion matrix in linear models with a common linear part. *Journal of American Statistical Association* **78**, 468–471.

McCulloch JH 1971 Measuring the term structure of interest rates. *Journal of Business* **44**, 19–31.

McCulloch JH 1975 The tax-adjusted yield curve. *Journal of Business*, **30**, 811–830.

McElroy FW 1967 A necessary and sufficient condition that ordinary least-squares estimators be best linear unbiased. *Journal of American Statistical Association* **62**, 1302–1304.

Mehta JS and Swamy PAVB 1976 Further evidence on the relative efficiencies of Zellner's seemingly unrelated regressions estimator. *Journal of American Statistical Association* **71**, 634–639.

Muirhead RJ 1982 *Aspects of Multivariate Statistical Theory*. John, Wiley & Sons.

Nawata K 2001 *Excel Toukei Kaiseki Box niyoru Data Kaiseki (in Japanese) (Data Analysis by Excel Statistical Analysis Box)*. Asakura Shoten.

Ng VM 2000 A note on predictive inference for multivariate elliptically contoured distributions. *Communications in Statistics: Theory and Methods* **29**, 477–483.

Ng VM 2002 Robust Baysian inference for seemingly unrelated regressions with elliptical errors. *Journal of Multivariate Analysis* **83**, 409–414.

Pan JX and Fang KT 2002 *Growth Curve Models with Statistical Diagnostics.* Springer-Verlag.

Perron F 1992 Equivariant estimator of the covariance matrix. *Canadian Journal of Statistics* **43**, 16–28.

Potthoff RF and Roy SN 1964 A generalized multivariate analysis of variance model useful especially for growth curve problems. *Biometrika* **51**, 313–326.

Puntanen S and Styan GPH 1989 The equality of the ordinary least squares estimator and the best linear unbiased estimator. *American Statistician* **43**, 153–161.

Rao CR 1967 Least squares theory using an estimated dispersion matrix and its application to measurement of signals. In *Proceedings of the Fifth Berkely Symposium on Mathematical Statistics and Probability* (eds. Lecam LM and Neyman J), Vol. 1, pp. 255–372. University of California Press.

Rao CR 1973 *Linear Statistical Inference and its Applications*, 2nd edn. John Wiley & Sons.

Rao CR and Mitra SK 1971 *Generalized Inverse of Matrices and its Applications.* John Wiley & Sons.

Rao CR and Rao MB 1998 *Matrix Algebra and its Applications to Statistics and Econometrics.* World Scientific Publishing.

Revankar NS 1974 Some finite sample results in the context of two seemingly unrelated regression equations. *Journal of American Statistical Association* **68**, 187–190.

Revankar NS 1976 Use of restricted residuals in SUR systems: some finite sample results. *Journal of American Statistical Association* **69**, 183–188.

Schoenberg IJ 1938 Metric spaces and completely monotone functions. *Annals of Mathematics* **39**, 811–841.

Sen A and Srivastava M 1990 *Regression Analysis: Theory, Methods, and Applications* Springer-Verlag.

Srivastava VK and Giles DEA 1987 *Seemingly Unrelated Regression Equation Models.* Marcel Dekker.

Srivastava VK and Maekawa K 1995 Efficiency properties of feasible generalized least squares estimators in SURE models under non-normal disturbances. *Journal of Econometrics* **66**, 99–121.

Sugiura N and Kubokawa T 1988 Estimating common parameters of growth curve models. *Annals of Institute of Statistical Mathematics* **40**, 119–135.

Swamy PAVB and Mehta JS 1979 Estimation of common coefficients in two regression equations. *Journal of Econometrics* **10**, 1–14.

Takemura A 1984 An orthogonally invariant minimax estimator of the covariance matrix of a multivariate normal population. *Tsukuba Journal of Mathematics* **8**, 367–376.

Taylor WE 1977 Small sample properties of a class of two stage Aitken estimators. *Econometrica* **45**, 497–508.

Taylor WE 1978 The heteroscedastic linear model: exact finite sample results. *Econometrica* **46**, 663–675.

Tong YL 1990 *The Multivariate Normal Distribution.* Springer-Verlag.

Toyooka Y 1987 An iterated version of the Gauss-Markov theorem in generalized least squares estimation. *Journal of Japan Statistical Society* **17**, 129–136.

Toyooka Y and Kariya T 1986 An approach to upper bound problems for risks of the generalized least squares estimators. *Annals of Statistics* **14**, 679–690.

Toyooka Y and Kariya T 1995 A note on sufficient condition for the existence of 2nd moments of generalized least squares estimator in a linear model. *Mathematica Japonica* **42**, 509–510.

Usami Y and Toyooka Y 1997a On the degeneracy of the distribution of a GLSE in a regression with a circularly distributed error. *Mathematica Japonica* **45**, 423–431.

Usami Y and Toyooka Y 1997b Errata of Kariya and Toyooka (1992), Bounds for normal approximations to the distributions of generalized least squares predictors and estimators. *Journal of Statistical Planning and Inference* **58**, 399–405.

von Rosen D 1991 The growth curve model: a review *Communications in Statistics: Theory and Methods* **20**, 2791–2822.

Wang SG and Shao J 1992 Constrained Kantorovich inequalities and relative efficiency of least squares. *Journal of Multivariate Analysis* **42**, 284–298.

Wang SG and Ip WC 1999 A matrix version of the Wielandt inequality and its applications to statistics. *Linear Algebra and its Applications* **296**, 171–181.

Wu L and Perlman MD 2000 Lattice conditional independence model for seemingly unrelated regressions. *Communications in Statistics: Simulation and Computation* **29**, 361–384.

Zellner A 1962 An efficient method of estimating seemingly unrelated regressions and tests for aggregation bias. *Journal of American Statistical Association* **57**, 348–368.

Zellner A 1963 Estimators for seemingly unrelated regressions: some finite sample results. *Journal of American Statistical Association* **58**, 977–992.

Zyskind G 1967 On canonical forms, non-negative covariance matrices and best and simple least squares estimators in linear models. *Annals of Mathematical Statistics* **38**, 1092–1109.

Zyskind G 1969 Parametric augmentations and error structures under which certain simple least squares and analysis of variance procedures are also best. *Journal of American Statistical Association* **64**, 1353–1368.

Index

Generalized Least Squares Takeaki Kariya and Hiroshi Kurata
© 2004 John Wiley & Sons, Ltd   ISBN: 0-470-86697-7 (PPC)

WILEY SERIES IN PROBABILITY AND STATISTICS

ESTABLISHED BY WALTER A. SHEWHART AND SAMUEL S. WILKS

Editors: *David J. Balding, Noel A. C. Cressie, Nicholas I. Fisher,*
Iain M. Johnstone, J. B. Kadane, Geert Molenberghs, Louise M. Ryan,
David W. Scott, Adrian F. M. Smith, Jozef L. Teugels
Editors Emeriti: *Vic Barnett, J. Stuart Hunter David G. Kendall*

The *Wiley Series in Probability and Statistics* is well established and authoritative. It covers many topics of current research interest in both pure and applied statistics and probability theory. Written by leading statisticians and institutions, the titles span both state-of-the-art developments in the field and classical methods.

Reflecting the wide range of current research in statistics, the series encompasses applied, methodological and theoretical statistics, ranging from applications and new techniques made possible by advances in computerized practice to rigorous treatment of theoretical approaches.

This series provides essential and invaluable reading for all statisticians, whether in academia, industry, government, or research.

ABRAHAM and LEDOLTER · Statistical Methods for Forecasting
AGRESTI · Analysis of Ordinal Categorical Data
AGRESTI · An Introduction to Categorical Data Analysis
AGRESTI · Categorical Data Analysis, *Second Edition*
ALTMAN, GILL, and McDONALD · Numerical Issues in Statistical Computing
    for the Social Scientist
ANDĚL · Mathematics of Chance
ANDERSON · An Introduction to Multivariate Statistical Analysis, *Third Edition*
*ANDERSON · The Statistical Analysis of Time Series
ANDERSON, AUQUIER, HAUCK, OAKES, VANDAELE, and WEISBERG ·
    Statistical Methods for Comparative Studies
ANDERSON and LOYNES · The Teaching of Practical Statistics
ARMITAGE and DAVID (editors) · Advances in Biometry
ARNOLD, BALAKRISHNAN, and NAGARAJA · Records
*ARTHANARI and DODGE · Mathematical Programming in Statistics
*BAILEY · The Elements of Stochastic Processes with Applications to the Natural
    Sciences
BALAKRISHNAN and KOUTRAS · Runs and Scans with Applications
BARNETT · Comparative Statistical Inference, *Third Edition*
BARNETT · Environmental Statistics: Methods & Applications
BARNETT and LEWIS · Outliers in Statistical Data, *Third Edition*
BARTOSZYNSKI and NIEWIADOMSKA-BUGAJ · Probability and Statistical Inference
BASILEVSKY · Statistical Factor Analysis and Related Methods: Theory and
    Applications
BASU and RIGDON · Statistical Methods for the Reliability of Repairable Systems
BATES and WATTS · Nonlinear Regression Analysis and Its Applications
BECHHOFER, SANTNER, and GOLDSMAN · Design and Analysis of Experiments for
    Statistical Selection, Screening, and Multiple Comparisons
BELSLEY · Conditioning Diagnostics: Collinearity and Weak Data in Regression
BELSLEY, KUH, and WELSCH · Regression Diagnostics: Identifying Influential
    Data and Sources of Collinearity
BENDAT and PIERSOL · Random Data: Analysis and Measurement Procedures,
    Third Edition

*Now available in a lower priced paperback edition in the Wiley Classics Library.

BERRY, CHALONER, and GEWEKE · Bayesian Analysis in Statistics and
  Econometrics: Essays in Honor of Arnold Zellner
BERNARDO and SMITH · Bayesian Theory
BHAT and MILLER · Elements of Applied Stochastic Processes, *Third Edition*
BHATTACHARYA and JOHNSON · Statistical Concepts and Methods
BHATTACHARYA and WAYMIRE · Stochastic Processes with Applications
BILLINGSLEY · Convergence of Probability Measures, *Second Edition*
BILLINGSLEY · Probability and Measure, *Third Edition*
BIRKES and DODGE · Alternative Methods of Regression
BLISCHKE AND MURTHY (editors) · Case Studies in Reliability and Maintenance
BLISCHKE AND MURTHY · Reliability: Modeling, Prediction, and Optimization
BLOOMFIELD · Fourier Analysis of Time Series: An Introduction, *Second Edition*
BOLLEN · Structural Equations with Latent Variables
BOROVKOV · Ergodicity and Stability of Stochastic Processes
BOULEAU · Numerical Methods for Stochastic Processes
BOX · Bayesian Inference in Statistical Analysis
BOX · R. A. Fisher, the Life of a Scientist
BOX and DRAPER · Empirical Model-Building and Response Surfaces
*BOX and DRAPER · Evolutionary Operation: A Statistical Method for Process
  Improvement
BOX, HUNTER, and HUNTER · Statistics for Experimenters: An Introduction to
  Design, Data Analysis, and Model Building
BOX and LUCEÑO · Statistical Control by Monitoring and Feedback Adjustment
BRANDIMARTE · Numerical Methods in Finance: A MATLAB-Based Introduction
BROWN and HOLLANDER · Statistics: A Biomedical Introduction
BRUNNER, DOMHOF, and LANGER · Nonparametric Analysis of Longitudinal
  Data in Factorial Experiments
BUCKLEW · Large Deviation Techniques in Decision, Simulation, and Estimation
CAIROLI and DALANG · Sequential Stochastic Optimization
CHAN · Time Series: Applications to Finance
CHATTERJEE and HADI · Sensitivity Analysis in Linear Regression
CHATTERJEE and PRICE · Regression Analysis by Example, *Third Edition*
CHERNICK · Bootstrap Methods: A Practitioner's Guide
CHERNICK and FRIIS · Introductory Biostatistics for the Health Sciences
CHILÈS and DELFINER · Geostatistics: Modeling Spatial Uncertainty
CHOW and LIU · Design and Analysis of Clinical Trials: Concepts and Methodologies, *Second Edition*
CLARKE and DISNEY · Probability and Random Processes: A First Course with
  Applications, *Second Edition*
*COCHRAN and COX · Experimental Designs, *Second Edition*
CONGDON · Applied Bayesian Modelling
CONGDON · Bayesian Statistical Modelling
CONOVER · Practical Nonparametric Statistics, *Second Edition*
COOK · Regression Graphics
COOK and WEISBERG · Applied Regression Including Computing and Graphics
COOK and WEISBERG · An Introduction to Regression Graphics
CORNELL · Experiments with Mixtures, Designs, Models, and the Analysis of Mixture
  Data, *Third Edition*
COVER and THOMAS · Elements of Information Theory
COX · A Handbook of Introductory Statistical Methods
*COX · Planning of Experiments
CRESSIE · Statistics for Spatial Data, *Revised Edition*
CSÖRGŐ and HORVÁTH · Limit Theorems in Change Point Analysis
DANIEL · Applications of Statistics to Industrial Experimentation
DANIEL · Biostatistics: A Foundation for Analysis in the Health Sciences, *Sixth Edition*
*DANIEL · Fitting Equations to Data: Computer Analysis of Multifactor Data,
  Second Edition

*Now available in a lower priced paperback edition in the Wiley Classics Library.

*Now available in a lower priced paperback edition in the Wiley Classics Library.

*Now available in a lower priced paperback edition in the Wiley Classics Library.

*Now available in a lower priced paperback edition in the Wiley Classics Library.

MURRAY · X-STAT 2.0 Statistical Experimentation, Design Data Analysis, and Nonlinear Optimization

MURTHY, XIE, and JIANG · Weibull Models

MYERS and MONTGOMERY · Response Surface Methodology: Process and Product Optimization Using Designed Experiments, *Second Edition*

MYERS, MONTGOMERY, and VINING · Generalized Linear Models. With Applications in Engineering and the Sciences

NELSON · Accelerated Testing, Statistical Models, Test Plans, and Data Analyses

NELSON · Applied Life Data Analysis

NEWMAN · Biostatistical Methods in Epidemiology

OCHI · Applied Probability and Stochastic Processes in Engineering and Physical Sciences

OKABE, BOOTS, SUGIHARA, and CHIU · Spatial Tesselations: Concepts and Applications of Voronoi Diagrams, *Second Edition*

OLIVER and SMITH · Influence Diagrams, Belief Nets and Decision Analysis

PALTA · Quantitative Methods in Population Health: Extensions of Ordinary Regressions

PANKRATZ · Forecasting with Dynamic Regression Models

PANKRATZ · Forecasting with Univariate Box-Jenkins Models: Concepts and Cases

*PARZEN · Modern Probability Theory and It's Applications

PEÑA, TIAO, and TSAY · A Course in Time Series Analysis

PIANTADOSI · Clinical Trials: A Methodologic Perspective

PORT · Theoretical Probability for Applications

POURAHMADI · Foundations of Time Series Analysis and Prediction Theory

PRESS · Bayesian Statistics: Principles, Models, and Applications

PRESS · Subjective and Objective Bayesian Statistics, *Second Edition*

PRESS and TANUR · The Subjectivity of Scientists and the Bayesian Approach

PUKELSHEIM · Optimal Experimental Design

PURI, VILAPLANA, and WERTZ · New Perspectives in Theoretical and Applied Statistics

PUTERMAN · Markov Decision Processes: Discrete Stochastic Dynamic Programming

*RAO · Linear Statistical Inference and Its Applications, *Second Edition*

RENCHER · Linear Models in Statistics

RENCHER · Methods of Multivariate Analysis, *Second Edition*

RENCHER · Multivariate Statistical Inference with Applications

RIPLEY · Spatial Statistics

RIPLEY · Stochastic Simulation

ROBINSON · Practical Strategies for Experimenting

ROHATGI and SALEH · An Introduction to Probability and Statistics, *Second Edition*

ROLSKI, SCHMIDLI, SCHMIDT, and TEUGELS · Stochastic Processes for Insurance and Finance

ROSENBERGER and LACHIN · Randomization in Clinical Trials: Theory and Practice

ROSS · Introduction to Probability and Statistics for Engineers and Scientists

ROUSSEEUW and LEROY · Robust Regression and Outlier Detection

RUBIN · Multiple Imputation for Nonresponse in Surveys

RUBINSTEIN · Simulation and the Monte Carlo Method

RUBINSTEIN and MELAMED · Modern Simulation and Modeling

RYAN · Modern Regression Methods

RYAN · Statistical Methods for Quality Improvement, *Second Edition*

SALTELLI, CHAN, and SCOTT (editors) · Sensitivity Analysis

*SCHEFFE · The Analysis of Variance

SCHIMEK · Smoothing and Regression: Approaches, Computation, and Application

SCHOTT · Matrix Analysis for Statistics

SCHUSS · Theory and Applications of Stochastic Differential Equations

SCOTT · Multivariate Density Estimation: Theory, Practice, and Visualization

*SEARLE · Linear Models

*Now available in a lower priced paperback edition in the Wiley Classics Library.

*Now available in a lower priced paperback edition in the Wiley Classics Library.